Biology and Management of Noxious Rangeland Weeds

Edited by Roger L. Sheley and Janet K. Petroff

Biology and Management of Noxious Rangeland Weeds

Edited by Roger L. Sheley and Janet K. Petroff

Oregon State University Press
Corvallis

Publication of this book was made possible in part
through the generous support of

Montana Department of Agriculture's Noxious Weed Trust Fund
and
Montana State University

The Oregon State University Press is grateful for their support

Front cover: Diffuse knapweed. Photograph by R.D. Richard, USDA-APHIS, Bozeman, Montana. *Back cover:* Yellow starthistle. Photograph by Cindy Talbott Roché.

The paper in this book meets the guidelines for permanence and durability of the Committee on Production Guidelines for Book Longevity of the Council on Library Resources and the minimum requirements of the American National Standard for Permanence of Paper for Printed Library Materials Z39.48-1984.

Library of Congress Cataloging-in-Publication Data
Biology and management of noxious rangeland weeds / edited by Roger L. Sheley and Janet K. Petroff.
   p. cm.
Includes bibliographical references and index.
ISBN 0-87071-461-9 (alk. paper)
1. Rangelands—Weed control—West (U.S.) 2. Weeds—West (U.S.) 3. Range plants—Control—West (U.S.) 4. Range management—West (U.S.) 5. Invasive plants—West (U.S.) I. Sheley, Roger L. II. Petroff, J.K. (Janet K.)
SB612.W47B564 1999
633.2'02—dc21

       98-52348
       CIP

Oregon State University Press
101 Waldo Hall
Corvallis OR 97331-6407
541-737-3166 • fax 541-737-3170
http://.osu.orst.edu/dept/press

Table of Contents

Section II-The Weeds

Section I

Theory and Practice of Weed Management

Introduction

By Roger L. Sheley, Janet K. Petroff, and Michael M. Borman

North American rangelands and forests are vast and complex. These phenomenal natural resources sustain food and fiber production, protect watersheds, provide wildlife habitat and recreation sites, maintain a healthy diversity of flora and fauna, and appeal to our aesthetic values and need for open space. Noxious weeds are the foreign organisms—the invaders—of forest and rangeland.

Legally, a noxious weed is any plant designated by a federal, state, or county government to be injurious to public health, agriculture, recreation, wildlife, or any public or private property. In this text, we also use the term "noxious" to refer to those weeds that have invasive characteristics, regardless of whether they have been legally designated "noxious" at some governmental level. Long-term ecosystem management is necessary to reduce all weeds' threat to healthy, diverse rangelands and forests.

Most of the weeds invading U.S. rangelands originated in Europe and Asia. Many were introduced to western rangelands during the nineteenth century and are rapidly spreading. New weeds are introduced continually. In their ecosystem of origin, these plants are not problems because they evolved with natural controls such as insect predators, plant pathogens, fungi, other competing plants, and grazing. Plant species that arrived in North America without those control agents have dominated locations ranging from small, well-defined areas to very extensive, heterogeneous areas. Outcompeting the native plant community, the weeds create a monoculture that can increase wind and water erosion; decrease capture, storage, and proper release of precipitation; and alter nutrient cycling. Weed monocultures further reduce biological diversity by displacing the macro- and microfauna that depend on native plants for habitat and food. The grazing capacity of weed-infested land can be reduced by up to 75%.

Weeds prefer highly disturbed sites such as river and stream banks, trailheads, roadsides, building sites, trails, wildlife bedgrounds, overgrazed areas, and campgrounds. Well-managed land is the best defense against the spread of weeds. However, even well-managed land in good condition is susceptible when natural disturbances (such as wind, water, and a wide variety of wildlife, including birds) open niches in the plant community and distribute plant parts and seeds. Once established, weeds are spread by many vectors, including vehicles, wind, recreationists, waterways, animals, and weed-contaminated hay.

Over the years, support for and application of weed management programs have been variable and inadequate. Following the development of phenoxy herbicides at the end of World War II, weed control via herbicides became practical. Initially, chemical control was mainly directed at woody species that increased on rangelands. However, chemical control was also attempted for plants such as Canada thistle (*Cirsium arvense*) and halogeton (*Halogeton glomeratus*). Subsequent generations of herbicides provided additional management possibilities. When combined with mechanical control efforts, changes in livestock management, and reseeding with site-adapted plant species, some successes were achieved. However, concern about pesticides' possible detrimental effects on human health and the environment led to increasing restrictions on herbicide use, particularly by government agencies. Federal agency concerns about re-seeding with competitive, non-indigenous species also became a constraint. Interest, budgets, and activity in range weed control all declined because of concern and restrictions. More recently, however, a commitment to addressing the rapid invasion of noxious weeds has been revived and intensified. This book is the result of the increasing interest in noxious weeds. It is meant to provide natural resource managers, and students who are interested in noxious weeds management, with a reference and text.

Section I deals with the theory, concepts, and principles of rangeland weed management. The section is organized in the same fashion as one might develop a weed management plan. An effective weed management plan first considers the impacts of the weeds (Chapter 1). Then managers must survey and map the management unit (Chapter 2). Any sustainable management will require coordinated planning (Chapter 3). The cost and benefits of effective management must be evaluated (Chapter 4). Research indicates that integrating management strategies holds the most promise for long-term success (Chapter 5). Prevention of a weed invasion is the most cost-effective management strategy (Chapter 6), followed by early detection and control of small infestations (Chapter 7). Three management techniques—grazing (Chapter 8), biological control (Chapter 9), and chemical control (Chapter 10)—will be major components of most management programs. And finally, revegetating rangelands is an emerging science (Chapter 11), central to establishing and maintaining weed-resistant plant communities that meet other land-use objectives.

Section II deals specifically with twenty-nine of the most serious noxious weed species in the western United States. Each chapter discusses the identification, origin, and potential invasion of the weed. The ecological and economic effects of each weed are also discussed. Understanding the biology and ecology of each weed provides insights into how to manage the infestation effectively within a dynamic plant community. The chapters conclude with a full discussion of practical management options. The goal is to provide enough information to allow for effective decision-making and weed management planning for the long-term.

A note on the compilation of this book: Each of the chapters reflects the knowledge, experience, and interests of its author(s). Therefore, readers will discover a different "voice," different word usage in some cases, and different emphases in each chapter. Just as diversity creates a strong plant community, it creates a strong scientific and educational community. We welcome the insights that each author offers. Overall, the format and language of this book is intended to help bridge the gap between weed science and weed management. To a great extent, weed management is a planning process, whereas weed science comprises the theory on which weed managers base their planning. The relationship between these bodies of complementary knowledge seems clear, but the language and application gap can be as wide as the gap between art and science. In this book, the authors have provided practical, science-based information needed for the art of weed management.

The weed distribution maps were created at Montana State University by Lee Faulkner under the direction of Diana Cooksey, Cooperative Agricultural Pest Survey (CAPS) program, funded by USDA Animal and Plant Health Inspection Service Plant Protection and Quarantine (APHIS-PPQ). Data were collected in 1996-97 from various authorities (state Departments of Agriculture, Extension specialists, university researchers, county weed supervisors, Agricultural Experiment Station personnel, etc.) surveyed in each state. The maps do not indicate density of infestations. Rather, the maps show where a given weed has appeared—in any amount—so land managers can be alerted to watch for potential invasions.

Sustainable weed management and land restoration are as complex as the land itself, and involve ecological processes that determine plant community dynamics. Knowledge of these processes, along with economic data, must be compiled into decision-support systems to allow land managers to make informed choices when faced with tough decisions that involve diverse natural resources, land uses, aesthetic values, and budgetary constraints. This book is a contribution to building a practical, long-term, decision-support system for land managers.

Acknowledgments

We wish to acknowledge, first, the many authors who contributed their best work to this book. These contributors are often called upon to share their expertise in weed management. They are busy people, and we greatly appreciate the time they have spent writing, peer-reviewing, rewriting, and proofreading during the past several years. It is rewarding to see this work in print. In addition, this book would not have been possible without support from Montana State University and the Montana Department of Agriculture's Noxious Weed Trust Fund.

Roger Sheley, Janet Petroff, Mike Borman
October 1998

Impacts of Noxious Weeds on Ecologic and Economic Systems

Bret E. Olson

Noxious weeds alter the structure, organization, or function of ecological systems. "Structure" refers to the type and abundance of organisms in the environment, as well as to its physical features. "Organization" refers to the relative abundance of species and their relationships. "Function" refers to the various processes that occur in ecosystems. By affecting ecological systems, noxious weeds also affect economic systems. This review will focus on how noxious weeds affect the structure, organization, and function of ecological and economic systems. Research has been limited in these areas; thus, at times causal mechanisms or relationships will be suggested, and examples from native or desirable introduced species will be used when necessary.

Impacts on Structure

• *Soils*

Noxious weeds may affect the structure of ecosystems by altering soil properties. Soil in areas dominated by noxious weeds may have lower amounts of organic matter and available nitrogen than areas supporting native grasslands. This may result from several factors. First, noxious weeds may increase soil erosion (Lacey et al. 1989). Even modest erosion of the soil surface layer will cause disproportionate losses of biologically active organic matter and nitrogen because most of the biologically active organic matter is concentrated in the top 1 to 4 inches (3 to 10 cm) (Woods 1989). Second, roots of noxious weeds tend to be deeper than roots of native grasses, so they will contribute less organic matter near the soil surface. Third, thick taproots of noxious weeds may decompose more slowly than fine roots of grasses, reducing the annual input of organic matter to the soil. Finally, some noxious weeds have secondary compounds throughout the plant (Locken and Kelsey 1987, Mahlberg et al. 1987, Olson and Kelsey 1997). These chemical compounds may hinder soil macro- and microfauna from feeding on living roots, and thus may slow decomposition of litter and root material.

Many non-native species, including noxious weeds, deplete soil nutrient reserves to very low levels. Potassium, nitrogen, and phosphorous levels were 44%, 62%, and 88% lower on soils from a spotted knapweed-infested site than from adjacent soils with a grass overstory (Harvey and Nowierski 1989). The success of cheatgrass in the intermountain sagebrush region can be partly attributed to its ability to exploit soil moisture and nutrients in early spring before the native grasses are actively growing (Miller et al. 1994). In a greenhouse study, nitrate levels were much lower in pots containing non-native species than in pots containing native species (Elliot and White 1989). Plants that reduce soil nutrient availability to very low levels have a competitive advantage over neighboring plants (Tilman and Wedin 1991). Thus the success of many non-native species may reflect two strategies. Many early successional non-native species take up soil nutrients rapidly, whereas late successional non-native species can often tolerate low levels of available soil nutrients. Presumably, high concentrations of nutrients in litter indicate high nutrient uptake rates by non-native plants. Litter of several native perennial grasses is often relatively nutrient poor, whereas litter of non-native perennial grasses is often nutrient rich (Wedin and Tilman 1990). When non-natives immobilize or "tie up" more than their share of nutrients, these nutrients are not readily available for neighboring species.

Many noxious weeds are early successional species that colonize recently disturbed sites (Baker 1986). Disturbances such as livestock trampling and construction expose subsurface organic matter and incorporate litter into the soil. By adding additional oxygen and water to organic matter, a disturbance increases soil microbial activity and subsequent mineralization, the release of inorganic molecules which are available for plant uptake. Many weeds rapidly exploit these newly available nutrients (Best et al. 1980, Belcher and Wilson 1989). This may be one reason that the density of weedy plants increases as the intensity of disturbance increases (Jensen 1995).

Although not documented, dense infestations of annual and perennial weeds may alter the physical properties of soil. Most noxious weeds have deep taproots, whereas the native perennial grasses they displace have extensive, fibrous roots. These extensive, fibrous root systems of grasses have high annual turnover rates. In some grasses, up to 50% of the root material is replaced annually. A high turnover rate of these fibrous grass roots promotes greater aggregation of soil particles compared with taprooted forbs.

• *Plants*

Noxious weeds may affect the structure of ecosystems by altering the composition of plant communities. Noxious weeds often displace native plant species. This is the most frequently measured and reported impact on ecosystems, e.g., cheatgrass (*Bromus tectorum*) displacing native bunchgrasses and shrubs in the Great Basin (Miller et al. 1994), tamarisk (*Tamarix parviflora*) displacing riparian species in the southwest (Hildebrandt and Ohmart 1982), and leafy spurge (*Euphorbia esula*) displacing many species in the Northern Great Plains (Belcher and Wilson 1989).

5

Altering the composition of plant communities also changes the physiognomy or physical structure of the plant community. In upland areas, perennial grasses with multiple stems and relatively high basal cover are replaced by forbs or shrubs with single or spreading stems and low basal cover. The structure of the grasses more readily dissipates energy associated with falling raindrops, covers soil to minimize erosion, and traps blowing soil particles and snow. Belowground, changes from a predominance of fibrous-rooted species to taprooted species will alter soil-holding capability and water infiltration. On slopes, surface runoff and sediment production increase when bunchgrasses are replaced by weedy forbs (Lacey et al. 1989).

• Animals

By altering the structure of plant communities, noxious weeds alter the structure of animal communities. Many animal species that co-evolved with a particular native plant community for forage, cover, and shelter cannot shift their requirements rapidly enough to adapt to rapid changes in the plant community. Either reproduction is compromised, reducing population growth, or the animals simply leave or avoid infested areas. When spotted knapweed was removed (with picloram) from an historic elk winter range in western Montana, elk use of the area increased dramatically, especially in the winter immediately following the herbicide treatment (Thompson 1996). This shift in distribution reduced elk densities on the remaining winter range. The effect was greatest the first winter following treatment because of the rapid flush of high-quality forage of the perennial grasses. In subsequent winters, the accumulation of dead grass material reduced the attractiveness of these grasses to elk. In Theodore Roosevelt National Park in North Dakota, bison, deer, and elk used areas infested with leafy spurge significantly less than uninfested areas (Trammell and Butler 1995). The difference in animal use patterns was attributed to the fact that the desired forage species were less productive in areas infested with leafy spurge, and the animals simply avoided infested areas.

Some weeds can have subtle direct effects on animals, including humans. Spotted knapweed (*Centaurea maculosa*) was introduced into the Bitterroot Valley in Montana around 1900. Beekeepers in the valley have valued spotted knapweed for the honey produced from its flowers. However, this weed produces a common airborne allergen (pollen) that peaks in August (Gillespie and Hedstrom 1979). Initially, *Centaurea* species were thought to be pollinated only by insects and not the wind. *Centaurea* pollen produces strongly positive skin tests, and it is a significant allergen in causing allergic rhinitis (D.N. Gillespie, personal observation).

Impacts on Organization

• *Soils*

Most noxious weeds have secondary compounds that may affect soil microorganisms, either positively or negatively. Most microbial populations can adapt to different types of plant materials containing secondary compounds to increase their population biomass, increasing the rate of breakdown of secondary compounds (Vaughan et al. 1983). For example, in the presence of high phenolic concentrations in plant material, the proportion of microbes in a population that can metabolize phenolics increased (Blum and Shafer 1988). Similarly, terpenes can be used as a carbon and energy source by soil microbes (Vokou et al. 1984).

Conversely, secondary compounds can negatively affect activity and growth of anaerobic rumen microorganisms (Olson and Kelsey 1997), which reduces digestibility of certain forages such as spotted knapweed. In some cases, these secondary compounds may limit activity and growth of aerobic soil microbial populations, resulting in thick litter layers and slowing rates of nutrient cycling. Many weeds are also high in lignin, which is more resistant to microbial breakdown than cellulose and hemicellulose. In grasses, nitrogen mineralization is slower with plant materials with high lignin concentrations (Wedin and Tilman 1990).

• *Plants*

Noxious weeds alter the relative abundance of native plant species for one (or often many) of the following reasons: noxious weeds often outcompete native species, can be allelopathic (Rice 1984), produce abundant seed, have fast growth rates, grow before native species initiate growth, may exploit the entire soil profile for water and nutrients, have no natural enemies, and are often avoided by large herbivores grazing native plants (Lym and Kirby 1987). Competition has been inferred from field studies and observations of weeds displacing native species. In one field study, spotted knapweed reduced seed germination and seedling establishment of a rare Montana endemic forb, *Arabis fecunda*, but did not affect adult growth and fecundity of this species (Lesica and Shelly 1996). This was attributed to spotted knapweed exploiting resources before the seedlings of *Arabis* could use them. Competition has also been inferred from controlled greenhouse studies (Lindquist et al. 1996). However, few studies have assessed the actual processes (for example, greater nutrient uptake rates, water use efficiencies, photosynthetic rates, etc.) which may confer a competitive advantage to the weed.

Allelopathy is the effect of one plant on germination and seedling establishment of another. This is often associated with plant secondary compounds, and has been documented for several noxious weeds (reviewed by Kelsey and Everett 1995). Allelopathy has been used to explain the presence of essentially pure monocultures of several non-native species. However, it is often difficult to separate the effects of allelopathy from competition.

7

Even though some weeds may contain high concentrations of secondary compounds, those weeds are not necessarily allelopathic. Spotted and diffuse knapweed (*Centaurea diffusa*) contain high concentrations of toxic secondary compounds (Muir and Majak 1983, Kelsey and Locken 1987), but soil concentrations of these compounds are usually below phytotoxic levels. Soil particles may deactivate the toxic compound or make it unavailable for plant uptake (Dao 1987, Dalton et al. 1989). Soil microbes may convert plant-derived secondary compounds from inactive to active forms, or conversely, reduce their effective toxicity by metabolizing secondary compounds (Vaughan et al. 1983).

Weed success can also be attributed to high rates of seed production and long-term seed viability. Diffuse and spotted knapweed can produce from 400 to 25,000 seeds per plant when growing on dry and wet soils, respectively (Watson and Renny 1974). The non-native oxeye daisy (*Chrysanthemum leucanthemum*), a showy forb, usually produces 1,300 to 4,000 fruits, but vigorous plants can produce 26,000 fruits (Dorph-Petersen 1925).

Long-term seed viability allows weed seeds to remain in an undisturbed plant community for decades, then germinate when the site is disturbed. A small percentage of spotted knapweed seeds remain viable in soil for at least seven years, and possibly much longer (Davis et al. 1993). This exceeds the phytotoxic residual period of picloram, which is often the most effective herbicide for this plant. Eighty-two percent of buried oxeye daisy seeds were still viable after six years, whereas 1% were still viable after 39 years (Toole and Brown 1946).

Partly because of abundant seed production and dispersal and long-term seed viability, spotted knapweed readily invades pristine grasslands in Glacier National Park (Tyser and Key 1988). Invasion epicenters in the park include elk wallows, badger mounds, and areas of ground squirrel activity. Invading early successional non-native species often grow quickly once the seed has germinated, exploiting soil water and nutrients before the associated natives begin growing (Miller at al. 1994). In addition, many late successional non-native forbs have deep taproots, and thus are able to explore the entire soil profile for water and nutrients, especially into mid-summer.

Many non-natives were introduced to North America without the complement of native insects and pathogens that help to keep them in check in their native countries. When scientists from North America visit these plants' native countries to collect potential biological control agents, these "weeds" are often found only in scattered patches, not dense monocultures. This partly reflects the checks and balances between these plants and their "enemies" in their native countries.

In many ecosystems, the displacement of native species by noxious weeds is accelerated by animal foraging preferences. Large herbivores such as cattle, horses, and bison inherently prefer grasses over shrubs and forbs. In addition, many shrubs and forbs contain secondary compounds that further deter feeding by large herbivores. Therefore, constant grazing of the preferred grasses while avoiding shrubs and forbs reduces the ability of these grasses to compete with shrubs and forbs. In contrast, sheep and goats prefer forbs and shrubs, respectively, and use grasses sparingly, which confers a competitive advantage to the grasses. On many western rangelands, cattle numbers are close to record highs, while sheep and goat numbers have declined significantly since the 1940s. That trend continues today. Many noxious weeds did not begin to increase exponentially until after the 1950s, as numbers of sheep and goats were beginning to decline.

The effect of a noxious weed on a native plant community would not be too devastating if the noxious weed species simply replaced one member of the community. However, severe infestations of noxious weeds usually reduce community productivity, species diversity (relative abundance of species), and species richness (number of species). For example, cheatgrass and other non-native annuals continue to dominate sites in southeastern Washington, almost to the total exclusion of native grasses (Rikard and Cline 1980). Spotted knapweed is reducing species richness and diversity in Glacier National Park (Tyser and Key 1988).

• Animals

Again, since animal communities often prefer certain plant communities, the relative abundance of animal species should change with an infestation of a noxious weed, although this has seldom been documented, especially for invertebrates. Domestic livestock and wildlife avoid areas with heavy infestations of noxious weeds (Lym and Kirby 1987, Bedunah 1992, Hein and Miller 1992, Trammel and Butler 1995, Thompson 1996). In the Southwest, tamarisk stands have 50% fewer small mammals than native riparian stands (Anderson and Ohmart 1977) and less species diversity of reptiles and amphibians (Jakle and Katz 1985).

Impacts on Function

• Soils

The amount of moisture in the soil may be altered in several ways. First, infiltration may be reduced and runoff increased on sites dominated by noxious weeds (Lacey et al. 1989) because of lower basal cover and crusting of exposed soil. Taprooted forbs may reduce infiltration because they do not have the dense, fine root systems of grasses, which contribute organic matter and enhance soil structure (Tisdall and Oades 1982).

Second, many noxious weeds have relatively sparse plant canopies, although some noxious weeds such as leafy spurge have dense plant canopies on favorable sites. Where canopies are sparse, evaporation from exposed soil surface is greater than where there is dense vegetative cover. Greater evaporation from the soil surface will reduce levels of soil water for uptake by fine roots of native grasses. Dry surface soils hinder the survival of seedlings that cannot access water in the deeper soil layers (Lauenroth et al. 1994). Where canopies are dense, high transpiration rates by plants may deplete soil water content, also hindering seedling establishment.

9

Third, water uptake is a function of biomass allocation, architectural plasticity of plant root systems (Fitter 1994, Hook and Lauenroth 1994), and plant water uptake kinetics in semiarid regions (Sala et al. 1982, Hook and Wraith 1995, Wraith et al. 1995a,b). Although never quantified, these traits likely differ between noxious weeds and perennial grasses, which will alter the availability of soil water in infested areas.

Finally, sites infested with certain noxious weeds may have more extreme soil temperature changes because of 1) lower soil water content; 2) poorer soil aggregation; and 3) greater exposure of bare soil to direct sunlight (Jones 1983, Monteith and Unsworth 1990). Water has a very high capacity to store heat. By reducing soil water content in surface soils, greater evaporation enhances rapid heating and cooling of near-surface layers. Soils in weed-infested areas are often low in organic matter and poorly aggregated. This will increase runoff but lower infiltration, again reducing thermal conductivity and capacity of the soil to store heat, resulting in greater temperature extremes at the soil surface. The sparser canopies of many weed species will allow more sunlight to reach the ground surface during the day and cause greater heat loss at night, increasing temperature extremes on a 24-hour basis.

Noxious weeds may reduce soil nutrient availability by having higher nutrient uptake rates, or by having roots that decompose more slowly than roots of neighboring plants, or both. Regardless of the mechanism, this reduces nutrient availability for native species. Many noxious weeds are associated with highly disturbed sites during the early stages of succession when nutrient availability is generally high (Tilman 1994). By preempting nutrients, invasive plants may greatly slow the rate of succession, or even create a new stable state (Laycock 1991, D'Antonio and Vitousek 1992). When a system has crossed a threshold into a new stable state dominated by introduced species, massive amounts of fossil fuels are needed to restore the native vegetation (Laycock 1991).

Nutrient cycles reflect how rapidly plants take up (immobilize) nutrients, and then how rapidly microorganisms decompose (mineralize) plant or animal materials. Although litter from some non-native species may slow decomposition, some non-natives actually increase the rate of nutrient cycling. Nitrogen cycled more rapidly under an invasive annual forb than under native perennial grasses on a shortgrass prairie (Vinton and Burke 1995). However, their study involved a site which had a fast-growing annual forb (*Kochia scoparia*) and had received additional inputs of fertilizer and water. Nutrients may also cycle faster in systems dominated by non-native annual grasses such as cheatgrass. The fine above- and belowground material of cheatgrass decomposes quickly. In addition, cheatgrass-dominated areas burn frequently, releasing minerals rapidly. Unfortunately, cheatgrass and another introduced annual, medusahead (*Taeniatherum caput-medusae*), germinate and grow more rapidly under cooler conditions than many native species. This allows them to exploit recently released minerals, perpetuating their dominance in the intermountain area.

• *Plants*

Litter accumulated from noxious weeds may reduce native grass seed germination and seedling establishment by leaching inhibitory compounds, altering the microclimate, and preventing shoot extension (Schlatterer and Tisdale 1969, Bosey and Reader 1996). Litter from the invasive leafy spurge can accumulate up to 4 inches (10 cm) thick (Olson and Wallander 1997, unpublished data, Bozeman, MT). Leachate from dried stems and leaves of leafy spurge and spotted knapweed inhibited seed germination, and especially root hair development, of two native perennial grasses, bluebunch wheatgrass (*Agropyron spicatum*) and western wheatgrass (*Agropyron smithii*) (Olson and Wallander 1997, unpublished data, Bozeman, MT). Leaves of tamarisk contain and exude salts. This produces a dense leaf litter that increases soil salinity. In turn, this may reduce soil microbial activity and populations of soil arthropods and other biota.

Altered soil water and nutrient dynamics on sites infested with noxious weeds may reflect the ability of forbs to preempt soil water and nutrients. Effective competition for water may involve greater use early in the growing season (Eissenstat and Caldwell 1988), rapid resumption of water uptake when soil water deficits are relieved by rainfall (Sala et al. 1982, Cruz et al. 1992, Nobel 1994, Wraith et al. 1995a), or more effective use of water deep in the soil profile. Further, because of their relatively deep root systems, many noxious weeds are active later in the summer than native cool-season grasses, which may affect soil water and nutrient availability the following growing season (Gerlach and Rice 1996).

On a watershed level, reduced infiltration and increased runoff associated with dense infestations of weeds will alter seasonal water flows. Less water infiltration means less water storage and subsequent interflow (subsurface) movement. Interflow helps maintain high stream levels through summer, providing adequate flows for fish populations.

Noxious weeds that did not evolve in a particular system may alter historic disturbance cycles, including fire and grazing. In Hawaii, the invasion of non-native warm- and cool-season grasses has greatly increased fire frequencies by providing abundant fine fuels to carry fire. Also, the introduced species are more fire-tolerant than the native species (D'Antonio and Vitousek 1992). In the intermountain region, the introduction of the highly flammable cheatgrass has increased the frequency of fires. Cheatgrass areas often burn every 3 to 5 years, compared with 60 to 110 years for the native plant communities displaced by cheatgrass (Whisenant 1990). The native species did not evolve with such frequent fires, and consequently are at a great disadvantage. Whether invasive forbs increase fire frequency like the annual cheatgrass has not been documented. At a minimum, the invasive leafy spurge and spotted knapweed are seldom impacted by fire (B. Olson 1992-1997, personal observation).

• Animals

Animal productivity can be reduced by noxious weeds through several mechanisms. First, animals may simply avoid infested areas (Hein and Miller 1992), or infested areas may act as physical barriers to water sources or preferred, uninfested areas (Lym and Kirby 1987). Second, secondary compounds may deter feeding because they impart an unsavory odor or a bitter taste. Third, even though animals may ingest noxious weeds, secondary compounds in the forage may affect rumen microbial activity (Olson and Kelsey 1997), thereby reducing forage intake, or cause general malaise resulting in aversive postingestive feedback (Provenza 1995).

Impacts on Economics

• Microeconomics

Many noxious weeds greatly reduce the land's carrying capacity for domestic livestock. Because of such reductions, the value of a ranch in North Dakota heavily infested with leafy spurge dropped from the full market value of $100 to $125 per acre in 1975 to $40 per acre in 1991 (Weiser 1995). The cost to control the infestation was estimated at $20 to $22 an acre. These are annual costs; the leafy spurge was not eradicated. In Klamath County, Oregon, the value of a 1,360-acre ranch was $170,000 to $204,000 (uninfested) in 1988, but because it was infested with leafy spurge, it was sold by the county for only $27,500 with the stipulation that the new owner had to try to control the leafy spurge infestation (Weiser 1995). In six years the new owner had spent $60,000 and had made little headway in controlling the leafy spurge.

The effect of noxious weeds on land value can be measured by either the sales approach or the income approach (Godfrey et al. 1988). The sales approach relies on market sales to estimate value; that is, it reflects what a purchaser is willing to pay, represented by the example above. As long as the buyer and seller are aware of the existence and economic impact of noxious weeds, property infested with noxious weeds should sell for less than property without noxious weeds. In some areas of the West, the seller is required to disclose the presence of noxious weeds to the buyer. In many desirable areas of the West, a buyer may purchase a large landholding primarily for its aesthetic appeal rather than its productive potential. In these cases, the value of the land often is not discounted by the presence of noxious weeds. Once the general value of grazing land has been estimated, the impact of noxious weeds on asset value can be estimated (Godfrey et al. 1988). The productive value of the land is decreased by weeds that detract from or limit its productivity, or increase operating and management costs. Any factor that reduces potential net returns from the land will diminish its value.

Net returns are calculated by subtracting total expenses from total returns. Noxious weeds reduce net returns by either increasing operating expenses, decreasing total returns, or both. Operating expenses increase when landowners implement strategies to limit weed invasion or manage current infestations. Total returns are directly affected when weight gains of livestock are reduced, or animals are poisoned by certain noxious weeds, such as common tansy (*Tanacetum vulgare*), houndstongue (*Cynoglossum officinale*), Russian knapweed (*Centaurea repens*), St. Johnswort (*Hypericum perforatum*), tansy mustard (*Descurainia pinnata*), and yellow starthistle (*Centaurea solstitialis*).

Economic losses from poisonous noxious weeds are relatively insignificant compared with losses from non-poisonous noxious weeds. Most noxious range weeds have lower forage value than the native plants, primarily because most large domestic livestock avoid them. Grazing capacities of leafy spurge-infested land in North Dakota, and knapweed-infested land in Montana for cattle have been reduced by 75% and 63%, respectively (Bucher 1984, Thompson et al. 1990).

Capacity for cattle grazing decreases proportionately with the loss of forage caused by weed infestation. For example, when an animal unit month (AUM) is valued at $10, the annual grazing value of a hectare of land rated at 1 AUM/ha is $10. A weed infestation which reduces forage capacity 75% will reduce stocking rate to 0.25 AUM/ha, and thus the value to $2.50 per hectare. In summary, the economic impact of a weed on livestock forage could be estimated by placing a value on forage, and then multiplying the reduction in desirable forage by the number of affected acres.

At the microeconomic or ranch level, the feasibility of controlling a few range weeds has been evaluated. Controlling scotch thistle was evaluated by incorporating diminishing returns, fixed and variable costs, optimality, marginality, opportunity costs, and time comparisons into a model. On wet meadows and wheatgrass pastures, the model indicated that scotch thistle could be effectively controlled with low rates of picloram (Hooper et al. 1970). Economic returns from controlling leafy spurge with four herbicides were evaluated in east-central South Dakota (Gylling and Arnold 1985). Forage yields did not differ among treatments which controlled 90% or more leafy spurge. Marginal net return over marginal cost from herbicide treatments ranged from $35/ha to $63/ha. In North Dakota, the most cost-effective treatment was picloram plus 2,4-D (Lym and Messersmith 1990). Annual spring applications provided a net return of $284 and $109/ha in eastern and western North Dakota, respectively.

The efficacy of controlling spotted knapweed with picloram was evaluated with an economic model (Griffith and Lacey 1991). On high-producing sites, the cost-benefit ratio for controlling the weed was positive when the value of an AUM ranged from $6 to $14. On the low-producing site, the cost-benefit ratio was positive only when the value of an AUM was $14. Thus, high value lands infested with weeds provide a better return than low value lands because the response of the forage species will be greater on high value lands. As the rate of knapweed spread to new acres increased and forage displacement increased, returns increased relative to treatment costs, because costs per acre would be less and expected forage responses would be greater.

Many noxious weeds are here to stay, and all we can hope for is to minimize their spread. Given this reality, an alternative philosophical approach would be to consider at least some of these weeds as a forage resource for certain grazing animals, especially small ruminants such as sheep and goats. Using a LOTUS spreadsheet model, returns over total costs were calculated to project the effect of incorporating sheep grazing into a typical Northern Great Plains cattle ranch infested with leafy spurge (Williams et al. 1996). Annual returns exceeded total costs by $4,675 when adding sheep grazing to this typical 10,800-acre cattle ranch infested with 1,150 acres of leafy spurge. For this particular scenario, the break-even lamb price was $.53/lb. Returns are higher when the leafy spurge is concentrated, because less field fence is needed to tack onto existing barb wire fence. Annual returns were positive when as little as 4% of the ranch was infested. Costs, returns, acres, level of leafy spurge infestation, number of pastures, etc., can be easily modified with this model to match any operation. Results from several scenarios indicate that adding sheep to a ranch infested with leafy spurge usually results in positive net returns.

Instead of constant, often annual, costs associated with chemical control, sheep grazing can provide real positive returns. A landowner would not have to buy sheep, but could allow a sheep producer to graze his land with no exchange of money. The producer benefits by having access to a free, nutritious forage; the landowner benefits by the sheep controlling the weed. There are examples of this mutually beneficial relationship throughout the western United States.

• *Macroeconomics*
Spotted knapweed could potentially invade 33 million acres of rangeland in Montana (Chicoine et al. 1986). In a recent study, the effects of three knapweed species (spotted, diffuse, Russian) on grazing lands (grazing capacity) and wildlands (wildlife habitat and watershed capacity) were assessed, based on a survey of county weed boards and an input-output model (Hirsch and Leitch 1996). From their analysis, they estimated that total direct and secondary annual economic impacts exceeded $42 million, which could support 518 jobs in the state's economy. Using the same approach, Leitch et al. (1996) estimated that the total direct and secondary annual economic impacts of leafy spurge in Montana, South Dakota, North Dakota, and Wyoming exceeded $129 million in 1993.

Conclusions
Noxious weeds affect the structure, organization, and function of ecological systems in various ways, from nutrient cycling to plant species displacement, to altered activity patterns of animals. Economically, many noxious range weeds: 1) reduce value of land and net returns by increasing operating costs, decreasing total returns, or both; 2) can be managed at the ranch level; and 3) have far-reaching impacts on regional economies.

14

The most disturbing conclusion regards the lack of scientifically based information on weed biology and ecology. In the past, too much emphasis has been placed on weed control, especially with herbicides and, more recently, biological control. When these treatments succeed, everyone is happy. When they fail, and they often fail, we cannot explain why. Until the effect of noxious weeds on the structure, organization, and function of our ecosystems and economy are fully known, future allocation of funding and legislation will continue to be based on anecdotes and observations (Lacey and Olson 1991).

Literature Cited

Anderson, B.W., and R.D. Ohmart. 1977. Wildlife Use and Densities of Birds and Mammals in the Lower Colorado River Valley. Report, U.S. Bureau of Reclamation, Lower Colorado Region.

Baker, H.G. 1986. Patterns of plant invasion in North America. In: H.A. Mooney and J.A. Drake (eds.), Ecology of biological invasions of North America and Hawaii. Springer-Verlag, Berlin, 44-57.

Bedunah, D.J. 1992. The complex ecology of weeds, grazing and wildlife. West. Wildlands. 18:6-11.

Belcher, J.W., and S.D. Wilson. 1989. Leafy spurge and species composition of a mixed-grass prairie. J. Range Manage. 42:172-75.

Best, K.F., G.G. Bowes, A.G. Thomas, and M.G. Shaw. 1980. The biology of Canadian weeds. 39. *Euphorbia esula* L. Can. J. Plant Sci. 60:651-63.

Blum, U., and S.R. Shafer. 1988. Microbial populations and phenolic acids in soil. Soil Biol. Biochem. 20:793-800.

Bosey, J.L., and R.J. Reader. 1996. Mechanisms underlying the suppression of forb seedling emergence by grass (*Poa pratensis*) litter. Funct. Ecol. 9:635-39.

Bucher, R.F. 1984. Potential spread and cost of spotted knapweed on range uses. Bull. 1316, CES Montana State Univ., Bozeman, MT.

Chicoine, T.K., P.K. Fay, and G.A. Nielsen. 1986. Predicting weed migration from soil and climate maps. Weed Sci. 34:57-61.

Cruz, R.T., W.R. Jordan, and M.C. Drew. 1992. Structural changes and associated reduction of hydraulic conductance in roots of *Sorghum bicolor* L. following exposure to water deficit. Plant Physiol. 99:203-12.

Dalton, B.R., U. Blum, and S.B. Weed. 1989. Differential sorption of exogenously applied ferulic, *p*-coumaric, *p*-hydroxybenzoic, and vanillic acids in soil. Soil Sci. Soc. Am. J. 53:757-62.

D'Antonio, C.M., and P.M. Vitousek. 1992. Biological invasions by exotic grasses, the grass/fire cycle, and global change. Ann. Rev. Ecol. Syst. 23:63-87.

Dao, T.H. 1987. Sorption and mineralization of plant phenolic acids in soil. *In*: G.R. Waller (ed.), Allelochemicals: Role in agriculture and forestry. ACS Symp. Series 330. Amer. Chem. Soc., Wash. D.C., 358-70.

Davis, E.S., P.K. Fay, T.K. Chicoine, and C.A. Lacey. 1993. Persistence of spotted knapweed (*Centaurea maculosa*) seed in soil. Weed Sci. 41:57-61.

Dorph-Petersen, K. 1925. Examinations of the occurrence and vitality of various weed seed species under different conditions, made at the Danish State Seed Testing Station during the years 1896-1923. Rep. 4th Int. Seed Test. Congr. 4:128-38.

Eissenstat, D.M., and M.M. Caldwell. 1988. Competitive ability is linked to rates of water extraction. A field study of two aridland tussock grasses. Oecologia 75:1-7.

Elliot, K.J., and A.S. White. 1989. Competitive effects of various grasses and forbs on ponderosa pine seedlings. For. Sci. 33:356-66.

Fitter, A.H. 1994. Architecture and biomass allocation as components of the plastic response of root systems to soil heterogeneity. *In*: M.M. Caldwell, R.W. Pearcy (eds.), Exploitation of environmental heterogeneity by plants. Academic Press, San Diego, CA, 305-23.

Gerlach, J.D., and K. Rice. 1996. *Centaurea solstitialis* threatens California's ecosystems. Bull. Ecol. Soc. Amer. 77:160.

Gillespie, D.N., and L. Hedstrom. 1979. Aeroallergens of western Montana. Rocky Mountain Medical J. 79-82.

Godfrey, E.B., D.B. Nielsen, and N.R. Rimbey. 1988. *In*: L.F. James, M.H. Ralphs, D.B. Nielsen (eds.), The ecology and economic impact of poisonous plants on livestock production. Westview Press, Boulder, CO, 17-25.

Griffith, D., and J.R. Lacey. 1991. Economic evaluation of spotted knapweed (*Centaurea maculosa*) control using picloram. J. Range Manage. 44:43-47.

Gylling, S.R., and W.E. Arnold. 1985. Efficacy and economics of leafy spurge (*Euphorbia esula* L.) control in pasture. Weed Sci. 33:381-85.

Harvey, S.J., and R.M. Nowierski. 1989. Spotted knapweed: Allelopathy or nutrient depletion. Proc. Knapweed Symposium. Montana State University, Bozeman, MT.

Hein, D.G., and S.D. Miller. 1992. Influence of leafy spurge on forage utilization by cattle. J. Range Manage. 45:405-407.

Hildebrandt, T.D., and R.D. Ohmart. 1982. Biological Resource Inventory (Vegetation and Wildlife) Pecos River Basin, New Mexico and Texas. Final Report, Bureau of Reclamation.

Hirsch, S.A., and J.A. Leitch. 1996. The impact of knapweed on Montana's economy. Ag. Econ. Rep. No. 355.

Hook, P.B., and W.K. Lauenroth. 1994. Root system response of a perennial bunchgrass to neighbourhood-scale soil water heterogeneity. Funct. Ecol. 8:738-45.

Hook, P.B., and J.M. Wraith. 1995. Modeling spatial and temporal patterns of soil water availability in semiarid grasslands: effects of plant spacing and lateral distribution of water uptake. Supp. Bull. Ecol. Soc. Amer. 76(3):342.

Hooper, J.F., J.A. Young and R.A. Evans. 1970. Economic evaluation of scotch thistle suppression. Weed Sci. 18:583-86.

Jakle, M.D., and T.A. Katz. 1985. Herpetofaunal use of four habitats on the middle Gila River drainage, Arizona, 355-58. Proc. First North American Riparian Conference, Tucson, AZ.

Jensen, P.K. 1995. Effect of light environment during soil disturbance on germination and emergence pattern of weeds. Ann. Appl. Biol. 127:561-71.

Jones, H.G. 1983. Plants and microclimate. Cambridge University Press, Cambridge, England.

Kelsey, R.G., and L.J. Locken. 1987. Phytotoxic properties of cnicin, a sesquiterpene lactone from *Centaurea maculosa* (spotted knapweed). J. Chem. Ecol. 13:19-33.

Kelsey, R.G., and R.L. Everett. 1995. Allelopathy. *In*: D.J. Bedunah and R.E. Sosebee (eds.) Wildland plants: Physiological ecology and developmental morphology. Soc. Range Manage. Denver, CO, 479-549.

Lacey, J.R., C.B. Marlow, and J.R. Lane. 1989. Influence of spotted knapweed (*Centaurea maculosa*) on surface runoff and sediment yield. Weed Tech. 3:627-31.

Lacey, J.R., and B.E. Olson. 1991. Economic and resource impacts of noxious range weeds. *In*: L.F. James, J.O. Evans, M.H. Ralphs, and R.D. Child (eds.), Noxious range weeds. Westview Press, Boulder, CO.

Lauenroth, W.K., O.E. Sala, D. Coffin, and T.B. Kirchner. 1994. The importance of soil water in the recruitment of *Bouteloua gracilis* in the shortgrass steppe. Ecol. Appl. 4: 741-49.

Laycock, W.A. 1991. Stable states and thresholds of range condition on North American rangelands: a viewpoint. J. Range Manage. 44:427-33.

Leitch, J.A., F.L. Leistritz, and D.A. Bangsund. 1996. Economic effect of leafy spurge in the upper Great Plains: Methods, models, and results. Impact Assessment 14:419-33.

Lesica, P., and J.S. Shelly. 1996. Competitive effects of *Centaurea maculosa* on the population dynamics of *Arabis fecunda*. Bull. Torrey Bot. Club. 123:111-21.

Lindquist, J.L., B.D. Maxwell, and T. Weaver. 1996. Potential for controlling the spread of *Centaurea maculosa* with grass competition. Great Basin Natur. 56:267-71.

Locken, L.J., and R.G. Kelsey. 1987. Cnicin concentrations in *Centaurea maculosa*, spotted knapweed. Biochem. Syst. Ecol. 15:313-20.

Lym, R.G., and D.R. Kirby. 1987. Cattle foraging behavior in leafy spurge (*Euphorbia esula*) -infested rangeland. Weed Tech. 1:314-18.

Lym, R.G., and C.G. Messersmith. 1990. Cost-effective long-term leafy spurge (*Euphorbia esula*) control with herbicides. Weed Tech. 4:635-41.

Mahlberg, P.G., D.G. Davis, D.S. Galitz, and G.D. Manners. 1987. Lacticifers and the classification of *Euphorbia*: The chemotaxonomy of *Euphorbia esula* L. Bot. J. Linn. Soc. 94:165-80.

Miller, R.F., T.J. Svejcar, and N.E. West. 1994. Implications of livestock grazing in the Intermountain sagebrush region: plant composition. *In*: M. Vavra, W.A. Laycock, R.D. Pieper (eds.), Ecological implications of livestock herbivory in the west. Soc. Range Manage. Denver, CO.

Monteith, J.L., and M.H. Unsworth. 1990. Principles of environmental biophysics. 2nd ed. Edward Arnold, London, England.

Muir, A.D., and W. Majak. 1983. Allelopathic potential of diffuse knapweed (*Centaurea diffusa*) extracts. Can. J. Plant Sci. 63:989-96.

Nobel, P.S. 1994. Root-soil response to water pulses in dry environments. *In:* M.M. Caldwell, R.W. Pearcy (eds.), Exploitation of environmental heterogeneity by plants. Academic Press, New York, NY, 285-304.

Olson, B.E., and R.G. Kelsey. 1997. Effect of *Centaurea maculosa* on sheep rumen microbial activity and mass. J. Chem. Ecol. 23:1131-44.

Provenza, F.D. 1995. Postingestive feedback as an elementary determinant of food selection and intake in ruminants. J. Range Manage. 48:2-17.

Rice, E.L. 1984. Allelopathy. Academic Press. New York, NY.

Rikard, W.H., and J.F. Cline. 1980. Cheatgrass communities: effects of plowing on species composition and productivity. Northwest Sci. 54:216-21.

Sala, O.E., W.K. Lauenroth, and W.J. Parton. 1982. Plant recovery following prolonged drought in a shortgrass steppe. Agric. Meteorol. 27:49-58.

Schlatterer, E.F., and E.W. Tisdale. 1969. Effects of litter of *Artemisia*, *Chrysothamnus* and *Tortula* on germination and growth of three perennial grasses. Ecol. 50:869-73.

Thompson, M.J. 1996. Winter foraging response of elk to spotted knapweed removal. Northwest Sci. 70:10-19.

Thompson, F., F.L. Leistritz, and J.A. Leitch. 1990. Economic impact of leafy spurge in North Dakota. Agric. Econ. Rep. No. 257, Agric. Exp. Stat., North Dakota State Univ., Fargo, ND.

Tilman, D. 1994. Competition and biodiversity in spatially structured habitats. Ecol. 75:2-16.

Tilman, D., and D. Wedin. 1991. Plant traits and resource reduction for five grasses growing on a nitrogen gradient. Ecol. 72:685-700.

Tisdall, J.M., and J.M. Oades. 1982. Organic matter and water-stable aggregates in soils. J. Soil Sci. 33:141-63.

Toole, E.H., and E. Brown. 1946. Final results of the Duvel buried seed experiment. J. Agri. Res. 72:201-10.

Trammel, M.A., and J.L. Butler. 1995. Effects of exotic plants on native ungulate use of habitat. J. Wildl. Manage. 59:808-16.

Tyser, R.W., and C.H. Key. 1988. Spotted knapweed in natural area fescue grassland: an ecological assessment. Northwest Sci. 62:151-60.

Vaughan, D., G.P. Sparling, and B.G. Ord. 1983. Amelioration of the phytotoxicity of phenolic acids by some soil microbes. Soil Biol. Biochem. 15:613-14.

Vinton, M.A., and I.C. Burke. 1995. Interactions between individual plant species and soil nutrient status in shortgrass steppe. Ecology 76:1116-33.

Vokou, D., N.S. Margaris, and J.M. Lynch. 1984. Effects of volatile oils from aromatic shrubs on soil microorganisms. Soil Biol. Biochem. 16:509-13.

Watson, A.K., and A.J. Renney. 1974. The biology of Canadian weeds. 6. *Centaurea diffusa* and *C. maculosa*. Can. J. Plant Sci. 54:687-701.

Wedin, D.A., and D. Tilman. 1990. Species effects on nitrogen cycling: a test with perennial grasses. Oecologia 84: 433-41.

Weiser, C. 1995. Economic effects of invasive weeds on land values from an agricultural bankers perspective. Proc. Symp. Alien plant invasions: Increasing deterioration of rangeland ecosystem health. BLM/OR/WA/PT-95/048.

Whisenant, S. 1990. Changing fire frequencies on Idaho's Snake River plains: ecological and management implications. Proc. Symp. Cheatgrass invasion, shrub dieoff and other aspects of shrub biology and management, 4-10. USFS Gen. Tech. Rep. INT-276.

Williams, K.E., J.R. Lacey, and B.E. Olson. 1996. Economic feasibility of grazing sheep on leafy spurge-infested rangelands in Montana. J. Range Manage. 49:372-74.

Woods, L.E. 1989. Active organic matter distribution in the surface 15 cm of undisturbed and cultivated soil. Biol. Fert. Soil 8:271-78.

Wraith, J.M., J.M. Baker, and T.K. Blake. 1995a. Water uptake resumption following soil drought: a comparison among four barley genotypes. J. Exp. Bot. 46:873-80.

Wraith, J.M., P.B. Hook, and D. Or. 1995b. Kalman filter improves analysis of temporal dynamics of soil water extraction by plants. Bull. Ecol. Soc. Amer. 76(3):405.

Surveying, Mapping, and Monitoring Noxious Weeds on Rangelands

Douglas E. Johnson

The primary objective of weed surveying is to accurately determine land areas with populations of unwanted plants, and to monitor effects of management. Surveys generally define the boundaries and surface area of infested lands. They may also catalog the density and pattern of weed populations. Weeds are surveyed so scientists and managers can:

1. Determine the location of weeds for eradication and future monitoring;
2. Plan, optimize, and facilitate management efforts;
3. Assess economic impacts of weed invasion and dominance on lands;
4. Determine which environments or sites are potentially subject to weed invasion;
5. Determine invasion dynamics and the means by which weeds spread;
6. Develop predictive models, so sites that may have been invaded can be more easily identified and search time for new invasions reduced;
7. Better understand the biology and ecology of weed species;
8. Increase public awareness, education, and weed management efforts.

Survey information is often compiled into maps showing weed distribution and the relative severity of the infestation across the landscape. Mapping simplifies communication among people working with weeds, especially those implementing control measures. It can also facilitate understanding of the invasion process, weed biology, and weed ecology.

Weed monitoring involves repetitive sampling or examination of historical documents, such as county weed records, herbaria records, or photographs. This provides a temporal perspective which leads to an appreciation of weed population dynamics. Sequential surveys can define expansion or contraction rates for both plant species and plant communities, permitting evaluation of control efforts and projections of future weed impacts. Non-indigenous plant populations are extremely dynamic, and studies documenting spread and dominance of ecosystems are some of the most interesting and important in ecological literature.

Setting the Objectives for Weed Survey

Weed sampling techniques vary considerably, depending upon the objective of the study. Studies generally fall into two types. The first are those designed to further scientific understanding of weed populations or weed biology. These studies attempt to obtain accurate estimates of population parameters, such as density of weeds in an area, weed distribution, and phytomass. Objectives of the second type of sampling are generally related to managerial decisions. In this case, we may simply want to know whether a noxious weed exists in an area; it is not necessary to know exactly how many plants are on the range. Managerial decisions, such as when to spot-spray a pasture, can be made using rapid ocular surveys that indicate when a weed problem is serious enough to have noticable economic impact, and when control actions are needed.

Sighting Reports

The easiest means of conducting weed surveys is to ask people to note and report any weeds observed during their normal daily activities. Commonly, county, state, or federal government employees, technicians, landowners, and interested private citizens are asked to report sightings of specific problem plants. Weeds encountered are reported on forms like the one shown in Figure 1.

There are several problems with this approach. First, the search for weeds is not systematic or uniform across the landscape. Coverage of lands is often spotty, with areas near roadways or work sites better documented than outlying lands. A second problem is that weeds may not be reported because surveyors do not actively search for weeds. The survey is sometimes perceived to be of relatively low priority and secondary to other jobs. The third problem is related to weed identification. Frequently weeds are not reported because they are not correctly identified. Surveyors may not receive sufficient training and refresher training to recognize all weeds of interest in all stages of development. Weeds that are new to an area may not be recognized.

Several examples illustrate these problems. Recently, federal workers brought an unknown plant to a district botanist for identification. It was Mediterranean sage (*Salvia aethiopis*), a noxious weed, requiring a sighting report. Fortunately, they were curious enough about its identity to bring the plant in for identification. In another case, landowners that had attended several informational meetings about distaff thistle (*Carthamus* sp.) invasion in Douglas County, Oregon, failed to detect the weed on their own property.

Sighting reports, therefore, are valuable for alerting managers to a weed problem and approximating weed distribution, but are not adequate for detailed or scientific work.

Noxious Weed Sighting Report Form

Noxious weed species _____ Date _____

Specific location: Section _____ Range _____

                       Township _____ Quarter _____

*** Important: Give as detailed location as possible, so the spot can be found again for followup (e.g., Otis Mountain, South Loop Road, etc.)

_____

_____

****

Growth form (circle one):   Basal rosette Upright growth

Growth stage (circle one):   Preflower   Flowering   Seeded

Number of plants _____ or size in acres _____

Relative frequency (circle one):   Spot   Scattered   Patch

Type of action taken (circle one): Pulled   Chopped   None

Land ownership: BLM   USFS   Private   Other   Unknown

General location: Roadside   Ditch   Rangeland   Riparian   Forest

                     Other _____

Note: Be sure to fill in at least the top half of the form.

Turn this form in to your supervisor or the Weed Coordinator!

Figure 1. Example of a weed sighting report form used by the Burns District of the Bureau of Land Management in Oregon. This information is passed to the Oregon Department of Agriculture and entered into its database of noxious weeds.

21

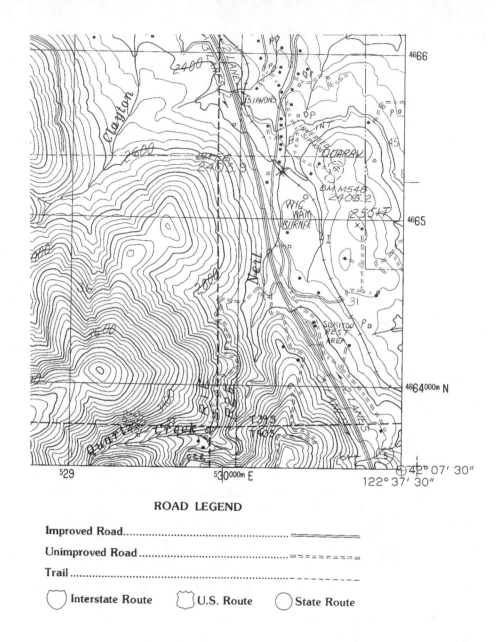

ROAD LEGEND

Improved Road.. ═══════

Unimproved Road .... ==========

Trail .... ─ ─ ─ ─ ─ ─

⬡ Interstate Route   ⬔ U.S. Route   ◯ State Route

Figure 2. A portion of a USGS 1:24,000 topographic quadrangle for the area near Ashland, Oregon. Random locations or cells can be determined from x and y coordinates or by using an overlain grid.

Systematic Weed Survey

Systematic surveys give much better information on the distribution and pattern of weeds across a landscape. If done correctly, they permit inferences to be drawn and predictions to be made regarding the distribution and impact of weeds. Systematic approaches to weed surveying require that areas being examined are divided, and specific portions of the area methodically examined for weeds. There are a wide range of strategies for weed sampling, ranging from random designs to systematic designs, to targeted sampling and geostatistically-based designs. If the area to be sampled is relatively small, random sampling locations or cells can be determined using an overlaid grid (Figure 2) and selecting random coordinates.

If the area to be sampled is large, it may be broken into smaller units on the basis of administrative boundaries, vegetation, or soil types, which are referred to as "strata." Sites to be surveyed are determined by randomly selecting a number of grid sections within each stratum. Transect lines within each randomly-chosen grid section can then be established to cross the landscape and sample it uniformly. Technicians walk transects and measure weed occurrences. Once areas with weeds are located, more intensive sampling can reveal the true extent of the infestation,

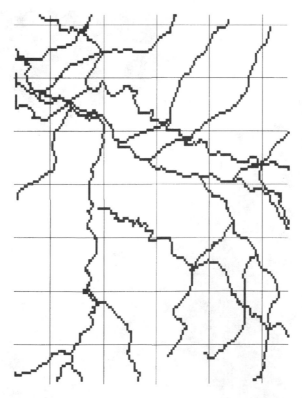

Figure 3. A search area for weeds associated with riparian communities based upon a 50-meter buffer zone around live streams. This zone was generated from USGS digital line graph (DLG) data on streams and overlaid on a digital elevation model (DEM). The total area of this map is 14,445 ha. The search area has been reduced to 963 ha.

relative abundance, and distribution. Surveys of this nature are much more time consuming, difficult, and expensive than are sighting surveys. However, weed patterns across the landscape will indicate factors that are important in controlling weed invasion, such as moisture regime, associated disturbance, soil type, etc.

Weeds that have specific known requirements, such as those that are closely associated with watercourses or roads, are more easily surveyed by following these linear landscape features. A systematic examination of selected portions of these linear features can be accomplished by subsampling. In the case of riparian plants, for example, selected sections of the riparian zones along watercourses would be sequentially examined. (In Figure 3, areas within 50 m of a permanent stream have been identified and outlined.)

Weeds that have specific requirements, such as those limited by soil type or elevation, can be located more quickly and efficiently by reducing the search area to those areas that meet specific criteria. For example, yellow starthistle (*Centaurea solstitialis*) dominates lower elevation foothills in southeastern Oregon and is expanding its range into higher zones. Delineation of its distribution is more efficient if surveyors concentrate on examining elevational zones where plants are at the

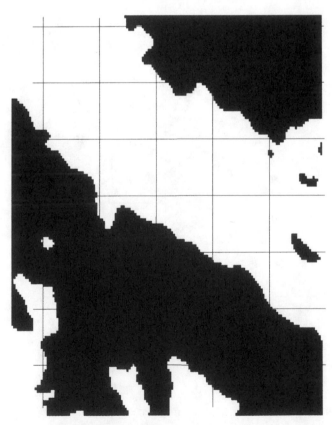

Figure 4. A search area for yellow starthistle in the 900-1,350 meter elevation range near Ashland, OR. This mask was generated from USGS digital elevation models (DEM). The total area of this map is 14,445 ha and the search area is 7,770 ha.

limits of their present range. An appropriate search zone would be between 900 and 1,350 meters (Figure 4). Records should be maintained that indicate the results of the survey, so that systematic examinations can continue. Periodic reexamination would allow calculation of expansion rate and appraisal of the economic or biological threat posed by the invading species.

Quantitative Measurements

One of the most useful methods for obtaining plant information involves the use of wire frames called quadrats or plots. Quadrats are generally between 0.1 m² and 1 m², and are laid upon the ground at random locations. Plants located within the quadrat are first identified and their presence recorded on data forms. The relative percentage of the quadrat that is covered by each species can be estimated, number of individuals counted, and above-ground phytomass clipped, dried, and weighed. Data collected from quadrats can be summarized by the following parameters:

• *Frequency*

Frequency is the relative number of quadrats or sampling units that contain the plant. The plant is recorded as either present or absent for each quadrat or sample unit. The tally of the number of samples that contained the plant is divided by the total number of samples, and multiplied by 100 to yield a percentage. Frequency is dependent upon the quadrat size, size of the plants being sampled, and the pattern of distribution of the species. Plants that have clumped distribution will have lower frequencies than species with dispersed distributional patterns. Frequency is therefore a non-absolute index, and the procedures used to obtain the values should be stated.

• *Density*

Density is the number of individuals per unit area. It is calculated by averaging the plant counts for all quadrats measured. Obtaining data for density calculation can be quite tedious, especially if you are measuring annual grasses. Small quadrats are most effective for small, very common weeds.

• *Above-Ground Phytomass*

Above-ground phytomass is the dry weight of plant material (exclusive of roots and below-ground plant parts) per unit area. It is usually measured by clipping sample quadrats or plots, placing the material in paper bags, and drying at 105˚ C until the weight stabilizes. Phytomass can then be averaged and its variability from plot to plot calculated. Phytomass is very dynamic, especially in spring, so we often try to obtain estimates when plants are near the peak value for the year.

• *Cover*

Cover is the vertical projection of above-ground plant parts onto the ground, or shadow with the sun at the zenith. It may be estimated as a percentage of the quadrat concealed, or as a cover class, such as 0% to 4%, 5% to 24%, 25% to 49%, etc. Numerous techniques have been developed to quantify cover, including point sampling, line intercept, and quadrat methods.

Importance Indices

Synthetic indices of importance can be generated from the above parameters by dividing the value of each species by the sum of values for all weed species, multiplying by 100, and summing as shown below:

$$\text{Relative Frequency} = \frac{\text{frequency of species j x 100}}{\text{sum of frequency values for all species}}$$

$$\text{Relative Density} = \frac{\text{density of species j x 100}}{\text{sum of density values for all species}}$$

$$\text{Relative cover} = \frac{\text{cover of species j x 100}}{\text{sum of cover values for all species}}$$

The three values are summed to obtain an importance index for each species:

$$\text{Importance Index} = \text{relative frequency} + \text{relative density} + \text{relative cover}$$

The importance index ranges from 0 to 300 because each of the individual values can range from 0 to 100. It is used to estimate which weedy plants have a greater impact on the systems in which they are found.

Core Infestations and New Populations

Core infestations of weeds or those areas in which weeds are prevalent are much easier to survey than the areas where weeds are represented by a few isolated and widely spaced individuals. Unfortunately, core infestations are often much more difficult and costly to control. Locating new invasion sites may, therefore, be far more important in the long run.

If something is known about the biology of the invading plant, including its site requirements for establishment, it may be possible to divide or stratify the landscape, which facilitates evaluation. If the area being examined is divided into subareas based upon land use, vegetation, geomorphology, microclimate, or ecological sites before the survey is undertaken, the areas most likely to have nascent populations can be examined in more detail.

In any case, the product of a weed survey should be a thorough, documented examination of the landscape, so that information can be easily shared. Locations of weeds, their relative abundance and other observations should be identified by map coordinates or map overlays on an accurate base map. Electronic navigational, remote sensing, and mapping technologies are revolutionizing the way spatial information is collected and processed.

Global Positioning Systems (GPS)

The Navigation Satellite Timing and Ranging (NAVSTAR) Global Positioning System (GPS) was developed by the U.S. Department of Defense to provide the military with 24-hour all-weather navigation, timing, and surveying information. This system consists of a network of 21 satellites in six orbital planes. Three additional satellites are positioned in orbits that permit their use as spares in case of malfunction of a primary satellite. Each satellite has a high-precision atomic clock and radio transmitters that broadcast an identifying code. Because the position of these satellites can be calculated very accurately at precise times, they can be used as reference points for calculating geographic positions around the globe.

The quality of the position being calculated by the GPS unit is influenced by buildings, trees, mountains, etc., since these objects can block satellite signals. Since signals are received on a line of sight from the satellite, ravines, steep-walled valleys and mountains are especially troublesome. In addition, the configuration of the constellation of satellites being tracked also affects the quality of position.

GPS software and hardware have developed rapidly. As of 1998, with sophisticated receivers equipped with data loggers, it is possible to obtain real-time locations accurate to a single meter and label them with weed information in the field. This electronic data can be exported to a Geographic Information System (GIS), generating an accurate map of weed infestations. GPS units equipped with data collectors can log positions at rates up to one per second, which facilitates continuous mapping by individuals either on foot, in all-terrain vehicles, or conventional vehicles.

It is probable that in the future weed specialists will obtain field data with integrated GPS/GIS data collectors, which will be downloaded and networked with centralized spatial databases at the state or national level. These could then output maps and summary statistics directly.

In addition to two-dimensional positioning of weeds on the landscape, GPS units also log elevational information. Elevational information is less accurate than horizontal positioning by roughly a factor of two. In spite of this limitation, elevational models can be extremely valuable for landscapes of particular interest.

Remote Sensing

Remote sensing of weeds can be as simple as having observers in aircraft ranking the relative invasion of areas along a flight line, or as complex as satellite-borne multispectral scanning instrumentation and their associated computer storage facilities. Remote sensing technology, like GPS technology, has rapidly evolved. Computer software packages that permit the manipulation of radiographic, photographic, and videographic images have provided unprecedented power to remote sensing specialists. Data can now be acquired, processed, and output in several formats at a variety of scales.

• Satellite Remote Sensing

Coarse scale (ground resolution of 1 to 10 km) has found only limited usefulness for weed surveying, but it has been used for predicting sites that may be subject to weed invasion (Oregon Department of Agriculture). Intermediate scale remotely sensed data (ground resolution of 15 to 50 m) are more useful for rangeland, environmental, and forestry monitoring. They are also very valuable for land and water resource analysis and land use planning. Detection of insect outbreaks, land degradation, vegetation mapping, and land use / cover classifications are also common applications of this data. The most conventional data sources at this scale are Landsat MSS, Landsat TM, and SPOT data. The Systeme Probatoire de l'Observation de la Terre or SPOT program constructed by the French Centre National d'Etudes Spatiales (CNES) provides information at a slightly finer scale than Landsat.

Unfortunately, Landsat and SPOT data are expensive, which has limited their use for weed surveys. Information at intermediate resolution requires relatively sophisticated hardware, since data sets can consume 50 megabytes or more of hard disk storage. Processor speed is also critical if tasks are to be done in a timely fashion. Equipment usually consists of a computer workstation with tape or CD backup and reading capabilities. Landsat data have been processed to identify specific weed spectral signatures with generally poor results. Landsat works best with plants that grow in close association and can therefore provide a uniform spectral response across a large area (several acres). Plants such as juniper that are self spacing or those that grow as an understory are often undetected.

Landsat and SPOT data are more useful for typing environments and vegetation and for predictive models. Both a "tasseled cap" transformation and a Transformed Vegetation Index (TVI) can be used for vegetation typing. The "tasseled cap" transformation utilizes six bands of reflected data, and converts them into three dimensions defining planes of soils, vegetation and a transition zone between them. The TVI index utilizes TM Bands 3 and 4, and produces an image in which cell values are proportional to the amount of green biomass present in each pixel. Calibration relationships are established for each vegetation type on the ground by referencing to standing green vegetation.

• Fine Scale Monitoring Using Remote Sensing

Fine scale monitoring of weeds can be executed by aerial photography, videography, or radiometry coupled with ground measurements. Areas of interest can be photographed in scales from 1:12,000 for general studies to 1:1,000 or 1:1,200 for detailed work. The small scale images are generally obtained from fixed-wing aircraft, while large scale images are obtained using either helicopter or blimp-borne photographic platforms. Both color and infrared photographic films are commonly used (Figure 5). Individual trees, shrubs, and infestations can be monitored over time by repetitive overflights spaced at yearly or seasonal intervals. Plant cover, usually

Figure 5. The rectified aerial photograph above of a site in Coos County, Oregon, was classified to yield the gorse map in the lower image. White areas represent gorse (Ulex europaeus). Bright yellow blossoms on gorse facilitated the classification.

expressed as a percentage, extent of bare soil, and standing crop estimates can also be produced. Detailed ground surveying and measurement of the sites increases the scientific value of photographs. Since data generated increase exponentially as the scale increases, only small areas are normally sampled at high resolution. These techniques are therefore only employed in critical areas.

Processing of air-photographs may involve either direct manual photointerpretation techniques in which a trained technician identifies areas of weed infestation by color, pattern, texture, site, size, shape, etc. Output consists of georeferenced maps of observed weed distribution and calculated surface area of infestation. This is obviously a tedious, time-consuming task if large areas are to be surveyed. Computer software has been developed to automate and speed the interpretation process. Photographs are converted directly to digital format or are scanned to produce computer images. Flatbed scanners are commonly used for this type of information. Slide scanners have also been developed to convert images directly from color slides or color negatives. If simultaneous photographs on both color and infrared films are obtained, up to four bands of digital data can be extracted, which aids interpretation of features. Once in electronic format, processing is similar to satellite-generated data.

Mapping Weed Distributions Through Time

Vegetational dynamics can be ascertained from image data by using satellite images or photographs taken over the course of time. Subsequent georeferenced images can be compared by setting a threshold and differencing or by regressing one image upon the other. In both techniques, points of change between the images are identified. Changes in the vegetative blanket, anthropogenic disturbance, fire and flooding effects, and vehicular trails can be identified using these techniques. Images can be used in conjunction with digital elevation, slope, and aspect models, which further enhances their usefulness for monitoring weed spread.

Weed Mapping

Maps are especially effective for recording and communicating information about the location, extent and severity of weed infestations. They can accurately portray complex spatial information in a graphic, easily understandable form. Maps indicate the distances and direction between occurrences of weeds so that valuable insights regarding mechanisms and patterns of dispersion can be determined. Arrangements of weed patterns can be examined visually and intuitively or quantitatively using geostatistics.

Weed maps at the farm or ranch scale can be made quickly by crossing the property on an all-terrain vehicle equipped with a continuously recording GPS unit attached to a data logger. Weed locations are tagged with the name of the plant and other pertinent information. Large infestations can be circled, which creates a map polygon showing position and area affected. Although time-consuming to create, these maps can substantially improve management efforts and suggest preventative managerial actions.

Weed occurrences can also be tagged with existing environmental, geological, climatic, and ownership information through the process of creating weed distribution overlays on existing maps. Base maps generally are either USGS topographic or county maps providing scale and links to information about other features such as roads, rivers, soils, property ownership, etc. This process facilitates understanding of the weed invasion process. For example, we may determine by looking at distribution overlays that south-facing slopes are more likely to be dominated by a particular weed than are the cooler north-facing slopes with denser perennial vegetation. Weed management strategies could be adjusted and priority given to lands with a southern aspect.

Geographic Information Systems (GIS)

Geographic Information Systems (GIS) consist of computer hardware and software programs that together can process spatially based data. GIS programs input, manipulate, retrieve, analyze, and display information. By coupling ground observations, remote sensing, GIS, GPS, and mathematical modeling technologies, scientists and managers have extremely powerful and sophisticated analytical tools at their command. These tools help identify land use problems such as weed spread, as well as indicate possible solutions. Data needs and specific objectives determine the kind of information that is remotely collected, resolution or scale of the data, and processing techniques utilized. They also largely determine the type of GIS analysis employed.

Information extracted from ground surveys, sighting reports, and remote sensing sources is generally mapped and combined with ancillary information to increase its usefulness. For example, we may need to relate remotely derived vegetational cover with factors such as land slope, aspect, and soil type in a model that predicts the location of weeds and their economic impact.

Traditionally, we would have combined these data by creating a series of map overlays made on transparent map sheets and, for example, areas with high probability for invasion would have been identified by visual inspection. With the development of high-speed computers came the capability to create a spatially registered database, in which a matrix of cells contain information about a specific area on the map. Systems designed to store, manipulate, and display this information are called Geographic Information Systems (GIS). Numerous GIS software packages exist. These systems operate on either workstations or personal computers. The power of a Geographic Information System is derived from its ability to model and manipulate georeferenced data.

In the previous example, areas at risk could be distinguished by either: 1) identifying map areas that have a specific combination of feature characteristics; or

31

2) mathematically combining factors using regression or other techniques. Obviously, the more sophisticated mathematical approach has many benefits.

To support various modeling functions, a GIS must be able to handle both locational data and attribute or descriptive data about features. Several types of supplementary information are encoded by digitization, so that the full power of remotely sensed data can be realized. The following area, linear, and point data are becoming available as GIS is accepted by more and more agencies and administrations:

Area Features
1. Terrain Elevation (USGS Digital Elevation Models DEM)
    a. 1:250,000 scale
    b. 1:24,000 scale
2. Soil Type (USDA Natural Resources Conservation Service) (1:20,000)
3. Land Cover / Land Use (USGS, U.S. Census Bureau Tiger Files)
4. Surface Hydrology (1:100,000)
5. Subsurface Hydrology
6. Surface Geology
7. Land Ownership or Administrative Regions (USGS, County Governments) (1:100,000)
    a. Federal Ownership
    b. State Ownership
    c. Private Ownership
    d. County Boundaries
    e. Soil & Water Conservation Districts

Linear Features
1. Surface Water (USGS 1:100,000 Digital Line Graphs)
    a. Rivers
    b. Streams
    c. Intermittent streams
2. Transportation Corridors (USGS 1:100,000 Digital Line Graphs)
    a. Primary Roads
    b. Secondary Roads
    c. Railroads

Point Features
1. Wells
2. Farmsteads

Each of these themes or map layers can be combined with others in various ways to yield new information about the process of weed invasion and its affect on the land.

Conclusions

Assessment of ecologic and economic impacts of weed invasions can be improved through the use of modern techniques of weed survey, mapping, and monitoring. If proper survey techniques are used, then valuable information to devise management alternatives is obtainable. Sampling methods must be compatible with management or study objectives to ensure efficient use of resources. Survey data can be combined within a Geographic Information System with ancillary data to increase their utility. Remotely sensed information is valuable for mapping distribution and pattern of weed invasions. Techniques in these fields are rapidly evolving, and future weed specialists will have even more sophisticated tools at their disposal.

Selected References

Anderson, G.L. 1996. The application of spatial technologies for rangeland research and management: State of Art. Geocarto Internat. 11:5-11.

Anderson, G.L., J.H. Everitt, A.J. Richardson, and D.E. Escobar. 1993. Using satellite data to map false broomweed (Ericameria austrotexana) infestations on south Texas rangelands. Weed Tech. 7:865-71.

Benoit, D.L., D.A. Derksen, and B. Panneton. 1992. Innovative approaches to seedbank studies. Weed Sci. 40:660-69.

Boutin, C., and J.L. Harper. 1991. A comparative study of the population dynamics of five species of *Veronica* in natural habitats. J. Ecol. 79:199-221.

Bryson, C.T. 1986. Aerial survey of bermudagrass in cotton and soybean fields in the Mississippi delta. Proc. South. Weed Sci. Soc. 39:486-94.

Buchanan, G.A., E.W. Hauser, A.S. Causey, and J. Adams. 1985. Weed survey of permanent pastures in three southern states. 28p. Ala. Agric. Exp. Stn. Bull. No. 572.

Bultsma, P.M., and R.G. Lym. 1985. Survey for spotted knapweed in North Dakota. North Dakota Farm. Res. 43:19-22, N.D. Agric. Exp. Stn. Fargo, N.D.

Callihan, R.H., and L.W. Lass. 1993. Simplified weed-mapping computer software for individual counties. Res Prog. Rep. West Soc. Weed Sci. [S.l.] IV/3-IV/6.

Chancellor, R.J., and R.J. Froud-Williams. 1984. A second survey of cereal weeds in central southern England. Weed Res. 24:29-36.

Cuthbertson, E.G. 1978. Advances in weed distribution mapping. Proc. Conf. Counc. Aust. Weed Sci. Soc. 1:273-87, Melbourne, Australia.

Dewey, S.A., K.P. Price, and D. Ramsey. 1991. Satellite remote sensing to predict potential distribution of dyer's woad (*Isatis tinctoria*). Weed Technol. 5:479-84.

Dunn, P.H. 1979. The distribution of leafy spurge (*Euphorbia esula*) and other weedy *Euphorbia* spp. in the United States: Density mapping to start a biological control program. Weed Sci. 27:509-16.

Evans, A.W., and C.W. Yates. 1985. Survey of the distribution and occurrence of weeds in herbage seed crops in England and Wales in 1983 and a comparison with 1973 and 1978 seasons already reported. J. Natl. Inst. Agric. Bot. 17:75-93.

Everitt, J.H. 1985. Using aerial photography for detecting blackbrush (*Acacia rigidula*) on south Texas rangelands. J. Range Manage. 38:228-31.

Everitt, J.H., and D.E. Escobar. 1996. Use of spatial information technologies for noxious plant detection and distribution on rangelands. Geocarto International 11:63-80.

33

Everitt, J.H., and C.J. Deloach. 1990. Remote sensing of Chinese tamarisk (*Tamarix chinensis*) and associated vegetation. Weed Sci. 38:273-78.

Everitt, J.H., D.E. Escobar, M.A. Alaniz, R. Villarreal, and M.R. Davis. 1992. Distinguishing brush and weeds on rangelands using video remote sensing. Weed Technol. 6:913-21.

Everitt, J.H., D.E. Escobar, and F.W. Judd. 1991. Evaluation of video imagery for distinguishing black mangrove (*Avicennia germinans*) on the lower Texas gulf coast. J. Coastal Res. 7:1169-73.

Everitt, J.H., D.E. Escobar, and P.R. Nixon. 1987. Near-real-time video systems for rangeland assessment. Remote Sensing of Environment. 23:291-311.

Everitt, J.H., D.E. Escobar, R. Villarreal, M.A. Alaniz, and M.R. Davis. 1993. Integration of airborne video, global positioning system and geographic information system technologies for detecting and mapping two woody legumes on rangelands. Weed Technol. 7:981-87.

Everitt, J.H., R.D. Pettit, and M.A. Alaniz. 1987. Remote sensing of broom snakeweed (*Gutierrezia sarothrae*) and spiny aster (*Aster spinosus*). Weed Sci. 35:295-302.

Everitt, J.H. and R. Villarreal. 1987. Detecting huisache (*Acacia farnesiana*) and Mexican palo-verde (*Parkinsonia aculeata*) by aerial photography. Weed Sci. 35:427-32.

Froud-Williams, R.J., and R.J. Chancellor. 1982. A survey of grass weeds in cereals in central southern England. Weed Res. 22:163-71.

Harris, N.R., D.E. Johnson, T.L. Righetti, and M.R. Barrington. 1996. A blimp borne camera system for monitoring rangelands, riparian zones, or critical areas. Geocarto International 11:99-104.

Hickman, M.V., J.H. Everitt, D.E. Escobar, and A.J. Richardson. 1991. Aerial photography and videography for detecting and mapping dicamba injury patterns. Weed Technol. 5:700-706.

Lass, L.W., and R.H. Callihan. 1993. GPS and GIS for weed surveys and management. Weed Technol. 7:249-54.

Loux, M.M., and M.A. Berry. 1991. Use of a grower survey for estimating weed problems. Weed Technol. 5:460-66.

Marshall, E.J.P. 1988. Field-scale estimates of grass weed populations in arable land. Weed Res. 28(3):191-98.

Martyn, R.D. 1985. Color infrared photography for determining the efficacy of grass carp in aquatic weed control. Proc. South. Weed Sci. Soc. 38:381-90.

McCully, K.V., M.G. Sampson, and A.K. Watson. 1991. Weed survey of Nova Scotia lowbush blueberry (*Vaccinium angustifolium*) fields. Weed Sci. 39:180-85.

Monks, D.W., M.A. Halcomb, and E.L. Ashburn. 1991. Survey and control of musk thistle (*Carduus nutans*) in Tennessee field nurseries. Weed Technol. 5:218-20.

Northam, F.E., R.R. Old, and R.H. Callihan. 1993. Little lovegrass (*Eragrostis minor*) distribution in Idaho and Washington. Weed Technol. 7:771-75.

Pammel, L.H., and C.M. King. 1930. A weed survey of Iowa. J. Am. Soc. Agron. 22:587-94.

Peters, A.J., B.C. Reed, M.D. Eve, and K.C. McDaniel. 1992. Remote sensing of broom snakeweed (*Gutierrezia sarothrae*) with NOAA-10 spectral Image Processing. Weed Technol. 6:1015-20.

Prather, T.S., and R.H. Callihan. 1993. Weed eradication using geographic information systems. Weed Technol. 7:265-69.

Richardson, A.J., R.M. Menges, and P.R. Nixon. 1985. Distinguishing weeds from crop plants using video remote sensing. Photogramm. Eng. Remote Sensing. 51:1785-90.

Senseman, G.M., C.F. Bagley, and S.A. Tweddale. 1996. Correlation of rangeland cover measurements to satellite-image-derived vegetation indices. Geocarto International 11:29-38.

Steven, M.D. 1993. Satellite remote sensing for agricultural management: opportunities and logistic constraints. J. Photogrammetry and Remote Sensing 48:29-34.

Tueller, P.T. 1996. Near-Earth monitoring of range condition and trend. Geocarto International 11:53-62.

Tucller, P.T. 1989. Remote sensing technology for rangeland management applications. J. Range Manage. 42:442-53.

Tueller, P.T., P.C. Lent, R.D. Stager, E.A. Jacobsen, and K.A. Patou. 1988. Rangeland vegetation changes measured from helicopter-borne 35mm aerial photography. PE&RS. 54:609-14.

Thomas, A.G. 1985. Weed survey system used in Saskatchewan for cereal and oilseed crops. Weed Sci. 33:34-43.

Thomas, A.G., and J.A. Ivany. 1990. The weed flora of Prince Edward Island cereal fields. Weed Sci. 38:119-24.

Wilson, J.P., W.P. Inskeep, P.R. Rubright, D. Cooksey, J.S. Jacobsen, and R.D. Snyder. 1993. Coupling geographic information systems and models for weed control and groundwater protection. Weed Technol. 7:255-64.

Coordinated Weed Management Planning

Tony Svejcar

The length and breadth of this book should make clear the fact that weeds have become a major problem on western rangelands. Weed invasions can reduce forage production, biodiversity, and water quality of associated stream systems. There is a great deal yet to learn about the biology and control of rangeland weeds. However, many of the problems encountered in developing and implementing weed management plans are not technical, but rather social in nature. Communication, motivation, conflict, and consensus are more likely to influence the success of a plan than herbicide selection, nozzle size, and drift. The technical questions are certainly important, but are generally quite easy to address. The focus of this chapter will be more on the social—rather than technical—aspects of developing a coordinated weed management plan.

There was a time when weed management was not a particularly controversial subject, but that time has passed. The public is concerned about food safety, water quality, soil erosion, biodiversity, and a host of other environmental issues. In the western United States, there has been an ever-increasing urban/rural interface, and an expansion in the areas occupied by semi-rural "ranchettes." In addition, about half of the land area in the western states is public land, much of which is administered by the Bureau of Land Management or the U.S. Forest Service. These agencies must comply with the National Environmental Policy Act (NEPA), and are required to assess the impacts of their activities on the environment. Any large-scale activities require the development of an Environmental Impact Statement (EIS) so the general public, environmental groups, and agricultural interests have the opportunity to comment on the proposed plan. If one or more of the individuals or organizations disagree with the plan, they can use a whole series of appeals and court maneuvers to slow or stop it.

Conflicts

The degree of conflict over most natural resource issues seems to have increased over the past 10 to 15 years. Examples and literature from other areas of natural resource management are worth considering during the development of coordinated weed management plans. An understanding of the underlying reasons for the conflicts is useful. Clearly, there is more interest in environmental issues now than there was in the past. In addition, members of the general public are demanding more of a say in policy development. Wagner (1994) has argued that policy-setting by resource professionals is being replaced by "constituency-based, multi-resource management." The role of the resource professional has changed, but in many cases neither the professionals nor their institutions are sure how to deal with the change. We have seen an emphasis on bringing together diverse groups of people to work on solutions to resource controversies (Svejcar 1996). Whaley (1993) lists three major trends that help explain the need for group participation: 1) conflicting social goals, specifically economic activity vs. maintaining environmental quality; 2) increasing democratization of decisions; and 3) excessive information. These trends have clearly had an impact on weed management activities as society has raised questions about herbicide impacts on the environment, has influenced policy toward herbicide use, and has had difficulty accessing and synthesizing the available information.

An analysis of past resource conflicts may provide some clues for avoiding future problems. Several authors have evaluated the public participation process used by the U.S. Forest Service in its national forest planning efforts. Both Wondolleck (1985) and Blahna and Yonts-Shepard (1989) cite the importance of the process in dealing with potential controversies. There tends to be frustration when the public feels left out of decision-making. The time and form of communication is critical. Blahna and Yonts-Shepherd (1989) stress the importance of personal forms of public involvement, as opposed to simply receiving written comments. They also suggest that the public should be made aware of how the input will be used in the planning process. In their analysis of the forest planning process, Blahna and Yonts-Shepherd (1989) concluded that the Forest Service did not solicit public input effectively during the early stages of the planning process. Rather, the Forest Service hoped to reach consensus without making substantial changes to their draft plans. The authors suggest that by trying to avoid controversy and using non-confrontational methods of public involvement, the Forest Service actually increased the level of controversy. The lack of public involvement in developing planning alternatives may have forced interested parties to take extreme stands, even though small changes would have been acceptable.

There are no guarantees that using the appropriate process will eliminate controversy, but the level of conflict may be reduced. Planning should include an emphasis on how decisions will be made. Wondolleck (1985) stresses that the first question a decision-maker should ask is not what the outcome should be, but rather how the outcome will be determined. Wondolleck (1985) goes on to say: "A process focus immediately raises questions such as: What information do we need? Who should be involved? Where can we get the information needed? What are the likely problems we will encounter, and how might we overcome them? A process focus encourages consensus-building and collaborative problem-solving among affected interests when the decisions to be made are complex and value-laden, and when there are limits to technical expertise in reaching solutions."

An emphasis on process probably seems very formal to many of us who will participate in coordinated weed management planning. The need for a more formal approach will depend on the complexity of the issues and the diversity of participants. What will be done if a major weed infestation occurs on public land and the agency involved is under court order not to spray? Or if a significant landowner in the area does not consider weed management cost-effective? Or if five-acre ranchettes are a major reservoir of weeds? These are but a few examples where conflict is likely. In some cases, a good education and outreach program may reduce the conflict. But what if it does not? Some county weed control ordinances allow for fines and penalties if a landowner does not make a good-faith effort to control noxious weeds.

Developing a Coordinated Weed Management Plan

How are weed managers of the future to deal with the social environment? Again, the three trends mentioned earlier dictate planning procedures: 1) Social conflicts have increased, especially with regard to economic activity vs. environmental protection; 2) Individuals and groups wish to be more involved in decision-making and policy formulation; and 3) The Information Age has arrived, and we often are inundated with information. Including a broad array of interests and viewpoints in the planning process should help address the first two points. The third point will require some synthesis of the available information into a format that can be used for planning and decision-making.

Developing a list of potential participants is probably a good starting point in the planning process. The list will vary from area to area, but the following will provide a starting point:

1. Landowners
Include those who want to manage weeds, and those who may be affected by weed management activities. Small parcels of land can be significant reservoirs of weed species.

2. Federal Agencies
   a) U.S. Department of Agriculture
   • Natural Resources Conservation Service. Provides technical assistance and site inventories.
   • Farm Services Agency. May provide cost-share support for some weed management activities.

• Agricultural Research Service. Conducts research, and may provide technical assistance and scientific input.

• Forest Service. Depending on the area, may be a significant landowner, and lands may contain weeds.

• Economic Research Service. May be able to provide information on the economic impacts of weeds.

• Animal and Plant Health Inspection Service. Is a major player in the importation of biological control agents.

b) U.S. Department of Interior

• Bureau of Land Management. May be a significant landowner, and lands may contain weeds.

• Fish and Wildlife Service. Has authority on issues pertaining to threatened and/or endangered animal species. In the case of Federal wildlife refuges, may also be a landowner with weeds.

• Park Service. Is a landowner in some areas, and lands may contain weeds.

c) Other Federal Agencies

• Environmental Protection Agency. Has the authority for enforcement of pesticide laws and the Clean Water Act. Enforcement authority is generally given to state Departments of Agriculture or Departments of Environmental Quality.

• Occupational Safety and Health Administration. Sets the standards for worker protection procedures associated with pesticides.

3. States

a) Department of Agriculture. May provide technical assistance and often employs individuals familiar with weed management planning. Generally has the responsibility for administering pesticide applicator licenses. In many states, the Departments of Agriculture designate the state weed list, have active weed management programs, and may have grant programs available. Many state Departments of Agriculture run weed-free forage programs.

b) Department of Environmental Quality. May be responsible for regulating pesticide handling, safety, and disposal. May also enforce water quality standards and toxic material regulations.

c) Department of Fish and Wildlife. May pose concerns over pesticide effects on wildlife, or effects of weed invasions on wildlife habitats. In some cases, will cost-share on wildlife habitat improvement projects. Issues wildlife licenses and tags, and thus may participate in programs aimed at reducing spread of weeds by hunters and other recreationists.

d) Department of Water Resources. May have some jurisdiction over water quality and water distribution. In some cases, waterways and irrigation canals are a significant source of weeds.

e) Department of Natural Resources. May have jurisdiction over state lands and natural resources.

f) State Universities and Extension Service. Provide technical assistance, conduct research, and in many cases can provide assistance with meeting facilitation.

g) Department of Transportation. Has jurisdiction over roadways, which are often a critical source of weed problems.

4. County

a) Soil and Water Conservation Districts. Are involved in conservation projects, some of which involve weed control.

b) County Weed Board. Can serve as a coordinating body for county-level weed management programs. Sometimes owns spray equipment that is available for rental. May employ a weed supervisor and weed control specialists. Designates noxious weeds for the county.

c) County Road Departments. Responsible for activity on county roads.

5. Town/City

a) City Government. May influence weed management activities within the city limits (parks, vacant lots, schools, etc.).

6. Private

a) Chemical, biological control, and equipment suppliers; custom applicators.

b) Consultants. Depending on the situation, many help formulate costs, provide input on availability of products, and may have knowledge of past successes and failures with weed control efforts in the immediate area.

c) Environmental Groups. Generally represent concerns related to environmental impacts of herbicides, but also favor preserving native populations and maintaining biodiversity. Often provide input on public land management. Are increasingly concerned about the impacts of invasive weeds.

There are many ways to reach the broad array of potential participants. Personal contacts are important, and may be essential for those individuals who are not comfortable with group settings. Mailing lists of landowners and/or agricultural producers may be available from the County Extension Service. The media are also a powerful tool for soliciting input. Many newspapers, radio stations, and cable TV systems provide free or low-cost public service announcements. Let people know that a coordinated weed management plan is being developed and that input is needed. The forest planning example used earlier indicates the importance of receiving and addressing concerns early in the planning process.

There are two primary reasons why all points of view should be included in the planning process: 1) The general public perceives open participation as necessary for fair decision-making; and 2) unrepresented interests can block or undermine decisions (Dale and Hahn 1994). Again, in the example of the Forest Service, not soliciting and addressing concerns during the early phases of planning forced critics into extreme positions.

Table 1. Some of the critical steps and important components of a coordinated weed management plan. In this instance the plan is adaptive over time. Funding is an important consideration under each heading

Meeting Details	Education	Management Plan	Information Collection and Management
• agenda	• workshops	• identify critical weeds and affected habitats	• weed status
• identify and invite participants	• visitor centers	• outline management measures	• management success with various methods
• facilitation	• county fairs	• mechanisms of dispersal	• acres treated
• notes and minutes	• roadside displays	• problem areas	• new information from other areas
• meeting announcements	• mailing lists	• proposed actions	• effects on native or seeded species
• facilities	• media	• goals	• resulting plant communities
		• new information from other areas	

The mechanics of the initial meetings should be worked out in advance. Will there be a formal facilitator? Will notes be taken? What is the agenda for the meeting? What audiovisual aids are necessary? Individuals with meeting facilitation training and experience can be of tremendous assistance in setting up and conducting meetings. Once the initial meeting is held, how will the planning be handled? Will a steering committee be selected? Will there be a chairperson, secretary, etc.? A committee may already be in place before any public meetings take place. What if a group feels unrepresented and would like to have a member on the steering committee? The term "steering" or "coordinating" committee implies that the group will have to do more than simply develop a plan and then disband. Once a plan has been written, there is a need for implementation, data-collection, and plan adjustment as conditions change. Subgroups may be best suited to handling individual tasks. For example, if data management and analysis are part of the weed management plan, a subgroup or committee may be assigned those tasks.

The chances for long-term success of a coordinated weed management plan improve if the plan includes an education component, and if the plan is adaptive in nature. To be adaptive, the plan must include some type of data collection to evaluate successes and failures. A simple scheme of the steps necessary to develop an adaptive weed management plan is presented in Table 1. The information collected locally can then be used to adjust the emphasis of the plan. Some biological control agents are effective only in specific areas, and thus local testing may be necessary to determine the appropriateness of an agent. Combinations of treatments (for example, grazing and herbicides) will be necessary in some cases.

Nearly everyone agrees that educational workshops and monitoring are good ideas. The primary stumbling block usually centers on time and/or money. Who will conduct the workshops and collect weed response data? If weeds are a significant problem to an area, then many of the agencies listed earlier in this chapter should have an interest in assisting with weed management planning. Generally, individuals in the land management, technical assistance, or research organizations should have expertise in vegetation monitoring. These same organizations, along with farm chemical and equipment representatives, should be able to assist with educational materials. There may be advantages to conducting workshops in different areas of the management area to improve turnout and focus on specific local problems.

Conclusions

This chapter is focused primarily on the social and organizational aspects of developing a Coordinated Weed Management Plan. The basic approach should draw heavily on the concepts of Integrated Weed Management , which is covered in another chapter of this book. Here are suggestions for developing a Coordinated Weed Management Plan:

1. Identify the problem: weed species, resource and economic impacts, other concerns.

2. Define the boundaries of the management area. This step will provide ownership patterns, and aid in identifying participants.

3. Identify potential participants, both public and private. Identify who is responsible for weed control along roads, what recreational activities should be considered, and so on.

4. Determine the most appropriate methods of communication: local radio and television, mailings, phone calls, and/or personal contacts.

5. Set up meetings to get the process started.

a) Establish an organizational structure, meeting guidelines, etc.

b) Initiate weed management planning.

c) Develop educational programs.

It is important to keep in mind that technical problems are generally easier to solve than social problems. Support for plans and weed management efforts will increase if individuals are kept informed and feel they can provide input into the process. Maintaining an open and informative approach is important, as is emphasizing progress and successes.

Literature Cited

Blahna, D.J., and S. Yonts-Shepard. 1989. Public involvement in resource planning: Toward bridging the gap between policy and implementation. Society and Nat. Res. 2:209-27.

Dale, D.D., and A.J. Hahn (eds). 1994. Public issues education: Increasing competence in resolving public issues. Univ. Wisc. Coop. Ext., Madison, WI.

Svejcar, T.J. 1996. What are working groups and why should scientists be involved? Weed Tech. 10:451-54.

Wagner, F.H. 1994. Changing institutional arrangements for setting natural-resources policy, 281-88. *In:* Martin Vavra, William A. Laycock, and Rex D. Pieper (eds.), Ecological Implications of Livestock Herbivory in the West. Soc. Range Manage., Denver, CO.

Whaley, R.S. 1993. Working partnerships: Elements for success. J. Forestry 91:10-11.

Wondolleck, J. 1985. The importance of process in resolving environmental disputes. Environ. Impact Assess. Rev. 5:341-56.

Economic Evaluation Procedures for Noxious Weed Management on Rangeland

Duane Griffith

Noxious weed management on rangeland has both short (one-year) and long-term economic implications. Possible short-term impacts are mostly negative. Short-term impacts can include the cost of treatment and the cost of replacing the forage treated if the land is "rested" during a treatment year. While short-term considerations are important, noxious weed management requires long-term planning. The protection of non-infested areas is a key element in determining the economic benefit of managing noxious weeds. If noxious weed management is implemented when infestations are small, benefits include the value of desirable herbage produced on infested acreage, and the value of desirable herbage not lost, through a management strategy that prevents the spread of noxious weeds to uninfested areas.

Integrated Weed Management (IWM) is a structured approach to ecosystem management based on a general understanding of the ecology, uses, and interactions of the plants and animals within the system. An IWM approach can be used to identify the negative and positive effects (actual or potential) of weeds on an ecosystem. The identified effects form the basis of an integrated approach to amelioration consistent with the ecological capabilities of the ecosystem.

The cost and benefits of effective management vary with current site conditions (competing vegetation, amount of bare ground, organic matter, and other soil characteristics) and the site's productive potential. The term "site" refers to a management unit. A block of land managed under a single weed plan constitutes a management unit. A management unit may consist of a few acres within a pasture, an entire pasture, a ranch, or several ranches. Because the protection of uninfested acres in a management unit is a key economic consideration, economic implications of noxious weed management must be evaluated in the context of the entire management unit, and not just the acreage actually infested.

Once a management unit is defined, a management strategy can be determined. Three distinct noxious weed management strategies are available: 1) no action or no treatment; 2) containment to prevent spread, but not eradication; and 3) complete eradication. Management strategies will vary for each management unit. Selection

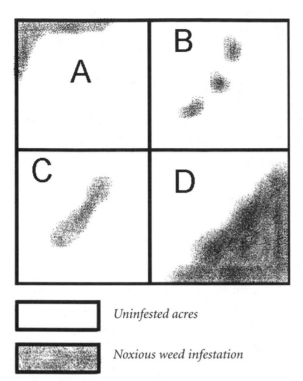

Figure 1. An illustration of possible types of infestations within a management unit. In addition to those shown, there could also be some combination such as that shown in blocks A and C.

Uninfested acres

Noxious weed infestation

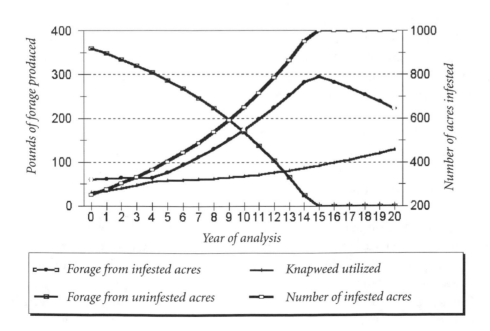

Figure 2. Estimated forage sources within a management unit infested with spotted knapweed, without implementation of a management strategy. A 20-year time frame is used in this example.

of a management strategy will depend on the type of infestation within the management unit. For example, four different management units, each with a distinct type of infestation, are shown in Figure 1. Eradication strategies may be logical for units B and C, as the infestations are rather isolated and relatively small. Because reinfestation from neighboring land is a problem for units A and D, a containment strategy, or a combination of containment and eradication, may be logical for these units. Combinations of these types of infestations are also possible: A and C, or A and B. Some combination of the three strategies may also be implemented, depending on the type of infestation.

Analyzing the costs and benefits of a selected management strategy for a specific management unit provides an estimate of the economic feasibility (benefits greater than costs) of controlling noxious weeds. Costs are assigned to resources used in managing the noxious weeds. Benefits of weed management accrue from two sources. The first is an increase in desirable herbage production that replaces the noxious weeds. The second is the protection of uninfested acres within the management unit. Benefits are measured as the value of increased forage and/or forage saved, over the planning horizon. Benefits are produced from three potential forage sources: 1) noxious weeds, 2) desirable herbage from infested acres, and 3) herbage from uninfested acres. Figure 2 illustrates potential forage from these three sources, in a spotted knapweed (*Centaurea maculosa*) infestation that is left uncontrolled (Bawtree and Mclean 1977, Bedunah 1989). Forage sources from other types of noxious weed infestations may look much different.

At first glance, Figure 2 may seem to be in error. The desirable herbage from infested acres continually rises until year 15. This increase in desirable herbage is not due to an increase in production, but an increase in the number of acres that are infested as the initial infestation spreads throughout the management unit. The increase in infested acres causes an increase in the amount of forage from infested acres. The key to understanding the graph in Figure 2 is the definition of infested acres. Infestation occurs in a two-step process. First, an individual plant or maybe several plants are established in a previously uninfested acre. The next step is the spread of the noxious weed throughout the remaining area. As the weed spreads, it increases in density until the noxious weed reaches its peak density on the infested acre. After year 15, all acres are infested and desirable herbage declines as knapweed continues to increase in density within the management unit through year 20. In this process, an acre is considered to be infested when it contains a seed source, one or more plants, that allows the spread to other areas within the infested acre. Figure 2 also shows projected growth in the number of acres infested.

The size and location of the infested areas determine the most cost-effective method of managing noxious weeds. Small patches may be permanently eliminated with persistent herbicide and/or cultural management programs. However, an IWM approach that uses all methods may be more appropriate for large infestations. The deliberate use of natural enemies (parasites, predators, or pathogens) to reduce weed densities is an appealing strategy, and can be economically feasible. Effective

Table 1. Production and economic information required for economic analysis of a management unit.

General Economic Information Required

Enter the Market Price of AUM's ($/AUM).	$10.00
Enter the current nominal interest rate.[1]	12.00%
Enter the expected annual inflation rate.	7.00%
Enter your marginal tax rate (percent).[2]	15.00%

Production Information Specifying Current Conditions Within the Management Unit

Enter the total acres in this management unit.	1,000
Enter the percent of the total acres infested.	25.00%
Enter your estimate on what percent of noxious weed is utilized as forage by livestock.	20.00%
Enter the average pounds of herbage currently being produced on each uninfested acre.	1000
Enter the minimum percent of total herbage left to maintain the quality of your pasture.	60.00%
Enter the maximum lbs. of herbage/acre this site is capable of producing under ideal conditions.	1500
Enter the pounds of forage required to support each animal unit month (AUM) of grazing.	660
Enter lbs. of herbage/acre produced by noxious weed during the initial year of infestation.	100

Production Information for Future Site Capacity

Enter annual rate total herbage production would increase with better pasture management.	3.00%
Enter annual rate forage production will decrease on infested acres if no control measures are taken.	15.00%
Enter annual rate which new acres will be infested if no control measures are taken.	10.00%
Calculated minimum lbs. of herbage/acre this site will produce with no treatment or control.	1080

1. The interest rate you would pay if you borrowed money.
2. The combined federal and state marginal tax bracket you fall into when calculating your income taxes. If your federal tax bracket is 15% and your state tax bracket is 6%, your combined marginal tax rate is 21%.

management must also include full cooperation and coordination of all parties involved in a management unit: private landowners, public land users, and government agencies. However, this approach does not suggest that a particular weed management strategy will be economically or financially feasible.

Determination of economic feasibility (profitability) is the first step, but financial feasibility must also be addressed. Financial feasibility addresses the question of a firm's ability actually to implement a proposed plan or action, that is, its ability to maintain cash flow during a project. If analysis shows a noxious weed management project is economically feasible, it does not necessarily follow that there are enough resources, cash, etc., to carry out the project. The analysis approach discussed in this chapter addresses the question of economic feasibility, not financial feasibility. There is one instance when economic feasibility and financial feasibility are the same. If all costs and returns used in the analysis represent actual cash outflows and inflows, then economic profitability equals financial feasibility. But, for example, if the "value" put on AUMs (animal unit months) is not an actual cash inflow, but only the value at which owned land could be rented to someone, then the analysis shows economic profitability rather than financial feasibility.

Economic Analysis Method

Economic criteria for controlling noxious weed infestations are simple. The benefits must be greater than the costs of any weed management strategy implemented. However, neither costs nor benefits occur instantaneously during the initial year of treatment. They accrue over time, so measuring the costs and benefits requires the use of net present value (NPV) analysis. In addition, costs and benefits of a selected management strategy do not necessarily parallel one another. That is, a dollar of cost spent at a given time does not necessarily generate a dollar's worth of benefits in the same time period. The majority of costs may be incurred during the initial phase of the program, or there may be an unequal stream of costs throughout the program. Benefits can also accrue in unequal amounts.

Net Present Value analysis measures the costs and benefits that accrue over time in current dollars. Economic analysis requires that future streams of benefits and costs are adjusted for inflation. Inflation adjustments simply state costs and benefits in the same unit of measure: current (today's) dollars. Comparing costs in current dollars to benefits measured in future dollars may result in the wrong decision about implementing a weed management plan. Comparison of costs and benefits adjusted to current dollars provides an accurate analysis of economic feasibility. A more precise definition of the economic decision criteria, then, is that the discounted (inflation-adjusted) stream of benefits received from noxious weed management through time must be greater than or equal to the discounted (inflation-adjusted) stream of costs incurred through time. If the discounted value of benefits is greater than or equal to the discounted value of costs, noxious weed management is economically feasible. However, while feasible, it may not be the most profitable use of available capital

Table 2. Required data, including time frame, to establish treatment costs and herbage response to treatment for noxious weeds and desirable herbage.

Years(s) Treated	Costs[1]	Infested Acres Treated in Each Year	Pounds of Herbage Produced Per Infested Acre From		Treated Acres Grazed	Untreated Acres Grazed
			Noxious Weeds	Desirable Herbage		
0	$14.00	250	600	600	0.00%	100.00%
1		50	50	975	100.00%	100.00%
2			0	1155	100.00%	100.00%
3	$14.00	150	100	1100	100.00%	100.00%
4			0	1200	100.00%	100.00%
5			0	1200	100.00%	100.00%
6	$14.00	75	100	1100	100.00%	100.00%
7			0	1200	100.00%	100.00%
8			0	1200	100.00%	100.00%
9	$14.00	40	100	1100	100.00%	100.00%
10			0	1200	100.00%	100.00%
11			0	1200	100.00%	100.00%
12	$14.00	15	50	1150	100.00%	100.00%
13			0	1200	100.00%	100.00%
14			0	1200	100.00%	100.00%
15	$14.00	15	50	1150	100.00%	100.00%
16			0	1200	100.00%	100.00%
17			0	1200	100.00%	100.00%
18	$14.00	15	50	1150	100.00%	100.00%
19			0	1200	100.00%	100.00%
20			0	1200	100.00%	100.00%

1. Treatment costs per infested acre are specified, in current dollars, for each year a treatment is applied throughout the planning horizon as shown above. The computerized analysis automatically adjusts for future inflation.

within an operation, and it may not be financially feasible. Keep in mind that while economic analysis addresses profitability (profit maximization), not all activities or projects undertaken have maximum profits as the primary goal.

Any analysis procedure used to evaluate the economic feasibility of eradicating or containing noxious weeds on rangeland should allow decision-makers to test combinations of: 1) longevity of control, 2) herbage response from desirable plants and noxious weeds after treatment, 3) economic variables affecting costs and returns through time, and 4) various treatment strategies. Such a model is presented below. This model allows the user to determine the economic feasibility of a noxious weed management plan. The user must specify relevant herbage and forage production data for the management unit, and costs and benefits of a particular management strategy.

A complete mathematical specification of this model is listed in the *Journal of Range Management* (Griffith and Lacey 1991). In addition, this model has been computerized and is available in four different spreadsheet formats: Lotus® for Windows® (v. 4.0 or greater), Lotus® for DOS (v. 2.x or greater), Excel® for Windows® (v. 5.0 or greater), and Quattro Pro® for Windows® (v. 5.0 or greater). A copy of these spreadsheets can be downloaded from the World Wide Web at: http://www.montana.edu/wwwextec/index.htm. The name of the spreadsheet is NOXSWEED. Because this chapter is not meant to provide exacting detail for an economic analysis, only the general approach and summary information of this model are presented.

The tables presented in this chapter are summary information from the computerized model developed to assist with economic analysis of managing a noxious weed infestation. Tables 1 and 2 show the information required to analyze the economic feasibility (economic profitability) of a noxious weed management plan. Current productive capacity of a management unit, future site productive capacity of the management unit, and general economic parameters are necessary to complete an economic analysis. This information is shown in Table 1. Because treatment costs and benefits are incurred through time, and these cost and benefits will vary with the management strategy implemented, the costs and benefits must be specified in the appropriate time frame. In addition, information about how the management unit is utilized during treatment and non-treatment years must be specified. This information is shown in Table 2. In column 2, treatment costs are specified in current dollars throughout the entire planning horizon. The number of acres treated in each year of the planning horizon is specified in column 3. The fourth and fifth columns specify the yield response of noxious weeds to the selected treatment strategy, column 4, and the yield response of desirable herbage, column 5. While this allows specification of a yield response, it also requires the user to know how a particular site will respond to a management strategy. The utilization of treated and untreated acres within the management unit are specified in columns 6 and 7. The model uses the economic and production information in Tables 1 and

2 to calculate the net present value of forage production from the management unit. This is accomplished in several steps.

Before economic costs and benefits can be determined, the production parameters specified in Tables 1 and 2 are used to estimate herbage production over the specified time frame. The planning horizon can be up to 20 years. The example used here is a 20-year analysis. Shorter periods can be analyzed. As previously mentioned, herbage production comes from uninfested acres, noxious weeds utilized as forage from infested acres, and other desirable herbage from infested acres. These sources of herbage production and the three treatment strategies (no action or no treatment; containment to prevent spread, but not eradication; and complete eradication) combine to define up to three "blocks" within a management unit. Herbage production from these blocks and the size of the blocks themselves are dynamic throughout the planning horizon. Figure 3 illustrates the three blocks for a particular type of infestation at a point in time. For example, with a no-treatment strategy, Block A will continually increase in size, Block B will not exist (there are no "treated" acres), and Block C will continually shrink until it disappears when the entire management unit is infested. Another weed management strategy may include only blocks B and C.

Using the information in Table 1, forage production for a no-treatment strategy can be estimated for all forage sources within the management unit over the planning horizon, Figure 2. Table 3 shows the detail of these production estimates on a management unit infested with spotted knapweed when no management strategy is implemented. Information presented in Table 3 is generated from the information in Tables 1 and 2. Remember that treatment not only affects those acres infested with a noxious weed, but also protects the value of forage produced on uninfested

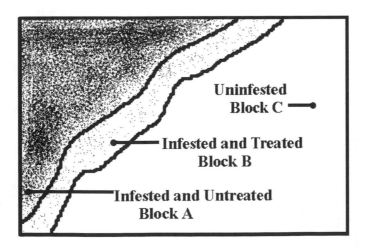

51

Figure 3. This figure illustrates the division of management blocks within a management unit. Each of the three blocks is dynamic through time as the infestation is either treated by a selected management strategy or left to spread.

Table 3. The body of this table contains the calculated pounds of forage produced from herbage other than knapweed on infested acres. The last three rows list the forage from knapweed utilization, the forage from desirable herbage on infested acres, and total forage from all sources.

Year of Analysis

Yr	Infested Acres[1]		UA[2]	0	1	2	3	4	5	6	7	8	9	10	11	12	13	14	15	16	17	18	19	20
0	250	0	750	600	510	407	287	151	120	120	120	120	120	120	120	120	120	120	120	120	120	120	120	120
1	275	25	725		1100	1085	1068	1048	1025	999	969	934	894	848	795	735	665	585	492	386	264	124	120	120
2	303	28	698			1100	1085	1068	1048	1025	999	969	934	894	848	795	735	665	585	492	386	264	124	120
3	333	30	667				1100	1085	1068	1048	1025	999	969	934	894	848	795	735	665	585	492	386	264	124
4	366	33	634					1100	1085	1068	1048	1025	999	969	934	894	848	795	735	665	585	492	386	264
5	403	37	597						1100	1085	1068	1048	1025	999	969	934	894	848	795	735	665	585	492	386
6	443	40	557							1100	1085	1068	1048	1025	999	969	934	894	848	795	735	665	585	492
7	487	44	513								1100	1085	1068	1048	1025	999	969	934	894	848	795	735	665	585
8	536	49	464									1100	1085	1068	1048	1025	999	969	934	894	848	795	735	665
9	589	54	411										1100	1085	1068	1048	1025	999	969	934	894	848	795	735
10	648	59	352											1100	1085	1068	1048	1025	999	969	934	894	848	795
11	713	65	287												1100	1085	1068	1048	1025	999	969	934	894	848
12	785	71	215													1100	1085	1068	1048	1025	999	969	934	894
13	863	78	137														1100	1085	1068	1048	1025	999	969	934
14	949	86	51															1100	1085	1068	1048	1025	999	969
15	1000	51	0																1100	1085	1068	1048	1025	999
16	1000	0	0																					
17	1000	0	0																					
18	1000	0	0																					
19	1000	0	0																					
20	1000	0	0																					
Forage from "other"				60	62	64	65	65	77	94	112	131	152	174	198	224	252	282	295	283	270	254	238	221
From noxious weeds				30	35	41	48	55	58	59	61	63	66	69	72	76	81	87	93	99	105	113	121	129
From uninfested				360	348	335	320	304	287	267	246	223	197	169	138	103	66	24	0	0	0	0	0	0
Total forage available				450	445	439	432	425	422	421	419	417	414	411	408	404	399	393	387	381	375	367	359	351

1. Column 1 = total acres; column 2 = marginal acres infested. 2. Uninfested acres in each year.

acres in future years. Future loss in forage productivity is a key element in determining the potential benefits of a weed management strategy. Loss of forage production and associated economic implications of not treating are listed in Table 4. General economic information, provided in Table 1 and combined with the production estimates in Table 3, allows calculation of the costs, measured as the current dollar value of forage lost, if an infestation is not treated. Table 4 shows an example calculation of the value of lost AUMs when a spotted knapweed infestation is left uncontrolled. This provides economic information about the cost of doing nothing to control a weed infestation. It does not indicate whether a specific weed management strategy would be economically feasible.

To answer that question, the model uses the treatment costs and herbage response (benefits) information specified in Table 2. Using information in Table 1, treatment costs (specified in Table 2) are inflated to an appropriate level for each year a treatment is applied to the management unit. Treatment costs should include any costs that are associated with the particular treatment strategy. For example, initial treatment costs might include the cost of a chemical application, labor costs for removal of noxious weeds near waterways, fencing costs, and the costs to introduce weed pathogens or insects. Costs in future years may include follow-up chemical applications on fewer acres as the weed infestation is slowly eradicated.

The benefits of the chosen treatments are determined by the yield of desirable herbage and noxious weeds to the treatment strategy. This yield response must be specified for each year of the planning horizon. The flexibility of specifying a yield response for a particular treatment strategy allows the model to handle any type of infestation: sparse, dense, concentrated, scattered, etc. It also requires the user to know what the yield response will be for a particular noxious weed within a given management unit.

With the information on cost and yield response to treatment by year, the model estimates the economic feasibility of the treatment strategy selected. The example used for illustration shows treatment every three years through the eighteenth year (Table 2). A smaller number of acres are treated through year 12, with the assumption that the infestation is completely eliminated by year 20. The NPV of costs and benefits for this strategy are listed in Table 5. In the example shown here, the present value of benefits is $8.42 per acre of the management unit, while the costs incurred are only $4.93 per acre. For this management unit, it is economically feasible to treat the weed infestation. The last two columns in Table 5 can be used to compare the cost/benefit ratio for any length of treatments, i.e., a 5-year versus a 15-year plan. These columns also provide information on when, if ever, benefits of treatment will exceed the costs of treatment. In the example shown, the cumulative benefits of treatment do not exceed the costs until the fourteenth year of the program. Resource managers interested in a complete specification of the economic analysis are referred to Griffith and Lacey (1991), or the Web site previously listed.

53

continued on page 56

Table 4. Summary data calculated for the specified production and economic parameters for a "No Treatment" management strategy.

Year(s) Treated	Number Infested Acres	Average lbs of Noxious Weed	Lbs of Noxious Weed Utilized	Total lbs Herbage Utilized	Acres per AUM	Change in Total Herbage Utilized (lbs)	Present Value of AUM Loss
0	250	150	30	450	1.47	0	$0.00
1	275	175	35	445	1.48	5	$0.05
2	303	204	41	439	1.50	11	$0.11
3	333	238	48	432	1.53	18	$0.18
4	366	277	55	425	1.55	25	$0.24
5	403	290	58	422	1.56	28	$0.26
6	443	297	59	421	1.57	29	$0.26
7	487	306	61	419	1.58	31	$0.26
8	536	316	63	417	1.58	33	$0.26
9	589	328	66	414	1.59	36	$0.27
10	648	343	69	411	1.60	39	$0.28
11	713	360	72	408	1.62	42	$0.29
12	785	381	76	404	1.63	46	$0.31
13	863	405	81	399	1.65	51	$0.32
14	949	434	87	393	1.68	57	$0.34
15	1000	464	93	387	1.70	63	$0.36
16	1000	493	99	381	1.73	69	$0.38
17	1000	526	105	375	1.76	75	$0.40
18	1000	564	113	367	1.80	83	$0.42
19	1000	605	121	359	1.84	91	$0.44
20	1000	646	129	351	1.88	99	$0.46

Per Acre Present Value of AUMs Lost to Weeds (in current dollars)	$5.88
Total Value Lost Given the Size of the Management Unit (in current dollars)	$5,879.33

Summary of information calculated assuming no management strategy is implemented. Values are average per acre prorated over the entire management unit.

Table 5. Summary information calculated given the parameters specified in Tables 1 and 2, with eradication as the management strategy.

Year(s) Treated	Year	Number Infested Acres	Average lbs of Noxious Weed	Lbs of Noxious Weed Utilized	Total lbs Herbage Utilized	Acres per AUM	Present Value of After Tax Treatment Costs	Present Value of Added AUMs With Treatment	Cumulative Costs of Treatment in Present Value	Cumulative Benefit With Treatment in Present Values
Treated	0	250	150.00	30.00	360	1.83	($2.53)	($0.99)	$2.53	$-0.99
	1	50	2.50	0.50	460	1.43	$0.00	$0.16	$2.53	$-0.83
	2	0	0.00	0.00	476	1.39	$0.00	$0.36	$2.53	$-0.47
Treated	3	150	15.00	3.00	475	1.39	($1.32)	$0.41	$3.85	$-0.06
	4	0	0.00	0.00	480	1.38	$0.00	$0.50	$3.85	$0.44
	5	0	0.00	0.00	480	1.38	$0.00	$0.51	$3.85	$0.95
Treated	6	75	7.50	1.50	475	1.39	($0.58)	$0.45	$4.43	$1.40
	7	0	0.00	0.00	480	1.38	$0.00	$0.49	$4.43	$1.89
	8	0	0.00	0.00	480	1.38	$0.00	$0.48	$4.43	$2.37
Treated	9	40	4.00	0.80	475	1.39	($0.27)	$0.44	$4.70	$2.81
	10	0	0.00	0.00	480	1.38	$0.00	$0.48	$4.70	$3.28
	11	0	0.00	0.00	480	1.38	$0.00	$0.48	$4.70	$3.76
Treated	12	15	0.75	0.15	478	1.38	($0.09)	$0.47	$4.78	$4.23
	13	0	0.00	0.00	480	1.38	$0.00	$0.49	$4.78	$4.72
	14	0	0.00	0.00	480	1.38	$0.00	$0.50	$4.78	$5.22
Treated	15	15	0.75	0.15	478	1.38	($0.08)	$0.50	$4.86	$5.72
	16	0	0.00	0.00	480	1.38	$0.00	$0.52	$4.86	$6.23
	17	0	0.00	0.00	480	1.38	$0.00	$0.53	$4.86	$6.76
Treated	18	15	0.75	0.15	478	1.38	($0.07)	$0.53	$4.93	$7.30
	19	0	0.00	0.00	480	1.38	$0.00	$0.56	$4.93	$7.85
	20	0	0.00	0.00	480	1.38	$0.00	$0.57	$4.93	$8.42

After Tax Treatment Costs Per Acre of Management Unit (Current Dollars)   ($4.93)
After Tax Benefit of Treatment (Current Dollars/Acre) $8.42

As a final note, the present values of both costs and benefits are adjusted for the effects of income taxes. Expenses incurred to treat an infestation are tax deductible. The allowed deduction effectively lowers the cost of treatment, hence the adjustment for income tax effects. Benefits are also adjusted for taxes, to keep both costs and benefits in after-tax current dollars.

Conclusions

Managing noxious weeds is a capital investment. Both costs and benefits of weed management occur through time. Because of the time element, economic evaluation requires the use of net present value analysis. The NPV analysis adjusts all costs and benefits to current dollars. This allows comparison of costs and benefits in the same unit of measure, current after-tax dollars.

In the example used for illustration, the results in Table 5 show that the eradication strategy is economically profitable. The NPV of benefits is greater than the NPV of costs. The eradication strategy selected for evaluation and the procedures used in this particular strategy are economically profitable. This is only an example, and does not guarantee that all other treatment strategies are less profitable. The analysis procedures summarized here do not address the financial feasibility of a weed management project.

Literature Cited

Bawtree, A., and A. Mclean. 1977. Rate of spread, potential range, losses and changes in pasture composition from diffuse and spotted knapweed in British Columbia. Proc. Knapweed Symp., Min. Agric., Victoria, BC Canada, Kamloops, BC.

Bedunah, D. 1989. Grass response following spotted knapweed control. Techline. Dow Chem. USA

Griffith, D., and J. Lacey. 1991. Economic evaluation of spotted knapweed (Centaurea maculosa) control using picloram. J. Range Manage. 44(1).

Integrated Weed Management on Rangeland

Roger L. Sheley, Susan Kedzie-Webb, and Bruce D. Maxwell

The pervasiveness and complexity of noxious rangeland weeds, combined with their cost of control, necessitates using integrated weed management (IWM). IWM on rangeland involves the use of several management techniques in a well-planned, coordinated, and organized program. Education, inventory, and impact assessment comprise the first phase of an IWM program. The second phase includes prioritizing weed problems, and choosing and strategically implementing management techniques for a particular area of land. The third phase is adopting proper grazing management practices. And the fourth phase is evaluating the management approach and techniques, and making adjustments to optimize the IWM program. A weed management program must fit into an overall range management plan.

Integrated weed management became an accepted and frequently used term by weed scientists in the 1970s. Walker and Buchanan (1982) defined IWM as "the application of many kinds of technologies in a mutually supportive manner. It involves the deliberate selection, integration, and implementation of effective weed control measures with due consideration of economic, ecological, and sociological consequences." It is a multi-disciplinary, ecological approach to managing weed infestations. For the purpose of this chapter, the overall goal of IWM will be to maintain or develop ecologically healthy plant communities that are relatively weed-resistant, while meeting other land-use objectives such as forage production, wildlife habitat development, or recreational land maintenance (Sheley et al. 1996).

A healthy, weed-resistant plant community consists of diverse species that occupy most of the available niches by capturing a large proportion of the resources in the system (Figure 1). Soil resources, particularly those associated with soil moisture, are limiting factors in the shortgrass prairies and intermountain regions. In these regions particularly, plant communities must be developed that effectively use the soil resources over time and space. A weed-resistant plant community may include an early emerging species, such as the shallow-rooted Sandberg's bluegrass (*Poa sandbergii*), which uses the resources available in the upper soil profile early in the growing season and during periods of light precipitation. As the season progresses, species that begin growing later in the season are needed to use the soil resources from moderate soil depths. Finally, the diverse plant community may include a deep-

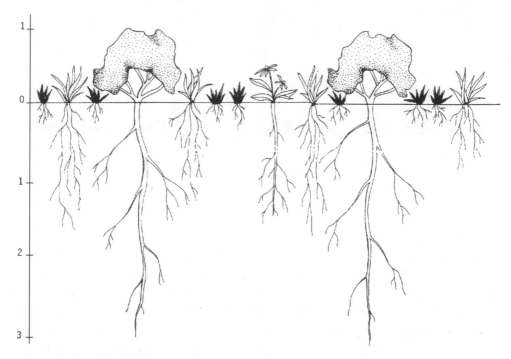

Figure 1. A healthy, weed-resistant plant community consists of diverse species that occupy most of the available niches and minimize weed invasion.

taprooted, very late-maturing species such as alfalfa (*Medicago sativa*) or big sagebrush (*Artemisia tridentata*). These species are capable of extracting resources from deep in the soil profile and throughout much of the growing season. Although little is known about the role of many species within the plant community, it has been proposed that maximum diversity provides for stability and resource capture over a wide range of unpredictable conditions (Tilman 1996, Chapin et al. 1992). Once a desired plant community has been determined, an ecologically based IWM system may be developed.

Phase I: Education, Inventory, and Impact Assessment

Correctly identifying weeds and knowing their most immanent characteristics is central to developing early detection and prevention strategies. Weeds may be native or non-indigenous—this determination will dictate different management approaches. Weed identification and management courses are offered locally and regionally in most states. Publications and computer programs that offer excellent management information and pictures of important weeds are available. Some states offer a certification program for weed managers. Also, the general public is more likely to support weed management efforts when it is educated about the threat of invasive weeds.

Taking a weed inventory is important. (See Chapter 2, Surveying, Mapping and Monitoring Noxious Weeds on Rangelands; and Chapter 7, Early Detection and Eradication of New Weed Infestations.) The goal of inventory is to determine and map the weed species present, area infested, density of the infestation, land under threat of invasion, soil types, and other site factors. Inventories can be conducted by field surveys, aerial photography, and geographic information systems. Identifying weeds and conducting an inventory may be complex, requiring considerable technical skills, or it may be as simple as systematically surveying an area on foot and sketching weed infestations on a map. In any case, inventorying is time-consuming. However, accurate mapping is important in developing a land use plan, setting priorities, and evaluating the success of an IWM program.

The ecological impacts of a weed on an ecosystem and agricultural production may be assessed generally by observing changes in the plant and animal communities in a given area over time. Change in land use is a factor to consider. An awareness of native plants of the area is helpful in arriving at general conclusions. However, a thorough, scientific assessment of ecological impacts may require four or five years of data. Indicators of ecosystem function are being developed for many ecosystems, but the technology is in its infancy (Spellerberg 1991, National Research Council 1994). Weed impacts on agricultural production may be measured by comparing biomass produced in a given area per unit time. Changes in forage production are good measures of agricultural impacts.

The decision to manage weeds is often an economic decision. Every piece of land has value. Sometimes that value is quantifiable, as when land is used for grazing or agricultural production. Sometimes the value is harder to quantify, as when land is valued for its aesthetic appeal, wildlife habitat, recreational use, or native plants. In any case, the cost of weed management, ideally, should be less than or equal to the value of the land.

Economic assessment methods can be developed to maximize returns based on ecological and economic thresholds. In some situations, the desired (possibly native) plant species may be displaced to the point where they are unlikely to return, even when the weeds are suppressed. This, in turn, may require additional effort by the land manager to create a desirable plant community. Knowing and predicting a plant community's response to weed management is critical for making wise decisions about management strategies. In some cases, weed management is not economically reasonable, despite legislative requirements for management. Careful economic assessments—including considerations of the ecology of the site, land use, and land value—should be conducted before large-scale weed management is implemented.

Improving the desirable plant community's competitiveness, durability, and resistance to weed invasion may reduce the economic loss from a piece of land. In this case, the density of the weed is allowed to increase until the value of the land is lowered enough to justify the cost of weed management. The healthier the desirable plant community, the denser the weeds can grow before causing damage and lowering the land value. Plant communities can be made more competitive, durable, and resistant by increasing species diversity, particularly in a way that fills the available resource niches and leaves little space, moisture, or light for weeds.

Phase II: Planning and Implementation

Weed management planning is the process by which problems and solutions are identified and prioritized in consideration of ecological, agricultural, economic, and sociological factors. An action plan is developed to provide direction for implementing the program.

Implementing an IWM plan includes: 1) preventing encroachment into uninfested rangeland; 2) detecting and eradicating new introductions; 3) containing large-scale infestations; 4) controlling large-scale infestations using an integrated approach; and, often, 5) revegetation. However, the key components of a successful IWM plan are sustained effort, constant evaluation, and the adoption of improved strategies.

Preventing the introduction of rangeland weeds is the most practical and cost-effective method for their management. Prevention programs include limiting weed seed dispersal, minimizing soil disturbance, and properly managing desirable vegetation.

Early detection and systematic eradication of new weeds are central to IWM. Weeds encroach by establishing small satellite infestations that are generally the spreading front of the large infestation. Eradication means totally removing the weed from the area. This is usually achievable on a small scale, and generally requires aggressive annual applications of herbicides. Revegetation of infested areas may be required in areas that lack an understory of desirable plant species to reoccupy the site after the weeds are eliminated. Eradication requires continual monitoring and evaluation to ensure the permanent removal of the weed.

Containment programs are generally used to restrict the encroachment of large-scale weed infestations. Studies have shown that containment is cost-effective because it preserves neighboring uninfected rangeland and enhances the success of future large-scale control programs. Containing a large-scale infestation requires using preventative techniques and spraying herbicides on the border of weed infestations to stop the encroachment.

Most successful, large-scale weed management programs are completed in a series of steps over a long period of time. Large infestations may be divided into smaller units to make them more manageable. Units should be selected and managed based on their similarity of range sites, weed species and density, desired understory species and density, and environmental constraints that determine management options. Weed management then may be carried out unit by unit, at a rate compatible with environmental and financial goals.

Weed Management Methods

Selection of a proper management program will depend on: weed species, effectiveness of the control technique, availability of control agents or grazing animals, use of the land, length of time required for control, environmental considerations, and relative cost of the control techniques.

• *Biological Weed Management*

Biological control involves using living organisms (such as insects, pathogens, and nematodes) to suppress the weed infestation to an acceptable level. Sheley and Jacobs (1997) defined an acceptable level of weed control as the level of weed suppression which shifts the competitive balance in favor of a desired species. Most noxious rangeland weeds have been recently introduced into North America and have few natural enemies. Classical biological control is the distribution of natural enemies collected from the weed's area of origin.

• *Chemical Weed Management*

Selective herbicides can effectively remove broadleaved weeds and shift the competitive balance in favor of desired grasses. In most cases, weed control is temporary and requires repeated applications. Large-scale weed management using a herbicide program is cost-effective only on highly productive areas with a substantial grass understory. For example, Davis (1990) found picloram at 0.25 pounds per acre provided nearly 100% spotted knapweed control for three to five years. Griffith and Lacey (1991) suggested this treatment applied over 20 years could be cost-effective, based on a typical livestock operation in Montana. On the other hand, continuous application of this chemical on leafy spurge does not appear cost-effective for most cattle ranches. In any case, long-term commitment and sustained effort are essential for effective weed management using herbicides.

• *Cultural Weed Management*

Cultural methods of noxious weed management are generally aimed at enhancing desirable vegetation to minimize weed invasion. Plant competition, grazing, and fertilization can favor desired species. Revegetation with desirable plants may be the best long-term alternative for controlling weeds on sites without an understory of desirable species. Establishing competitive grasses can minimize the re-invasion of rangeland weeds and provide excellent forage production (Lym and Tober 1997). On appropriate sites, a herbicide application after weeds have emerged, followed by plowing or disking and drill-seeding, is most effective for establishing desirable species.

Most animals prefer certain forages. Selectivity by herbivores can shift the competitive balance of plant communities (Crawley 1983, Lukan 1990). For example, in some situations leafy spurge (*Euphorbia esula*) can be controlled by sheep or goat grazing (Bowes and Thomas 1978, Lacey et al. 1984, Larson and McInnis 1989).

Appropriate grazing by animals preferring weeds can shift the plant community toward more desired grasses (Lacey et al. 1989). On the other hand, cattle grazing can selectively reduce grass competitiveness, shifting the community in favor of weeds (Svejcar and Tausch 1991).

Fertilization can be useful in some situations. In a two-year study, Sheley and Jacobs (1997) found that picloram plus fertilizer did not interact to affect either spotted knapweed density or grass yield. However, fertilizer applied at 32+40 kg/ha of nitrogen and phosphorous, respectively, increased grass yield on those sites that had a substantial grass understory.

• *Physical and Mechanical Management*

Physical weed control methods can be effective on small infestations. Hand-pulling, hoeing, tilling, mulching, burning, and mowing are all commonly used to control noxious weeds. Hand-pulling and hoeing are most successful under conditions where complete crowns can be removed. Shallow-rooted weeds can be removed by hand-pulling where the soil is loose or moist. Tillage can be successfully used on level areas. It is best used on cropland, and is usually a portion of a revegetation procedure on rangeland. Canada thistle (*Cirsium arvense*) has been effectively controlled by tilling every 21 days during the growing season. However, some rhizomatous plants such as leafy spurge can be spread by tillage.

A single, low-intensity fire does not effectively control most weeds because it is not hot enough to prevent resprouting from crowns or re-establishment from seeds in the soil (Sheley et al. 1998). A single fire increased the cover and density of diffuse knapweed (*Centaurea diffusa*) in northern Washington without enhancing desired species (Sheley and Roché 1982). Similarly, spotted knapweed increased about six-fold within two years after a controlled fire on a forested site in Montana. Fires may create the type of disturbance that promotes the colonization of many weeds. However, herbicide efficacy has been shown to increase when applied post-burn (Sheley and Roché 1982, Lacey et al. 1992).

Mowing reduces seed production in some plants, especially annuals. However, the stage of growth and weather after mowing appears critical. Plants mowed during late-bud or flower stages produce fewer seeds than those mowed later. If soil moisture is adequate or replenished, species such as diffuse knapweed can actually produce more seeds after mowing than without mowing. Repeated mowing in late-bud and early flower stages, and again in the fall, minimized seed production of spotted knapweed.

• *Integrated Weed Management*

Researchers are in the process of determining whether combining treatments will provide a synergistic response in controlling weeds. Preliminary evidence suggests most control techniques are compatible. Experimenting with combinations of control techniques may provide better and longer-term control than any single treatment.

For example, in areas with adequate precipitation, combining picloram with fertilizer can increase the longevity of spotted knapweed control and triple forage production more effectively than either treatment applied alone (Sheley and Roché 1982).

Phase III: Proper Grazing Management

Proper grazing management is crucial to a weed management program. Follow-up management determines the longevity of weed control. Proper grazing is essential to maintain competitive desirable plants, which will help prevent weed re-invasion after control measures are completed. A grazing plan should be developed for any management unit involved in a weed management program. (See Chapter 8, Grazing and Weeds.) The plan should include altering the season of use and stocking rates, to achieve moderate grass utilization. Grazing systems should rotate animals to allow desirable plants to recover before being regrazed, and to promote litter accumulation. A proper grazing system will benefit the desirable plants and help minimize the weeds.

Phase IV: Evaluation

Range monitoring and annual evaluations should be conducted to determine the adequacy of existing management plans, and to determine when plans need to be changed. Monitoring involves making observations, gathering data, and keeping records on the range condition and trend. Monitoring must be designed to detect changes in weeds and desirable plants, biological control agents, and surface conditions including litter accumulation, exposed soil, erosion, and soil compaction. Management practices (e.g., grazing utilization patterns) and climatic factors should be monitored as well. Data may be compared with data from earlier years, so that weed management programs can be adjusted according to the predetermined management objectives.

Ecologically Based Weed Management: Two Models

Integrated weed management must have an ecologically sound conceptual basis from which strategies can be developed and tested. These conceptual models are used to direct data collection, assemble and process information, and predict the outcome of management decisions (Westoby et al. 1989). Two ecologically-based models are proposed: 1) the successional rangeland weed model (Sheley et al. 1996); and 2) a model based on the life-history of key plants within the community (Sheley and Larson 1994, Maxwell and Sheley 1997).

• *Successional Rangeland Weed Management Model*

Managing plant succession requires knowledge of the three general causes of succession: disturbance, colonization, and species performance (Sheley et al. 1996; Figure 2). Within the limits of our knowledge about the conditions, mechanisms, and processes controlling plant community dynamics, these three components can be modified to allow predictable successional transitions. We can design the disturbance regime and attempt to control colonization and species performance through management. Successional management must be viewed as an ongoing process, moving from one successional component to the next, or repeating a single component through time (Figure 2). This model is driven by both naturally occurring and human-induced processes, and thus is robust enough to allow incorporation of virtually any management decision.

Disturbance plays a central role in initiating and altering successional pathways, although a unified disturbance theory has not been developed (Provenza 1991). Natural disturbances (such as landslides, fire, and severe climatic conditions) initiate, retard, or accelerate succession, or alter successional pathways. Designed disturbances are activities that are initiated to create or eliminate site availability, and are aimed at initiating and controlling succession (Figure 3). Weed management strategies have included designed disturbances such as cultivation, burning, and herbicide applications for decades. However, in successional management, designed disturbance is used to alter successional trajectories and to minimize the need for continuous high-energy inputs.

Controlled colonization is the intentional alteration of availability and establishment of various plant species. Colonization may be influenced in a positive or negative manner, depending on the species and successional goals (Figure 3). Controlled colonization efforts are directed toward influencing seed banks, propagule pools, and regulation of safe sites for germination and establishment of desirable species. Weed seed banks can be depleted through attrition if seed production is prevented or significantly reduced.

Controlled species performance involves manipulating the relative growth and reproduction of plant species, in an attempt to shift community dynamics in a desirable direction (Figure 3). Biological and chemical weed control, grazing, plant

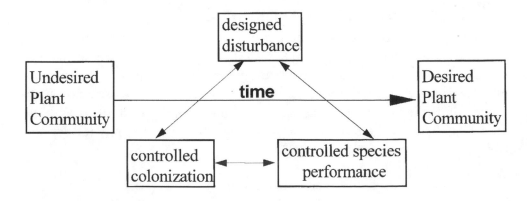

Figure 2. The three general causes of plant succession are disturbance, colonization, and species performance. These factors can be manipulated to change outcomes.

Plant community before weed management	Designed disturbance	Controlled colonization	Controlled species performance	Plant community after weed management
	Broadleaf herbicide	Biological control	Biological control	
Composition, density, and frequency of plant community before weed management	Cultivation	Mowing	Mowing	Composition, density, and frequency of plant community after weed management
	Nonselective herbicide	Drill seeding	Early spring grazing	
	Flooding and draining	Broadcast seeding	Fertilization	
	Grazing	Fertilization	Sheep grazing	
	Irrigation	Grazing	Broadleaf herbicide	
	Burning	Broadleaf herbicide	Reduce soil fertility	
		Irrigation	Irrigation	
		Burning		
		Preventing weed introductions		

Figure 3. The three components of succession (disturbance, colonization, and species performance) can be used to guide the integration of various management methods.

and plant-part removal, altering resource availability, and competitive plantings are techniques to change species performance.

In Figure 4, several schematics using successional weed management are shown for spotted knapweed-infested rangeland (Laycock 1991, Sheley and Larson 1994). In these examples, the plant community prior to weed management is composed of: 1) 98% spotted knapweed with a very suppressed understory of cheatgrass (*Bromus tectorum*) or Kentucky bluegrass (*Poa pratensis*); or 2) 50% spotted knapweed, 30% suppressed native species, and 20% cheatgrass and/or bluegrass. Two successional weed management systems are shown for each situation. Weed management treatments are integrated to direct succession. In the examples, the resulting plant community is influenced by both the weed management system and the initial plant community. Weed management actions should be selectively integrated to ensure that the three components of the successional management model are addressed in a complementary manner, based on the composition of the existing plant community. Climatic variation introduces a random element that can influence the short-term outcome.

The successional weed management model presented in this chapter allows for integration of currently available tools. Unfortunately, with conceptual models of this type, there are few large, comprehensive research projects that have tested all possible options for a particular plant community. Development of future

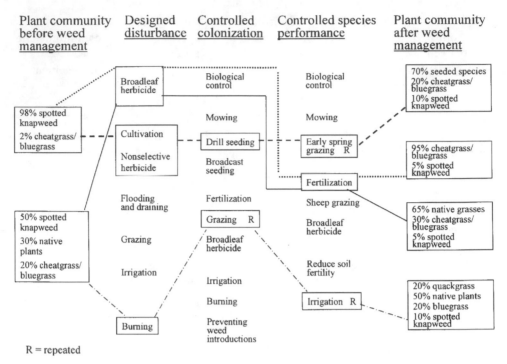

Plant community before weed management | Designed disturbance | Controlled colonization | Controlled species performance | Plant community after weed management

Plant community before weed management

98% spotted knapweed
2% cheatgrass/bluegrass

50% spotted knapweed
30% native plants
20% cheatgrass/bluegrass

Designed disturbance

Broadleaf herbicide
Cultivation
Nonselective herbicide
Flooding and draining
Grazing
Irrigation
Burning

Controlled colonization

Biological control
Mowing
Drill seeding
Broadcast seeding
Fertilization
Grazing   R
Broadleaf herbicide
Irrigation
Burning
Preventing weed introductions

Controlled species performance

Biological control
Mowing
Early spring grazing   R
Fertilization
Sheep grazing
Broadleaf herbicide
Reduce soil fertility
Irrigation   R

Plant community after weed management

70% seeded species
20% cheatgrass/bluegrass
10% spotted knapweed

95% cheatgrass/bluegrass
5% spotted knapweed

65% native grasses
30% cheatgrass/bluegrass
5% spotted knapweed

20% quackgrass
50% native plants
20% bluegrass
10% spotted knapweed

R = repeated

Figure 4. Shown are several schematics using successional weed management for spotted knapweed-infested rangeland. Multiple weed management tools are integrated to provide desired outcomes.

successional weed management plans will require use of existing research information, management experience, and monitoring of successes and failures.

• Life-History Model

A copy of a Noxious Weed Population Dynamics Education Model can be obtained on the Internet at ftp://weedeco.msu.montana.edu. This model is based on a conceptual diagram of a weed population, including life-history stages of spring and fall seed banks, spring and fall seedlings, rosettes, flowering plants, and seeds produced which return to the fall seed bank. The demographic processes that regulate the rate of transition between state variables over a one-year period are: fall and spring germination, summer and winter weed mortality and transition to flowering plants, flowering plant seed production and winter mortality, and seed migration away from the population. Intra- and interspecific density-dependent regulation of seed production per plant, for the weed and for single-species competitive grass, are included in the model. The model predicts the number of individual weed and grass plants per unit area at each life-history stage every year for a selected number of years. Once baseline life-history and interference information is available for specific weeds and grasses, management alternatives can be incorporated into the model to predict the outcome of various integrated weed management strategies.

Literature Cited

Bowes, G.C., and A.G. Thomas. 1978. Longevity of leafy spurge seeds in the soil following various control programs. J. Range Manage. 31:137-40.

Crawley, J.J. 1983. Herbivory: The Dynamics of Animal-Plant Interactions. Univ. of Calif. Press, Berkeley, CA.

Davis, E.S. 1990. Spotted knapweed (*Centaurea maculosa* L.) seed longevity, chemical control, and seed morphology. M.S. thesis, Montana State Univ., Bozeman, MT.

Griffith, D., and J.R. Lacey. 1991. Economic evaluation of spotted knapweed (*Centaurea maculosa*) control using picloram. J. Range Manage. 44:43-47.

Lacey, C.A., R.W. Kott, and P.K. Fay. 1984. Ranchers control leafy spurge. Rangelands 6:202-204.

Lacey, J.R., C.B. Marlow, and J.R. Lane. 1989. Influence of spotted knapweed (*Centaurea maculosa*) on surface runoff and sediment yield. Weed Technol. 3:627-31.

Lacey, J.R., R. Wallander, and K. Olson-Rutz. 1992. Recovery, germinability, and viability of leafy spurge (*Euphorbia esula*) seeds ingested by sheep and goats. Weed Technol. 6:559-602.

Larson, L.L., and M.L. McInnis. 1989. Impact of grass seedlings on establishment and density of diffuse knapweed and yellow starthistle. Northwest Sci. 63:162-66.

Laycock, W.A. 1991. Stable states and thresholds of range condition on North American rangelands: A viewpoint. J. Range Manage. 44:427-33.

Lukan, J.O. 1990. Directing Ecological Succession. Chapman and Hall, London.

Lym, G.R., and D.A. Tober. 1997. Competitive grasses for leafy spurge (*Euphorbia esula*) reduction. Weed Technol. 11:787-92.

Maxwell, B.D., and R.L. Sheley. 1997. Noxious weed population dynamics model. Weed Technol. 11:182-88.

Mortensen, D. 1997. Integrated pest management reduces reliance on "the big hammer." *In:* Weeds as Teachers, Proc. AERO's 1995 Weed Management Alternatives Conference, AERO, Helena, MT, 14-15.

National Research Council, 1994. Rangeland Health. National Academy Press, Wash., DC.

Pedigo, L.P. and H.G. Higle. 1992. The economic injury level concept and environmental quality: a new perspective. Am. Entomologist 38:12-21.

Provenza, F.D. 1991. Range management and range science. Rangelands 13:101-103.

Sheley, R.L., and J.S. Jacobs. 1997. Response of spotted knapweed and grass to picloram and fertilizer combinations. J. Range Manage. 50:263-67.

Sheley, R.L., and J.S. Jacobs. 1997. Acceptable levels of spotted knapweed (*Centaurea maculosa*). Weed Technol. 11:363-68.

Sheley, R.L. and L.L. Larson. 1994. Observation: Comparative life-histories of cheatgrass and yellow starthistle. J. Range Manage. 47:450-56.

Sheley, R.L., and B.F. Roché, Jr. 1982. Rehabilitation of spotted knapweed infested rangeland in northeastern Washington. Abstr. of papers, W. Soc. Weed Sci. Denver, CO.

Sheley, R.L., T.J. Svejcar, and B.D. Maxwell. 1996. A theoretical framework for developing successional weed management strategies on rangeland. Weed Technol. (10)7:766-73.

Sheley, R.L., J.S. Jacobs, and M.F. Carpinelli. 1998. Distribution, biology, and management of diffuse (*Centaurea diffusa*) and spotted knapweed (*C. maculosa*). Weed Technol. 12(2):353-62.

Spellerberg, I.F. 1991. Monitoring Ecological Change. Cambridge University Press.

Svejcar, T. and R. Tausch. 1991. Anaho Island, Nevada: A relict area dominated by annual invader species. Rangelands 13:233-36.

Tilman, D. 1996. Biodiversity: population vs. ecosystem stability. Ecology 77:350-63.

Walker, R.H., and G.A. Buchanan. 1982. Crop manipulation in integrated weed management systems. Weed Sci. 30:17-24.

Westoby, M., B. Walker, and I. Noy-Meir. 1989. Opportunistic management for rangelands not at equilibrium. J. Range Manage. 42:266-73.

Preventing Noxious Weed Invasion

Roger Sheley, Mark Manoukian, and Gerald Marks

The most effective method for managing noxious weeds is to prevent their invasion. There is a serious lack of information regarding the prevention of noxious weed spread. This chapter primarily addresses strategies that are intuitive rather than scientifically proven.

Developing a noxious weed prevention program requires using a combination of methods to limit weed encroachment. Methods of preventing noxious weeds from spreading include:
* Limiting weed seed dispersal;
* Containing neighboring weed infestations;
* Minimizing soil disturbances;
* Detecting and eradicating weed introductions early;
* Establishing competitive grasses;
* Properly managing grasses.

Limiting Weed Seed Dispersal

Noxious weed seeds can be carried along roadways in the undercarriage of vehicles. A Montana State University study showed that a vehicle driven several feet through a spotted knapweed infestation could pick up about 2,000 seeds. These seeds are then dispersed along highways. In the same study, only 10% of the weed seeds remained on the vehicle 10 miles from the infestation. Similarly, weed seeds are dispersed by machinery. Limit noxious weed seed dispersal by refraining from driving vehicles and machinery through weed-infested areas during the seeding period. Wash the undercarriage of vehicles after driving through an area infested with seed-producing noxious weeds. Be sure to control emerging weeds in the wash-up area.

Wildlife and livestock disperse seeds two ways. First, animals ingest noxious weed seeds which may pass through some digestive systems unaffected, introducing seeds into new areas as the animals roam. Second, many weed seeds can become tangled in the coat of animals, and may fall to the ground when animals move into weed-free areas. Little can be done to limit weed seed dispersal by wildlife. However, livestock should not graze weed-infested areas during flowering and seeding. If they do, they should be transported to a holding area for about 14 days before being moved to weed-free ranges.

Noxious weeds can be dispersed in feed. This is especially true on lands where recreational horseback riding and hunting is permitted, and can be a problem for ranchers as well. Using only feed that is certified free of noxious weed seeds is one method of preventing the introduction of noxious weeds. The best weed seed-free feed is produced by grinding and pelleting forage or grain that has been certified as weed-free. Noxious weeds can be dispersed in grass and crop seeds. Whenever possible, purchase certified weed-free seeds.

Hikers, campers, and recreationists can spread noxious weed seeds on their clothing. Recreationists disperse weed seeds when they pick the flowers and discard the wilted parts along trails and recreational access sites. Clothing and camping equipment should be brushed, and the seeds and plant parts placed into a hot fire before leaving an area.

Seed- and flower-feeding biological control agents can play a role in limiting the spread of weed seeds. Seed-feeding gall flies (*Urophora affinis* and *U. quadrifasciata*) reduce seed production of diffuse knapweed (*Centaurea diffusa*) and spotted knapweed (*C. maculosa*). These flies have been reported to reduce knapweed seed production up to 80%, thereby reducing the potential for seed dispersal. The thistle head weevil *Rhinocyllus conicus* attacks the seed-producing tissue of musk thistle, and can effectively reduce the seed production and dispersal of this weed. In any noxious weed prevention program, it is important to work with university Extension specialists and local weed management officials to insure that biological controls are well established on neighboring infestations.

Containing Neighboring Weed Infestations

An integral part of any weed prevention program is to contain neighboring weed infestations. Containment practices are designed to restrict the encroachment of noxious weeds onto adjacent lands. The most effective method of containment is to spray the borders of infested areas with a herbicide. This approach is designed to concentrate efforts on the advancing edge of the weed infestation. Containment programs typically require a long-term commitment to herbicide application because they are not designed to eliminate or reduce the infestation level, only to limit its spread. Roadways, railways, and waterways, where weed infestations often begin, should be under a constant prevention and containment program.

Minimizing Soil Disturbances

Most noxious weeds on North American rangelands are non-indigenous, and have evolved under abusive grazing practices that cause soil disturbance and erosion. Non-indigenous weeds have developed many characteristics—such as rapid growth rates, high seed production, and extended growing periods—that give them an advantage over native North American plants in occupying disturbed soil. Minimizing soil disturbance by vehicles, machinery, wildlife, and livestock is central to preventing noxious weed establishment.

Detecting and Eradicating Weed Introductions Early

Preventing and controlling noxious weed encroachment depends on early detection. One successful method for preventing the invasion of weeds is to survey the area, removing any individual weed plants before they become well established. A survey plan should be developed for each management unit to include inventory techniques (vehicle, horseback, motorcycle, foot), area surveyed, and survey time periods. At least three surveys should be conducted each year: a spring survey to detect weeds early enough to allow effective chemical control, a second survey in early summer, and the last survey in early fall. At each survey, individual noxious weed plants should be hand-removed or sprayed with the appropriate herbicide. Preventing weed seed production is critical. Late-season chemical applications generally do not prevent seed production, and hand removal is usually necessary. Hand-pulled plants should be burned. The weed infestation should be identified on a map, marked or flagged in the field, continually monitored, and controlled during subsequent surveys.

Establishing Competitive Grasses

Another method for preventing encroachment is to establish competitive, desirable grasses in areas susceptible to invasion. Competitive grasses limit the establishment and growth of weed populations by using resources needed by weeds. Well-established grass stands are central to limiting weed encroachment along roadways. Specific establishment techniques depend upon the weed/grass complex and environmental characteristics of the site. In areas with a good residual (suppressed) perennial grass stand, chemical weed control may reduce weed competition to allow grass to reoccupy the site.

Severe weed infestations may require revegetation. Where a heavy residual weed stand exists, it may be necessary to burn in the fall to remove old stems before revegetation procedures can be implemented. In areas without a heavy residual weed stand or areas that have been burned, the soil should be chisel-plowed in the fall. Plowing will create a quality seedbed, bury some weed seeds, and turn up others. In areas dominated solely by broadleaved weeds, a broadleaf herbicide such as picloram may be applied immediately after plowing. If broadleaved and grass weeds (e.g., cheatgrass [*Bromus tectorum*] and/or medusahead [*Taeniataerum caput-medusae* subsp. *asperum*]) co-dominate, the area should be sprayed with a non-selective, short-residual herbicide such as glyphosate following spring plowing. The application should be applied as early in the spring as possible, but after the majority of the weeds emerge. Do not spray desired trees and shrubs.

71

After a spring herbicide application, the area may be drill-seeded with the proper mixture of perennial grasses. Grass mixtures and rates depend on the specific range site. A local Soil and Range Conservationist can recommend a good seed mixture for the site. A follow-up herbicide treatment may be necessary to control weeds emerging in the seeded grass stand. After three years, a strong grass stand should be able to limit invasion or re-invasion by noxious weeds.

Other revegetation methods are also useful. Hydroseeding, plugging, or broadcasting and covering seeds with a layer of straw may be more effective on steep slopes or rough terrain.

Properly Managing Grasses

On areas with a competitive grass stand, proper land management insures that the grasses remain strong and vigorous, minimizing noxious weed encroachment. In most cases, grasses require defoliation every two to four years to remove old stems that shade plants and hinder growth. Mowing, burning, and grazing are the primary methods for defoliating grasses. Grasses are generally mowed in the summer or fall. Burning is conducted in the fall or early spring, before the grasses resume growth. Defoliation stimulates grass growth and enhances its competitive ability.

Proper livestock grazing can effectively maintain competitive grass plants. A grazing management plan should be developed for any management unit involved in a noxious weed prevention program. This plan should include a grazing system that outlines the movement of livestock throughout the year. Grazing systems should include altering the season of use, rotating livestock to allow desirable plants to recover before being regrazed, and promoting litter accumulation. Grazing in this manner enhances the vigor and strength of the grasses, which limits weed germination and promotes early mortality of weed seedlings and rosettes. The grazing management plan should include a monitoring program to determine the efficacy of the grazing system in protecting grasses and limiting weed invasion.

Conclusions

Many rangeland areas throughout the western United States are being invaded by noxious weeds. The most economical and ecologically sound method for managing noxious weeds is to prevent their invasion in the first place. To prevent invasion, limit noxious weed dispersal and contain neighboring weed infestations. Soil disturbances must be minimized. New weed introductions must be detected early and eradicated. Finally, grasses and other desirable plants must be properly established and managed.

Early Detection and Eradication of New Weed Infestations

David L. Zamora and Donald C. Thill

Eradication is the destruction of every individual of a species from an area surrounded by natural or human-made barriers sufficiently effective to prevent reinvasion except by human intervention (Newsome and Noble 1986). The economic threshold for a plant pest targeted for eradication is zero (Dahlsten et al. 1989). In contrast to eradication, control is simply containment or prevention of spread, possibly combined with reduction of the population below its economic threshold. To avoid confusion and controversy, management plans must clearly define the goal for a specific weed on a given land management unit.

The choice between eradication, containment, or no action after detecting an invading plant species depends on the plant's potential impact. If the plant is listed on federal, state, or county noxious weed lists, then established management plans may dictate the appropriate action. If the plant is unlisted, then other criteria may have to be used. Unfortunately, reliable criteria to determine the noxious potential of plants have not been established (Mack 1996). This is a very weak aspect of federal, state, and county noxious weed management policies. The lack of criteria probably reflects the lack of understanding of the complex interaction between the plant and its environment, rather than a failure of policymakers. General guidelines, such as the climatic characteristics associated with endemic populations, a plant's history in other countries, and its similarities to existing noxious weeds, may give some indication of its potential impact (Zamora et al. 1989); however, there are many problems with these generalizations. For example, climate-matching ignores dissimilar biotic factors among different regions (Mack 1996). How a plant species entering new regions interacts with other species (Bazzaz 1986) partly determines its noxious potential. A plant is noxious when the interaction threatens human activities or the native flora or fauna. Unfortunately, waiting to see how a plant interacts with endemic plant species may allow the invading species to become established. Once non-indigenous pests become established, they are usually permanent in ecological time and very difficult to manage (Coblentz 1990).

Eradication can be a confusing concept when considering the size of the eradication zone and the time period given to eliminate the pest population, which both depend on political and economic factors (Dahlsten 1986). The maximum size of an infestation that can be eradicated is an issue that should be decided after an infestation is found and the costs and benefits are analyzed. The size will be limited by financial resources, available technology, and social and geographical constraints. A 0.1-hectare infestation may be too large, if limited resources prevent even one appraisal survey. In contrast, the USDA Animal and Plant Health Inspection Service (APHIS) is attempting to eradicate witchweed (*Striga asiatica*) on thousands of hectares in the southeastern United States.

Dahlsten et al. (1989) identified the following factors that should be included in an economic analysis: 1) conflicting impacts, 2) relative potential danger, 3) ease of eradication, 4) ease of recolonization, 5) ease of reintroduction, 6) availability of alternative control strategies, and 5) action guidelines. Human health and potential environmental disruptions from the eradication treatments (Dahlsten 1986), creation of new pests and elimination of competing pests, and public and industrial support (DeBach 1964, Lelliott and Aitkenhead 1979) are other considerations. In the short-term, all of these factors may seem overwhelming; however, a long-term view of managing a pest weed must be adopted. An economic analysis of control programs for serrated tussock (*Nassella trichotoma*) indicated that the low annual cost of containment (annual herbicidal control of light infestations) was more expensive in the long term than several other options, including eradication (Auld et al. 1982). Initial costs of eradication were estimated to be high, but the net return was almost twice that of any other strategy.

If eradication is an option in a noxious weed management plan, then two major components of the plan must be established before any attempt at eradication will be successful. They are: 1) a method to detect new invaders before they have time to become established; and 2) an eradication plan that includes population dynamics, eradication technology, and an effective strategy.

Early Detection

Early detection of invading plant species can minimize spread (Navaratnam and Catley 1986). Early detection is required at multiple levels of organization. On a national level, the United States operates a cooperative pest survey and detection program through the USDA Animal and Plant Health and Inspection Service. This program can serve as an early reporting program to alert cooperators of newly found pests.

The most effective early detection efforts probably occur at the local level by land managers, pest management professionals, and property owners. Detection at this level can be improved by making people aware of the importance of detecting new invaders early, by increasing the public's awareness of potential invaders, and by training weed management professionals to use statistically valid sampling techniques

(Navaratnam and Catley 1986). Systematic sampling methods insure thorough and frequent surveys. A county, state or national forest, or other region could be divided into areas with similar features. Each type of area then would be assigned a priority for being surveyed, based on its probability of being invaded with weeds. For example, roadsides are disturbed habitats that are more susceptible to invasion (Forcella and Harvey 1983) and, therefore, require more frequent surveys.

Another method of detecting new species is through botanical surveys required for permits, licenses, and assessments by federal, state, and county agencies. For example, botanical surveys are required as part of the process to license hydroelectric projects. Species new to an area may be detected but not reported to weed managers. Educating people about weeds that are involved in environmental assessments may bring reports of new weed species to land managers.

Population Dynamics

Population dynamics are changes in a plant's distribution and abundance in response to the environment. Relating population dynamics to control will increase treatment effectiveness (Auld et al. 1978/1979). Population dynamics indicate the life cycle stage most vulnerable to eradication treatments, the time to deplete viable propagules from the seed bank, and the strategy to stop spread.

• Life Cycle

The life cycle of a plant is a continuum composed of mature plant, reproduction, dispersal, propagule, seedling, and vegetative growth stages (Sagar and Mortimer 1976). These stages, and the probability of surviving to each successive stage, indicate the point at which a plant's population may be managed most effectively (Radosevich and Holt 1984). The probability of surviving each stage and the transition between stages is determined by following changes in unmanaged populations. An eradication strategy based on these probabilities consists of treatments to augment high mortality periods. To stop spread quickly may require relying on established weed control practices, such as considering a seedling the most vulnerable stage. This may provide acceptable control, but not the most efficient and effective eradication strategy.

• Seed Bank

An invading plant species is not eradicated until its viable propagules are depleted from the soil. The persistence of viable propagules in the soil can be predicted from losses and replenishment by natural causes and eradication treatments. Cohen (1966) developed a model that predicts seed numbers in soil for annual plants after one year under natural conditions. The model includes the number of seeds after one year, the initial number of seeds in the soil, the fraction of seeds that germinate, the fraction of seeds that decay (and also are lost through predation), and the number of seeds produced per germinated seedling dependent on the probability of successful reproduction. This model does not consider that the seed bank often is composed of several year's contribution, and that germination, dormancy, and longevity of these propagules can vary by year of production (Argel and Humphreys 1983, Silvertown 1984).

Cohen's (1966) model can predict the years required in an eradication to deplete viable propagules from the soil by incorporating the probability of escaping control. Seed yield would then depend on natural factors affecting reproduction, and the probability of escaping eradication. Applying this modified model to common crupina (*Crupina vulgaris*) illustrates the model's predictive ability (Zamora et al. 1989). When 99.9% of common crupina is controlled, less than 1 seed/m^2 remains in the soil after two years. When 95% of common crupina is controlled, it takes 138 years before the seed bank declines to less than 1 seed/m^2. As long as control is less than 100%, there will always be viable seeds in the soil, emphasizing that reproduction must be stopped completely if eradication is to succeed.

• *Spread*

Spread of an invading plant species determines the area requiring survey, quarantine, or treatment. If biological control is a treatment in an eradication strategy, spread rate of an invading plant species determines whether the control organism will spread fast enough to be an effective biological control agent (Roughgarden 1986). An eradication plan may use biological controls temporarily to stop spread from a portion of the infestation. The important components of spread to an eradication program are population growth rate, spread rate, dispersal vectors, and the probability of establishment after dispersal.

Population growth rate is the rate of change in the number of individuals, and spread rate is the rate of change in the distribution of individuals. These rates of change are indicated by documented invasions and predictive models. *Echium plantagieum* spread in Australia at an exponential rate (Forcella et al. 1986). Downy brome (*Bromus tectorum*) initially invaded western North America slowly, but increased rapidly and culminated with limited growth (Mack 1981). Modeling growth rate is extremely complicated (Auld and Coote 1980, Roughgarden 1986). For example, a model for an isolated population may not predict growth in mixed populations (Harper 1977). Auld and Coote (1980) developed a model of a single species that shows exponential population growth and linear spread rate.

The model of Auld and Coote (1980) shows the importance of mobility and multiple foci. Greater mobility increased total population growth rate, and multiple foci increased spread and population growth rates compared to a centralized population. As the number of foci increase, the area occupied per unit time increases, and as the area increases, the rate of spread increases (Auld et al. 1982). Multiple foci are important before and after applying eradication treatments. Remnant populations escaping eradication treatments can increase at a faster rate than a single initial focus (Mack 1985). Invading plants that are highly mobile and have scattered foci must be eradicated quickly to prevent rapid spread.

Spread rate and pattern of plant species are affected by the number and type of dispersal vectors. Copious seed producers attract many different animals capable of dispersing seeds to numerous, widely scattered, diverse habitats (Howe and Estabrook 1977). However, for invading plant species, the type of dispersal vector may be most important (Carter and Prince 1981, Forcella 1985). Rate of spread is higher for weeds in cereal and forage seed than in lawn seed, livestock, or as garden and agricultural escapees (Forcella 1985).

Distant spread by livestock may be relatively slow (Forcella 1985), but local spread throughout the animals' range along game or cattle trails may be rapid. Even dispersal by rodents has distinctive spread patterns (Kraus 1983) that can affect an eradication program. For plant species with heavy seeds, the dissemination pattern is a solid front, compared to a diverse front of scattered individuals for lighter, wind dispersed seeds (Harper 1977).

The spread pattern also is a characteristic of the habitat. Forcella and Harvey (1983) discovered that most non-indigenous species are found along roadsides, and that their frequency decreases as distance from the roadside increases. Roadsides and plant communities that lack a closed tree canopy have comparatively more non-indigenous species. Fewer non-indigenous plants are found in plant communities at higher elevations. Spread in a disturbed habitat is by many isolated small populations, while spread in established vegetation is by an expanding front (Bazzaz 1986).

An invading plant species will spread only if it encounters susceptible habitats. The susceptibility of adjacent habitats to invasion may be the most important determinant of an invader's success in a new habitat (Newsom 1978). A susceptible habitat has microsites or safe sites (Harper 1977) with environmental conditions sufficient to promote growth. Safe site density and distribution in a habitat determine the probability that propagules of an invading plant species will encounter safe sites (Green 1983) and, therefore, determine the site's susceptibility to invasion.

The probability of establishment in a susceptible habitat also depends on the plant. Plant species with a wide ecological amplitude can invade a greater variety of habitats, increasing the probability that the species will spread. To become established, a plant also must disperse enough propagules to a habitat for the birth rate to exceed the death rate (Roughgarden 1986). In relatively intact vegetation, successful establishment requires a single introduction of many propagules or repeated introductions (Bazzaz 1986).

Eradication Technology

Because the goal of eradication is different from the goal of weed management, eradication treatments must have higher efficacy than conventional weed control. Demanding so much of standard weed control methods or newly developed eradication treatments requires careful consideration of the potential effect on human health and the environment (Dahlsten 1986). Eradication technology should be developed for different infestation locations and different stages in the plant's life cycle.

Eradication treatments must be tailored to each specific location that a plant inhabits, for maximum efficacy and safety. Nonchemical treatments such as grazing, mowing, burning, or physical removal may be used near residences, water, or other situations where pesticides may be inappropriate. Chemical treatments include herbicides, fumigants, and germination stimulants. The best eradication technology will probably be a combination of treatment methods. The eradication technology should include alternate treatments, in case the primary treatments fail. Alternate treatments should be developed with the same rigorous standards used in developing primary treatments.

Chemical treatments are best applied during the most susceptible stage of plant growth, but the timing also may depend on non-target vegetation. Dense, canopy vegetation may decrease the amount of spray solution contacting seedling plants, requiring application when the canopy is not present, a high delivery volume, or granular herbicide formulations. Application timing also can select phenotypes that germinate before or after the optimal control period (Radosevich and Holt 1984). This may require changing application timing, making multiple applications, or using an alternate treatment.

Eradication Strategy

An eradication strategy for an invading plant species is the step-by-step procedure to stop spread, to prevent reproduction, and to deplete viable propagules from the soil. The components of a comprehensive strategy are treatment location and schedule, quarantine measures, appraisal criteria, a revegetation plan, and a cost estimate.

• *Treatment Location*

Eradication treatments must be located strategically, and must cover sufficient area to stop spread. Strategically locating treatments requires assigning priorities to specific populations within the total infested area, based on the probability of that population spreading the infestation. The first priority is to eradicate satellite infestations, since they cause the fastest spread (Mack 1985, Auld and Tisdell 1986). The second priority is to eradicate the borders of large infestations, and the final priority is to control population growth in noneradication areas, such as the core of large infestations. An eradication plan for a specific weed may include noneradication areas, if resources or other factors require treating the infestation in stages. Noneradication areas need to be considered, since dispersal from them increases as population density increases (Menz et al. 1980/1981). Before priorities can be assigned to specific populations, a delimiting survey that identifies the border of the infested area must be undertaken.

Surveys usually work progressively outward from known foci of the infestation and along roads and railways (Lelliott and Aitkenhead 1979). The search area can be narrowed by identifying the invading plant's habitat-type and plant associations. Using these characteristics will improve roadside surveys (Forcella and Harvey 1983). Knowing where to survey can focus initial delimiting surveys on high-probability areas to establish major boundaries; later surveys can be more thorough. The physical characteristics of invaded sites and species characteristics, such as native latitudinal range and past invasion success, were used to predict sites likely to be invaded by cordgrass (*Spartina* spp.) in coastal areas (Daehler and Strong 1996). A survey should be expanded beyond the last occurrence to be reasonably sure that no plants were missed. The amount to expand the survey depends on the plant's potential dispersal distance. A useful rule-of-thumb is to survey to a geographical border or the limits of the habitat.

The importance of delimiting surveys is demonstrated by the experience of land managers attempting to eradicate common crupina in California. Common crupina's first discovery in California was on less than 0.01 ha in 1975 (Davis and Sherman 1991). The infestation was treated with a herbicide and considered eradicated until more than 100 ha were discovered 10 years later, located not more than 100 meters from the original site (Zamora, personal observation). This larger infestation probably already existed, but as a result of nonexistent or inadequate delimiting surveys was not found until later.

Experience and training will help surveyors to use multiple search images, such as the plant's various polymorphs, distinctive colors or silhouettes, habitat(s), and plant associations, to locate even single plants. Careful, thorough surveys are as important to an eradication effort as are effective treatments. Surveyors have been observed standing within arm's reach of a large target weed when they declared the area weed-free (Zamora, personal observation). It may be necessary to survey even small areas two or three times to find all of the weeds that survived the eradication treatment. Essential data to collect on the survey include location, plant size, density, and growth stage, habitat-type, plant associations, land use, and a geographical description of the infested area.

A minimal area must be treated to stop spread and decrease the initial infested area, whether an infestation is eradicated instantaneously or in portions. The minimal area that must be treated is an uninfested buffer zone on the periphery of the infestation, large enough to encompass the area plants would have invaded without treatment (Auld et al. 1982). The area, a, into which the plant spreads in the yth year, and thus the buffer zone to treat, is

$$a = \pi d^2(2y-1)$$

where d is the maximum distance to which 95% of the seeds disperse and y is the years of spread (Auld and Coote 1980). To decrease the initial infested area, treatments must extend the radius, d, into the infestation, or the area of the annulus calculated from this equation.

79

This strategy also can be applied to individual populations within the total infestation. The distance, *d*, does not account for the potentially longer dispersal distance that a minority of propagules may travel. The eradication procedure for this longer distance is an extended period of frequent detection surveys to insure that no plants escape.

• *Treatment Schedule*

Initial and follow-up treatments must be scheduled to prevent reproduction and to deplete viable propagules in the soil. Initial treatments eliminate the majority of plants, and follow-up treatments eradicate escaped plants. Small infestations may only require spot applications for follow-up treatments. Large infestations require frequent, costly surveys to find escaped plants. Single or multiple annual herbicide applications to the entire infestation for two or three years may decrease population density sufficiently so that fewer surveys will suffice to find escaped plants.

• *Quarantine*

Quarantine is the regulation of the production, movement, or existence of plants, plant products, animals, animal products, or any other article or material, or the normal activity of persons to prevent or limit introduction or spread of a pest (USDA-APHIS 1983). Quarantine and strategically located eradication treatments are an effective combination to stop spread. Strategically located treatments were only able to account for the distance (d in the equation above) to which 95% of seeds would disperse. Detection surveys and quarantine should stop spread from the potentially longer distance that the remaining seeds would disperse.

• *Revegetation*

Eliminating a weed leaves environmental resources available for exploitation and consequent reinvasion by the same or different undesirable plants, requiring regular retreatment (Auld and Tisdell 1986). Reinvasion can be stopped by increasing species diversity or improving groundcover, making a habitat undesirable to an invader or desirable to its natural enemies (Dahlsten 1986). Seeding may be necessary, depending on the desirability and seed longevity of the original, supplanted species.

Revegetation can prevent reinvasion, and can also lead to the extinction of remnant weed populations. A remnant population may become extinct if its density declines below the minimum level needed for birth rate to exceed death rate (May 1977). Revegetation can stop a weed population from reaching the minimum density required to sustain existence. Remnant populations of large weed infestations are extremely difficult and costly to find. In large eradication programs, relying on revegetation rather than surveys may be necessary to eradicate remnant populations.

• *Appraisal*

Appraisal surveys are needed to evaluate progress in eradicating a plant. Carefully surveying the entire area is most reliable. If the infestation is too large for a complete survey, and if there are many escaped plants that are evenly distributed, carefully examine a random sample of many small areas. If the escaped plants are aggregated, rapidly scanning the treated area will decrease the survey effort (Lelliott and Aitkenhead 1979). Appraisal surveys should extend to the distance from an infestation's periphery that escaped plants may have dispersed. Escaped plants are difficult to find, so appraisal surveys often are conducted several times per year. Presence of escaped populations, reproductive potential, and propagules in the soil are criteria used to assess eradication progress. When no propagules are found after sufficient time to deplete viable propagules from the seed bank, an infestation can be declared eradicated. If an infestation cannot be completely surveyed, points ranking eradication progress may be assigned to individual infestations, based on treatments applied and expected efficacy (USDA-APHIS 1983). Minimum points for release from eradication are based on probability of absence. Studies of treatment efficacy establish probability of absence based on expected control.

• *Cost*
Eradicating an invading plant species can vary in cost, depending on the strategy. For strategies other than instantaneous eradication, initial costs should be defined in terms of treatment area and rate of spread (Menz et al. 1980/1981). Costs generally increase as rate of spread increases, because both the area to be surveyed and the population increases. Costs per unit area in an eradication program are likely to increase as population density declines. As density falls, the cost of finding and killing each remaining weed will increase (Auld et al. 1978/1979). Compared to the cost of materials for an eradication program, administrative functions, including quarantine, are often the major cost (Menz and Auld 1977).

Conclusion

Eradication is a desirable, feasible choice while a plant's population is small. Large infestations require more knowledge, money, and effort to eradicate successfully. A comprehensive eradication plan is needed to eliminate invading plant species successfully. An established eradication plan facilitates quick action to stop an invasion from growing into a large infestation. A comprehensive eradication plan has provisions to detect and assess the threat of invasions by new plant species, to develop eradication technology, and to synthesize this information into a cohesive strategy with a high probability of success. Extended, detailed studies are not required for every invading plant species. The attention given to each part of the plan depends on the availability of information, extent of the infestation, and funding. If the infestation is small, less detailed information will suffice to design an effective eradication strategy for quick, limited action (Lelliott and Aitkenhead 1979).

General recommendations about size and cost of eradication programs are difficult to make. The complex biological, technological, social, geographical, and financial interactions allow no easy generalizations. A private landowner, working alone, might be able to eradicate a rangeland weed on less than two hectares. A county weed crew with a normal, limited budget and many other responsibilities probably could eradicate a rangeland weed on three to six hectares. Larger infestations require significant, committed resources. Any eradication requires persistent, dedicated personnel willing to conduct many careful, thorough surveys each year, for as many years as it takes to deplete all viable propagules from the soil.

The underlying theme of an eradication plan is the necessity for quick, informed action. When a new plant species is detected, alert local, state, and regional authorities. Gather enough information to make a timely, informed decision on the necessity and safety of eradication. Gather information from botanical, taxonomic, and weed control references. Specific information on biology and treatment efficacy may require a season of research. Synthesize this information into a strategy that will eradicate the invader quickly and successfully. Only quick, informed action will keep local, occasional plant invasions from becoming widespread, permanent weed problems.

Literature Cited

Argel, P.J., and L.R. Humphreys. 1983. Environmental effects on seed development and hardseededness in *Stylosanthes hamata* cv. Veran. I. Temperature. Aust. J. Agric. Res. 34:261-70.

Auld, B.A., and B.G. Coote. 1980. A model of a spreading plant population. Oikos 34:287-92.

Auld, B.A., and C.A. Tisdell. 1986. Impact assessment of biological invasions. *In:* R.H. Groves and J.J. Burdon (eds.), The Ecology of Biological Invasions. Cambridge University Press, Cambridge and New York, 79-88.

Auld, B.A., K.M. Menz, and N.M. Monaghan. 1978/1979. Dynamics of weed spread: implications for policies of public control. Prot. Ecol. 1:141-48.

Auld, B.A., D.T. Vere, and B.G. Coote. 1982. Evaluation of control policies for the grassland weed, *Nassella trichotoma*, in south-east Australia. Prot. Ecol. 4:331-38.

Bazzaz, F.A. 1986. Life history of colonizing plants: some demographic, genetic, and physiological features. *In* H.A. Mooney and J.A. Drake (eds.), Ecology of Biological Invasions of North America and Hawaii. Springer-Verlag, New York, 96-110.

Carter, R.N., and S.D. Prince. 1981. Epidemic models used to explain biogeographical distribution limits. Nature 293:644-45.

Coblentz, B.E. 1990. Exotic organisms: a dilemma for conservation biology. Bio. Cons. 4:261-65.

Cohen, D. 1966. Optimizing reproduction in a randomly varying environment. J. Theor. Biol. 12:119-29.

Daehler, C.C., and D.R. Strong. 1996. Status, prediction and prevention of introduced cordgrass *Spartina* spp. invasions in Pacific estuaries. USA. Bio. Cons. 78:51-58.

Dahlsten, D.L. 1986. Control of invaders. *In:* H.A. Mooney and J.A. Drake (eds.), Ecology of Biological Invasions of North America and Hawaii. Springer-Verlag, New York, 275-302.

Dahlsten, D.L., R. Garcia, and H. Lorraine. 1989. Eradication as a pest management tool: concepts and contexts. *In:* D.L. Dahlsten, R. Garcia, and H. Lorraine (eds.), Eradication of exotic pests: analysis with case histories. Yale University Press, New Haven, CT, 3-15.

Davis, L.H., and R.J. Sherman. 1991. *Crupina vulgaris* Cass. (Asteraceae: Cynareae), established in Sonoma County, California at Annadel State Park. Madrono 38:296.

DeBach, P. 1964. Some ecological aspects of insect eradication. Bull. Entomol. Soc. Am. 10:221-24.

Forcella, F. 1985. Final distribution is related to spread in alien weeds. Weed Res. 25:181-91.

Forcella, F., and S.J. Harvey. 1983. Eurasian weed infestation in western Montana in relation to vegetation and disturbance. Madrono 30:102-109.

Forcella, F., J.T. Wood, and S.P. Dillon. 1986. Characteristics distinguishing invasive weeds within *Echium* (bugloss). Weed Res. 26:351-64.

Green, D.S. 1983. The efficacy of dispersal in relation to safe site density. Oecologia (Berl.) 56:356-58.

Harper, J L. 1977. Population Biology of Plants. Academic Press, New York, NY.

Howe, H.F., and G.F. Estabrook. 1977. On intraspecific competition for avian dispersal in tropical trees. Am. Nat. 111:817-32.

Kraus, B. 1983. A test of the optimal-density model for seed scatterhoarding. Ecology 64:608-10.

Lelliott, R.A., and P. Aitkenhead. 1979. The eradication of diseases and pests from the United Kingdom: its practice and management. *In:* D.L. Ebbels and J.E. King (eds.), Plant Health, The Scientific Basis for Administrative Control of Plant Diseases and Pests. Blackwell Scientific Publications, Oxford, 185-97.

Mack, R.N. 1981. Invasion of *Bromus tectorum* L. into western North America: An ecological chronicle. Agro-Ecosystems 7:145-65.

Mack, R.N. 1985. Invading plants: their potential contribution to population biology. *In:* J. White (ed.), Studies on Plant Demography. Academic Press, Inc., London, 127-42.

Mack, R.N. 1996. Predicting the identity and fate of plant invaders: emergent and emerging approaches. Bio. Cons. 78:51-58.

May, R.M. 1977. Thresholds and breakpoints in ecosystems with a multiplicity of stable states. Nature 269:471-77.

Menz, K.M., and B.A. Auld. 1977. Galvanised burr control, and public policy toward weeds. Search 8:281-87.

Menz, K.M., B.G. Coote, and B.A. Auld. 1980/1981. Spatial aspects of weed control. Agric. Syst. 6:67-75.

Navaratnam, S.J., and A. Catley. 1986. Quarantine measures to exclude plant pests. *In:* R.H. Groves and J.J. Burdon (eds.), The Ecology of Biological Invasion. Cambridge University Press, Cambridge and New York, 106-12.

Newsom, L.D. 1978. Eradication of plant pests. Bull. Entomol. Soc. Am. 24:35-40.

Newsome, A E., and I.R. Noble. 1986. Ecological and physiological characters of invading species. *In:* R.H. Groves and J.J. Burdon (eds.), The Ecology of Biological Invasions. Cambridge University Press, Cambridge, 1-20.

Radosevich, S.R., and J.S. Holt. 1984. Weed Ecology. John Wiley and Sons, Inc., New York.

Roughgarden, J. 1986. Invasions and rates of spread. *In:* H.A. Mooney and J.A. Drake (eds.), Ecology of Biological Invasions of North America and Hawaii. Springer-Verlag, New York, 179-88.

Sagar, G.R., and A.M. Mortimer. 1976. An approach to the study of the population dynamics of plants with special reference to weeds. *In:* T.H. Coaker (ed.), Applied Biology Vol. I, 1-47. Academic Press, Inc., New York.

Silvertown, J.W. 1984. Phenotypic variety in seed germination behavior: the ontogeny and evolution of somatic polymorphism in seeds. Am. Nat. 124:1-16.

U.S. Department of Agriculture, Animal and Plant Health Inspection Service, Plant Protection and Quarantine. 1983. Domestic Program Manual: Witchweed (*Striga asiatica* Lour.). U.S. Government Printing Office, Washington, D.C.

Zamora, D.L., D.C. Thill, and R.E. Eplee. 1989. An eradication plan for plant invasions. Weed Technol. 3:2-12.

Grazing and Weeds

Bret E. Olson

Most weeds in western North America came from Eurasia. Their spread is often unchecked, because most weeds are highly aggressive, their native pathogens and invertebrate feeders were not introduced simultaneously with the plants, and most large herbivores in North America avoid weeds while grazing other plants. On range and some pasture lands, controlling weeds with herbicides is not always cost-effective (Griffith and Lacey 1991). Potentially, appropriate grazing management can be used to minimize the spread of certain weeds and to control large weed infestations.

The spread of many weeds can be related to two major factors: inherent preferences of grazing and browsing ungulates, and plant community succession. First, certain types of animals such as domestic sheep prefer forbs (broadleaved plants) over grasses and shrubs, whereas cattle prefer grasses and mule deer prefer shrubs. Most weeds, such as spotted knapweed (*Centaurea maculosa*), Dalmatian toadflax (*Linaria genistifolia* subsp. *dalmatica*), sulfur cinquefoil (*Potentilla recta*), and leafy spurge (*Euphorbia esula*) are forbs.

In grazed systems, these inherent diet preferences have been a major force in shifting species composition of native plant communities (Bowns and Bagley 1986). For example, in western North America the climate supported a perennial bunchgrass-shrub mosaic before it was settled. After cattle were introduced into this area in the late 1860s, selective grazing of bunchgrasses by cattle and effective wildfire suppression altered the balance of this system. In the late 1800s, non-native annual grasses, such as cheatgrass (*Bromus tectorum*) and medusahead (*Taeniatherum caput-medusae*), were introduced into this disturbed system, established quickly, and now dominate millions of acres (Mack 1981). Because cattle, the dominant grazer in this system, selectively graze native grasses and have low impact on these non-indigenous grasses and weeds, native species are at a disadvantage in competing for limited soil water and nutrients. This leads to a preponderance of weedy, undesirable species.

The second, interrelated factor is succession. Traditional grazing management on rangelands is based on plant succession. In concept, plant succession is the natural tendency for rangeland systems to undergo a directional, linear change through a series of plant communities (seral stages), without human intervention. One plant community replaces another until an endpoint, a stable state, is reached. This endpoint or climax for an area is determined by the climate and the soils of the area (Clements 1916). Change in plant communities away from this climax stage is considered retrogression. In most western rangeland systems, especially those west of the Continental Divide, this climax state would be dominated by perennial grasses with a minor component of native forbs and shrubs (Miller et al. 1994).

Cattle tend to avoid shrubs and forbs. Their constant grazing of native grasses has caused many rangeland ecosystems to retrogress from the climax or late seral stage of succession. There was little counteracting grazing pressure on the shrubs and forbs. In contrast, sheep prefer forbs and apply pressure on native and non-indigenous forbs, resulting in succession toward a climax stage.

However, this linear concept of succession may be too simplistic for some rangeland systems (Westoby et al. 1989). For example, the dominance of cheatgrass and medusahead in the intermountain West, partly caused by extensive overgrazing in the late 1800s and early 1900s, will not be rectified by simply removing cattle or by reducing their numbers. The new cheatgrass-dominated "steady state" will require massive levels of fossil fuel input via herbicides, seeding, fertilizing, etc., to restore the presettlement flora (Laycock 1991). In some instances, the dominance of rangeland ecosystems by weeds such as yellow starthistle (*Centaurea solstitialis*), spotted knapweed (*Centaurea maculosa*), whitetop (*Cardaria draba*), and leafy spurge may also represent new steady states, especially if these rangeland systems continue to be grazed only by cattle and horses, which avoid these plants. A potential solution is to introduce or reintroduce small ruminants to these disturbed systems, restoring a balance by using grazers that prefer non-indigenous forbs.

In the context of weeds and livestock grazing, different levels of weed infestation will lead to three different management strategies. These scenarios are: preventing weed infestations, minimizing the spread of weed infestations, and controlling weed infestations.

Preventing Weed Infestations

Often infestations of weeds are not noticed until they have become a major problem. Ideally, the best way to avoid this situation is to prevent weeds from infesting areas that do not have the weed. Preventing weed infestations means preventing dispersal of seeds into uninfested areas. Vehicles, humans, wind, water, birds, and grazing ungulates can spread weed seeds. Animals may disperse seeds by picking up seeds in their coat, fur, or between the pads of their feet. Cattle readily pick up burrs of houndstongue (*Cynoglossum officinale*) when grazing forested range (De Clerke-Floate 1997). Three of six sheep grazing a leafy spurge-infested area in southwestern Montana had significant numbers of seeds embedded in the fleece (Olson et al. 1997a). However, considering the barb-like nature of the wool follicle, these seeds would likely remain in the fleece until it was processed.

Ruminants also ingest weed seeds in the field (Olson et al. 1997a). Most weed seeds are destroyed within the gastrointestinal tract; however, 5% to 15% of leafy spurge and spotted knapweed seeds will pass through sheep, goats, and mule deer, and some of those seeds remain viable (Lacey et al. 1992, Wallander et al. 1995, Olson et al. 1997a). All viable leafy spurge seeds passed through sheep and goats within five days (Lacey et al. 1992); mule deer were still passing viable seeds when fecal collection ended on day 10 (Wallander et al. 1995). Given that land managers control movements of domestic ruminants, these grazers should not be moved from infested areas to uninfested areas when viable seed is present on the stems. Alternatively, the animals could be brought into a drylot for five to seven days to allow any viable weed seeds to pass. Partly because ungulates will consume seedheads or hay infested with weed seeds and then defecate viable seeds in uninfested areas, federal public land agencies are now requiring the use of certified weed seed-free forage on public lands in most western states.

Minimizing the Spread of Weeds

Cattle grazing has been and will continue to be an important economic force in western North America. Proper grazing management of beef cattle can minimize the spread of weeds. This may include one of three strategies: 1) moderate grazing levels, to minimize the physiological impact on native plants and to minimize soil disturbance; 2) intensive grazing to counteract inherent dietary preferences of cattle, resulting in equal impacts on all forage species including weeds; or 3) multispecies grazing.

Many native plants, especially west of the Continental Divide, did not evolve with herds of larger herbivores (Mack and Thompson 1982). Thus, these native plants are relatively intolerant of high grazing levels, especially during the growing season. Severe grazing reduces a plant's ability to fix carbon, and stops root growth (Crider 1955, Richards 1984). This can place a grazed plant at a great disadvantage when competing with an ungrazed weed for soil water and nutrients. In contrast to the effects of heavy grazing, moderate grazing has minimal effect on root growth (Crider 1955). In eastern Washington, the establishment of diffuse knapweed (*Centaurea diffusa*) was enhanced only when defoliation of the native bluebunch wheatgrass (*Agropyron spicatum*) exceeded 60%, suggesting defoliation above this level reduced competitiveness of the grass (Sheley et al. 1997).

Intensive, time-controlled grazing may be considered an alternative to traditional, moderate, season-long grazing systems. Conceptually, time-controlled grazing will minimize the grazers' ability to avoid "undesirable" plants, resulting in uniform use of pastures and balanced competitive relationships among native and non-indigenous species. Intensive cattle grazing reduced the number of seedlings and rosettes of oxeye daisy (*Chrysanthemum leucanthemum*), a non-indigenous forb, but the impact was attributed more to trampling than to cattle actually consuming significant amounts of the forb (Olson et al. 1997b). Sheep or goats would have

87

grazed this weed more readily than cattle (Howarth and Williams 1968). On shortgrass prairie in Wyoming, undesirable forbs increased more with time-controlled grazing than with season-long or rotational deferred grazing (Manley et al. 1997).

Whichever type of grazing system is used, managers should alternate the season of use on each pasture. Most native range plants have a period during their rapid growth phase when they are highly susceptible to defoliation. This usually occurs when the plant is bolting, or sending up the developing flowerhead (Blaisdell and Pechanec 1949, Olson and Richards 1988, Kennett et al. 1992). To promote a healthy range ecosystem, a land manager should not allow animals to graze the same plants at the same time year after year. Most plants can withstand one year of heavy grazing during this active growth period, but it is often the cumulative effect of grazing at the same time year after year that increases the probability of plant mortality.

Finally, a land manager could introduce multispecies grazing, for example adding sheep or goats or both to a cattle ranch infested with leafy spurge (Williams et al. 1996). Because of the inherent grazing preferences of different classes of livestock, multispecies grazing distributes the impact of livestock more uniformly across the pasture and among plant species, including weeds (Walker 1994). On moderately-stocked rangelands, one ewe can be added per cow without reducing cattle production (Glimp 1988). Besides direct positive effects on individual animal performance and production per unit area (Walker 1994), grazing sheep or goats among cattle may even help to reduce losses associated with predation (Baker 1985).

Controlling Weed Infestations

At times, the land manager's objective will be to control the spread or density of a weed infestation. An important part of controlling weed infestations will be to use the appropriate animal, considering the inherent dietary preferences of different types of ruminants and the effect on plant succession. Small ruminants are the most likely animal to control most, but not all, weed infestations. Some weeds such as bitterweed (*Hymenoxys odorata*) and larkspur (*Delphinium* spp.) are just too toxic (Manners et al. 1992, Taylor and Ralphs 1992) for ruminants to ingest enough of the plant to control dense populations.

Small ruminants may avoid extremely dense infestations of weeds, even when the weeds are not toxic. Sheep with considerable experience grazing leafy spurge went "off feed" when confined to dense patches of leafy spurge, but resumed consuming this weed when allowed to graze less dense areas (personal observation). All ruminants require a diverse diet. For example, even excessive consumption of preferred foods resulting from limited availability of other foods will reduce forage intake (Provenza 1996). Dense infestations may also physically deter movement into the weed infestation, or may provide excellent hiding cover for predators.

Selecting the appropriate grazer is important. Cattle grazing was compared with a herbicide to control redroot pigweed (*Amaranthus retroflexus*), velvetleaf (*Abutilon theophrasti*), common ragweed (*Ambrosia artemisiafolia*), and several foxtail species (*Setaria* spp.), while establishing a stand of big bluestem (*Andropogon gerardii* var. *gerardii*) (Lawrence et al. 1995). Not surprisingly, cattle did not use the redroot pigweed (because of high nitrate levels) or the other forbs, so grazing to control weeds in this trial was unsuccessful. Presumably, sheep would have grazed these plants more readily than cattle.

Similarly, cattle avoid grazing areas infested with leafy spurge (Lym and Kirby 1987), but sheep readily graze leafy spurge (Landgraf et al. 1984), and thereby can impact the plant community. In Saskatchewan, continuous summer-long sheep grazing reduced the number of leafy spurge seeds in the soil from more than 3,500 seeds/m^2 to 15 seeds/m^2 after eight years (Bowes and Thomas 1978). Seedbanks and perennial shoot densities should be reduced concurrently to control any competitive species over the long term (Bowes and Thomas 1978, Marlette and Anderson 1986). Often, aboveground disturbances such as mowing, fire, and grazing will at least initially increase stem densities by removing apical dominance and stimulating growth of root buds of certain weed species (Selleck et al. 1962, Olson and Wallander 1998). Over time, continuous sheep grazing will reduce stem densities of leafy spurge. In Saskatchewan, summerlong continuous sheep grazing had no effect on stem densities for the first three years, after which densities declined dramatically (Bowes and Thomas 1978).

Once an appropriate animal species has been selected, the land manager needs to decide when, how much, and how often to graze the animals to have maximum impact on the weed with minimum impact on the desirable species. The ideal would be to graze the infestation when the weed is most susceptible to defoliation, as the flowerhead is elevating or "bolting" (Kennett et al. 1992). However, if the flowerheads of the weed and preferred vegetation are elevating at the same time, an alternative grazing period may need to be identified to minimize the impact on the desirable vegetation. For example, in western North America, the development of many weeds and desirable species occur simultaneously in late May and early June, so grazing should be fairly light at this time. In southcentral Montana, several bands of sheep are moved quickly in early summer through foothill rangelands covered with leafy spurge (Wayne Pearson 1994, personal communication). The land managers want the sheep to graze the upper leaves and developing flowerheads, thereby removing the canopy that shades the desirable grasses, and then quickly move onto the next pasture. Grass forage has increased dramatically on these pastures, which cattle graze later in summer.

In western North America, many weeds remain green and physiologically active later in the summer when other species are dormant. This may reflect the ability of deep taproots of most weeds to use soil moisture deep in the soil profile. Besides remaining greener longer into the summer, the weeds are also more nutritious than the dormant cool-season grasses at this time (Olson 1997, unpublished data). Thus, small ruminants are even more attracted to these weeds. Stocking rates can be higher at this time to maximize impact on the weed while minimizing impact on the desirable, dormant cool-season native grasses and forbs.

Unfortunately, land managers cannot control how much each plant is grazed in a pasture. Yet by selecting the appropriate livestock species, a manager increases the probability that the undesirable plant species will be grazed. Some livestock producers have claimed that livestock and their young which are "experienced" with a weed will graze the weed more readily than "inexperienced" animals of the same species. In a field study, yearling sheep exposed to dense infestations of leafy spurge as lambs grazed more leafy spurge in early summer than yearlings which had never been exposed to this weed (Olson et al. 1996). However, this difference in use patterns was not biologically significant in affecting leafy spurge; neither group ate leafy spurge much in early summer. Further, the difference in use of leafy spurge was short-lived. Within three to four weeks, inexperienced yearlings were grazing as much leafy spurge as experienced yearlings. In that study, the yearlings were not grazing with their mothers. Presumably, the experienced yearlings would have grazed more leafy spurge in early summer if their mothers had been present (Mirza and Provenza 1990, Nolte et al. 1990).

All animals need variety in their diet, so promoting more intense grazing during the active growing season increases the probability that desired species will be grazed. Light to moderate (up to 60%) consumption of the desirable species should not be exceeded during the active growing season. When the selected animal species prefers the target weed species, weed consumption should be considerably higher than consumption of the desirable species. Grazing must be monitored closely to ensure that the desirable species are not used excessively. Potentially, weed species can be more intensely grazed when desirable species are dormant.

A land manager also needs to decide how often weed-infested pastures should be grazed by a selective grazer. Many weeds deter grazing by secondary compounds, tolerate grazing when defoliated because of their deep taproots, or are successful at both. Usually, a single defoliation of these weeds, even when they are most susceptible during bolting, has minimal (if any) impact (Kennett et al. 1992). This suggests that repeated grazing is often needed to reduce the competitiveness of these weeds. Most small ruminants are herded, so it would be relatively easy to incorporate repeated grazing into a management plan. Regrowth on previously grazed plants is often very high in crude protein and digestibility, and is preferred by most animals (Olson 1997, unpublished data).

Based on the traditional forage quality measures—crude protein and digestibility—many weeds are highly nutritious and readily digested during the growing season (Bosworth et al. 1985, 1990). However, most North American large herbivores avoid these plants, at least partly because of defensive secondary compounds in the plant (Locken and Kelsey 1987). Ruminants may avoid grazing plants containing secondary compounds because the compounds deter herbivory by taste or smell, or are toxic to the ruminant directly or to rumen microbial populations (Striby et al. 1987, Kronberg and Walker 1993, Provenza 1995, Olson and Kelsey 1997). This reduces forage intake and/or causes general malaise, resulting in aversive postingestive feedback (Provenza 1996). Small ruminants, especially goats, appear to have several mechanisms that detoxify certain secondary compounds (Hofmann 1989, Kronberg and Walker 1993).

Many grazing systems in the western United States are based on rest-rotation grazing, whereby a pasture is grazed relatively heavily two out of every three years, and then not used for one year. Grazing is not repeated within a growing season. Most weed species are well adapted to germinate and grow quickly on heavily grazed areas. Resting a pasture after two relatively heavy years of grazing may allow the weed to gain a foothold. Thus, in some cases rest-rotation grazing may actually be undesirable in terms of controlling weeds.

Although grazing with appropriate livestock can be used to control weed infestations, grazing will not in most cases eradicate a mature infestation of weeds. Many weeds are clonal, short- to long-lived perennials, and reproduce vegetatively soon (6 to 12 months) after a disturbance, such as fire, mowing, grazing, or even herbicide use. However, when small ruminants graze mature plants, in time the ruminants reduce the ability of these weeds to compete with neighboring plants. Further, many weeds produce seeds that remain viable for 10 to 20 years or more. Small ruminants will remove seedheads, minimizing seed production for that year (Olson et al. 1997c). Small ruminants also readily graze seedlings (Olson et al. 1997c). The seedling stage is probably the least resistant stage of many weed species to disturbances such as frost heaving or grazing (Sheley and Larson 1994, Olson et al. 1997c) because of the small root systems, and because livestock prefer seedlings over mature plants.

A long-term commitment to small ruminant grazing is needed because many weeds can compensate quickly after the grazing pressure is removed. They compensate quickly because of their long-lived seeds in the soil, and because they can rapidly increase flower stem production once grazing pressure is removed (Olson et al. 1997c).

Small ruminant grazing should be considered an integral part of a grazing management plan. The landowner or manager can consider the weed as a resource rather than simply as a problem, and thus make a long-term commitment to managing weeds by grazing small ruminants.

Most weeds were introduced into North America in the early 1900s, but the weed populations did not begin to increase exponentially until the 1940s and 1950s. In the United States, the number of sheep peaked in the mid-1940s and has declined steadily since for social and economic reasons. Cattle numbers peaked in the mid-1970s and remain high. Thus, the trend is away from using the type of animal that will graze these weeds, and toward animals that avoid these weeds. Today, there are too few small ruminants to begin to have an effect on these weeds (Fay 1991).

Integrated Management

Combining small ruminant grazing with other weed management tools has considerable promise. For example, grazing leafy spurge with sheep or goats during the spring and summer will remove excess canopy and stimulate root buds to grow in the fall. Followed by a fall application of an appropriate herbicide, like picloram, there is less dead material to intercept the herbicide and the herbicide will act on rapidly growing material. In the 1990s, various combinations of grazing goats and fall or spring applications of herbicides were compared at two sites in North Dakota (Lym et al. 1997). Averaged across two locations, the most effective treatments were picloram applied in the fall alone or preceded by spring grazing. Although the combination of herbicide and grazing was not the most effective treatment, the grazing animals represented a product from the system, whereas using herbicide alone represents an ongoing cost with no additional product.

The combined effects of small ruminant grazing and insect biological control agents have been assessed in small-scale field trials (Hansen 1993). After three years, sheep grazing and the flea beetle (*Aphthona nigriscutis*) reduced densities of leafy spurge more than sheep grazing or the flea beetles did alone.

Economics

Fay (1991) characterized three phases of infestations of leafy spurge and other range weeds. Phase One is characterized by 20 plants per ranch, Phase Two by 20 patches per ranch, and Phase Three by 20,000 acres (8,000 hectares) per ranch. Each phase has a certain weed management tool or combination of tools which is most cost-effective for that particular phase. For example, herbicides are most cost-effective for Phase One infestations, whereas small ruminant grazing is most cost-effective for Phase Three infestations. Phase Two infestations are problematic. They may be uneconomical to control with herbicides, yet they may not be large enough to warrant bringing in a different livestock species to control the weed.

When the management objective is to control a weed infestation (Phase Three), incorporating appropriate livestock species into the management scheme can be economically and ecologically effective. Admittedly, there will be some additional costs, but these costs need to be compared with other possible tools for that level of infestation. In most cases, land managers cannot afford to spend $15 per acre to control spotted knapweed, or $20 per acre to control leafy spurge, when the immediate forage increase per acre will not pay for that treatment cost. Yet land managers must control the spread of weed infestations, or the land becomes worthless (Weiser 1995), at least for cattle grazing. In addition, grazing with small ruminants results in a product (meat, fiber, or both) to the land manager.

Adding sheep to a cattle enterprise for control of weeds can potentially be very profitable. The economic feasibility of grazing sheep on a leafy spurge-infested cattle ranch was assessed with a simple spreadsheet model, whereby inputs and outputs could be easily modified to fit any enterprise (Williams et al. 1997). In a typical scenario, on a 12,260-acre (4,905-ha) cattle ranch infested with 1,300 acres (520 ha) of leafy spurge, annual returns of adding sheep to graze leafy spurge exceeded total costs by $4,675. The break-even lamb price was $0.53/lb ($1.16/kg). Returns were higher when leafy spurge was concentrated in fewer pastures, because less sheep fence had to be tacked onto barbed wire fences. Although not in the model, a major benefit from adding sheep to a cattle ranch infested with leafy spurge is the positive return of a product from the land, compared with the cost of repeated herbicide applications.

A sheep or goat producer can be allowed access to infested areas, or a cattle operation can diversify by owning small ruminants. Historically, prices paid for lambs and calves have been offset. Diversifying helps a producer manage through the low price cycles of a particular animal species.

Predation

Loss of small ruminants to predators is often cited as the reason why more land managers do not use small ruminants to control weeds. Usually predation losses are extreme for a few producers, not all producers. Potential tools to minimize losses to predation include herders, guard animals, portable power fencing, and possibly multispecies grazing. Unfortunately, good herders are not always available. Guard dogs, llamas, and burros are being used with generally good results. The best results occur where these guard animals have minimal contact with humans and remain protective of the flock or band rather than wanting to socialize with humans. On a small scale, certain types of power fencing are effective at keeping small ruminants in dense infestations while deterring coyotes, bears, and domestic and feral dogs. In some cases, cattle will deter coyotes and dogs, as long as the sheep are bonded to the cattle and move toward them when they sense a disturbance.

Some losses to predators can be tolerated, because the loss of the ruminant products (meat and fiber) to the predator is offset by the animals providing a service, weed control. In addition, the livestock producer often does not have to compensate the landowner for the forage consumed, and thus can more readily justify some losses to predators.

Conclusion

Most small ruminants can be valued for their ability to provide a service—weed control—as well as traditional fiber and meat products. Today, sheep are being used to control leafy spurge along several major rivers in Montana. Money is not exchanged between landowner and livestock producer. The sheep control the weeds while grazing a nutritious forage resource, and release forage preferred by cattle.

Although many weeds are a resource for small ruminants, weeds still alter ecosystem structure and function (Olson, Chapter 1, this volume), and by their nature will always have the potential to spread. Yet weed infestations are not a problem that arrived overnight, and will not be cured overnight. Too often our most common weed control techniques address the symptom of a land management problem, not the cause. If the cause of the problem is not addressed, weeds will continue to spread. Or, if we successfully eradicate a weed, another weed will fill that niche because we have not solved the underlying problem. Reintroducing small ruminants to many of our western systems while managing large ungulates properly will help to restore that balance to the ecosystem which existed before European settlement. It will also begin to address the cause of the problem.

Literature Cited

Baker, F.H. 1985. Multispecies grazing: the state of the science. Rangelands 7:266-69.

Blaisdell, J.P., and J.F. Pechanec. 1949. Effects of herbage removal at various dates on vigor of bluebunch wheatgrass and arrowleaf balsamroot. Ecol. 30:298-305.

Bosworth, S.C., C.S. Hoveland, and G.A. Buchanan. 1985. Forage quality of selected cool-season weed species. Weed Sci. 34:150-54.

Bosworth, S.C., C.S. Hoveland, G.A. Buchanan, and W.B. Anthony. 1990. Forage quality of selected warm-season weed species. Agron. J. 72:1050-54.

Bowes, G.G., and A.G. Thomas. 1978. Longevity of leafy spurge seeds in the soil following various control programs. J. Range Manage. 31:137-40.

Bowns, J.E., and C.F. Bagley. 1986. Vegetation responses to long-term sheep grazing on mountain ranges. J. Range Manage. 39:431-34.

Clements, F.E. 1916. Plant succession. Carnegie Institute of Washington, Pub. 242.

Crider, F.J. 1955. Root growth stoppage resulting from defoliation of grass. USDA. Tech. Bull. 1102.

DeClerck-Floate, R. 1997. Cattle as dispersers of hound's tongue on rangeland in southeastern British Columbia. J. Range Manage. 50:239-43.

Fay, P.K. 1991. Controlling leafy spurge with grazing animals. In: L.F. James, J.O. Evans, M.H. Ralphs, and R.D. Child (eds.), Noxious range weeds. Westview Press, Boulder, CO, 193-99.

Glimp, H.A. 1988. Multi-species grazing and marketing. Rangelands 10:275-78.

Griffith, D., and J.R. Lacey. 1991. Economic evaluation of spotted knapweed (*Centaurea maculosa*) control using picloram. J. Range Manage. 44:43-47.

Hansen, R. 1993. Effects of *Aphthona* flea beetles and sheep grazing in leafy spurge stands. Proc. 1993 Leafy Spurge Symposium, Silvercreek, CO.

Hofmann, R.R. 1989. Evolutionary steps of ecophysiological adaptation and diversification of ruminants: a comparative view of their digestive system. Oecologia 78:443-57.

Howarth, S.E., and J.T. Williams. 1968. Biological flora of the British Isles. J. Ecol. 56:585-95.

Kennett, G.A., J.R. Lacey, C.A. Butt, K.M. Olson-Rutz, and M.R. Haferkamp. 1992. Effects of defoliation, shading and competition on spotted knapweed and bluebunch wheatgrass. J. Range Manage. 45:363-69.

Kronberg, S.L., and J.W. Walker. 1993. Ruminal metabolism of leafy spurge in sheep and goats: a potential explanation for differential foraging on spurge by sheep, goats, and cattle. J. Chem. Ecol. 19:2007-17.

Lacey, J.R., R. Wallander, and K. Olson-Rutz. 1992. Recovery, germinability, and viability of leafy spurge (*Euphorbia esula*) seeds ingested by sheep and goats. Weed Technol. 6:599-602.

Landgraf, B.K., P.L. Fay, and K.M. Havstad. 1984. Utilization of leafy spurge (*Euphorbia esula*) by sheep. Weed Sci. 32:348.

Lawrence, B.K., S.S. Waller, L.E. Moser, B.E. Anderson, and L.L. Larson. 1995. Weed suppression with grazing or atrazine during big bluestem establishment. J. Range Manage. 48:376-79.

Laycock, W.A. 1991. Stable states and thresholds of range condition on North American rangelands: a viewpoint. J. Range Manage. 44:427-33.

Locken, L.J., and R.G. Kelsey. 1987. Cnicin concentrations in *Centaurea maculosa*, spotted knapweed. Biochem. Syst. Ecol. 15:313-20.

Lym, R.G., and D.R. Kirby. 1987. Cattle foraging behavior in leafy spurge (*Euphorbia esula*)-infested rangeland. Weed Technol. 1:314-18.

Lym, R.G., K.K. Sedivec, and D.R. Kirby. 1997. Leafy spurge control with angora goats and herbicides. J. Range Manage. 50:123-28.

Mack, R.N. 1981. The invasion of *Bromus tectorum* L. into western North America: an ecological chronicle. Agro-ecosystems 7:145-65.

Mack, R.N., and J.N. Thompson. 1982. Evolution in steppe with few large, hoofed mammals. Amer. Natur. 119:757-73.

Manley, W.A., R.H. Hart, M.J. Samuel, M.A. Smith, J.W. Waggoner, and J.T. Manley. 1997. Vegetation, cattle, and economic responses to grazing strategies and pressures. J. Range Manage. 50:638-46.

Manners, G.D., J.A. Pfister, M.H. Ralphs, K.E. Panter, and J.D. Olsen. 1992. Larkspur chemistry: Toxic alkaloids in tall larkspurs. J. Range Manage. 45:63-67.

Marlette, G.M., and J.E. Anderson. 1986. Seed banks and propagule dispersal in crested wheatgrass stands. J. Appl. Ecol. 23:161-75.

Miller, R.F., T.J. Svejcar, and N.E. West. 1994. Implication of livestock grazing in the intermountain sagebrush region: plant composition. *In*: M. Vavra, W.A. Laycock, R.D. Pieper (eds.), Ecological implications of livestock herbivory in the west. Soc. Range Manage. Denver, CO.

Mirza, S.N., and F.D. Provenza. 1990. Preference of the mother affects selection and avoidance of foods by lambs differing in age. Appl. Anim. Behav. Sci. 28:255-63.

Nolte, D.L., F.D. Provenza, and D.F. Balph. 1990. The establishment and persistence of food preferences in lambs exposed to selected foods. J. Anim. Sci. 68:998-1002.

Olson, B.E., and J.H. Richards. 1988. Annual replacement of the tillers of *Agropyron desertorum* following grazing. Oecologia 76:1-6.

Olson, B.E., and R.G. Kelsey. 1997. Effect of *Centaurea maculosa* on sheep rumen microbial activity and mass in vitro. J. Chem. Ecol. 23:1131-44.

Olson, B.E., and R.T. Wallander. 1998. Effect of sheep grazing on a leafy spurge-infested Idaho fescue community. J. Range Manage. 51:247-52.

Olson, B.E., R.T. Wallander, V.M. Thomas, and R.W. Kott. 1996. Effect of previous experience on sheep grazing leafy spurge. Appl. Anim. Behav. Sci. 50:161-76.

Olson, B.E., R.T. Wallander, and R.W. Kott. 1997a. Recovery of leafy spurge seed from sheep. J. Range Manage. 50:10-15.

Olson, B.E., R.T. Wallander, and P.K. Fay. 1997b. Intensive cattle grazing of oxeye daisy. Weed Technol. 11:176-81.

95

Olson, B.E., R.T. Wallander, and J.R. Lacey. 1997c. Effects of sheep grazing on a spotted knapweed-infested Idaho fescue community. J. Range Manage. 50:386-90.

Provenza, F.D. 1995. Postingestive feedback as an elementary determinant of food selection and intake in ruminants. J. Range Manage. 48:2-17.

Provenza, F.D. 1996. Acquired aversions as the basis for varied diets of ruminants foraging on rangelands. J. Anim. Sci. 74:2010-20.

Richards, J.H. 1984. Root growth response to defoliation in two *Agropyron* bunchgrasses: field observation with an improved root periscope. Oecologia 64:21-25.

Selleck, G.W., R.T. Coupland, and C. Frankton. 1962. Leafy spurge in Saskatchewan. Ecol. Monogr. 32:1-29.

Sheley, R.L., and L.L. Larson. 1994. Observation: Comparative life-history of cheatgrass and yellow starthistle. J. Range Manage. 47:450-56.

Sheley, R.L., B.E. Olson, and L.L. Larson. 1997. Effect of weed seed rate and grass defoliation level on diffuse knapweed. J. Range Manage. 50:33-37.

Striby, K.D., C.L. Wambolt, R.G. Kelsey, and K.M. Havstad. 1987. Crude terpenoid influence on in vitro digestibility of sagebrush. J. Range Manage. 40:244-48.

Taylor, C.A., and M.H. Ralphs. 1992. Reducing livestock losses from poisonous plants through grazing management. J. Range Manage. 45:9-12.

Walker, J.W. 1994. Multispecies grazing: the ecological advantage. Sheep Research J. Special Issue 52-64.

Wallander, R.T., B.E. Olson, and J.R. Lacey. 1995. Spotted knapweed seed viability after passing through sheep and mule deer. J. Range Manage. 48:145-49.

Weiser, C. 1995. Economic effects of invasive weeds on land values from an agricultural banker's perspective. Proc. Symp. Alien plant invasions: Increasing deterioration of rangeland ecosystem health. BLM/OR/WA/PT-95.

Westoby, M., B. Walker, and I. Noy-Meir. 1989. Opportunistic management for rangelands not at equilibrium. J. Range Manage. 42:266-74.

Williams, K.E., J.R. Lacey, and B.E. Olson. 1996. Economic feasibility of grazing sheep on leafy spurge-infested rangelands in Montana. J. Range Manage. 49:372-74.

Biological Control of Noxious Rangeland Weeds

Linda M. Wilson and Joseph P. McCaffrey

Management of rangeland weeds traditionally has relied heavily on conventional weed control technologies, yet new weeds continue to emerge, causing severe forage losses and threatening native plant diversity. Faced with fewer and more expensive herbicides, increased weed resistance to herbicides, environmental health concerns, biodiversity awareness, water quality issues, and a declining forage base, rangeland users and managers are impelled to look toward alternative management strategies to broaden their arsenal of management tools. One useful and proven rangeland weed management tool is biological control.

The reader may refer to a large body of literature on biological weed control, most notably DeBach (1964), Huffaker and Messenger (1976), Van Driesche and Bellows (1996), and Charudattan and Walker (1982). Reviews by Frick (1974), Andres (1982), Harris (1991), and McFadyen (1998) provide useful background information, theoretical overview, and practical guidelines for applied biological control of weeds. The purpose of this chapter is to review the application of biological control of weeds on rangelands in the United States, discuss the ecological, practical, and political considerations in developing and implementing a biological weed control program, and propose alternative viewpoints on how to evaluate biological control programs within the context of ecologically sound, long-term rangeland rehabilitation.

Definition of Biological Control

Biological control can be defined as the deliberate introduction or manipulation of a pest's natural enemies, with the goal of suppressing the pest population (Wilson and Huffaker 1976). Since the late eighteenth century, biological control has become a useful and important tool in pest management. Applicable to a broad array of pests, biological control has been used against invertebrates (insects, mites, and snails), vertebrates (rabbits and birds), plant pathogens (fungi, bacteria, and viruses), and weeds (terrestrial and aquatic).

The first record of an intentional biological control of weeds program was the introduction of a cochineal insect (*Dactylopius ceylonicus*) to India in 1795 from Brazil, to control an introduced cactus (*Opuntia vulgaris*) (Moran and Zimmerman 1984). Successful, large-scale biological control of weeds began in the late 1800s with the importation of the moth *Cactoblastis cactorum* from Argentina for the control of prickly pear cactus (*Opuntia* spp.) in Australia. The first use of biological weed control in the United States was in 1903 with the importation of natural enemies into Hawaii for the control of lantana (*Lantana camara*) (Goeden 1988). The next major program of biological weed control started in 1939 with the importation of two leaf-feeding beetles (*Chrysolina* spp.) for the control of St. Johnswort (*Hypericum perforatum*). In the United States, other programs followed, largely encouraged by the success of the St. Johnswort program and expanding cooperation between the United States (USDA Agricultural Research Service) and Canada (Agriculture Canada). Between 1950 and 1990, programs were initiated for 23 additional species or species groups of terrestrial weeds in North America, so that by 1997, 165 species of natural enemies (insects, mites, nematodes, and phytopathogens) were imported and released in the continental United States and Canada (Goeden 1993, Van Driesche and Bellows 1996).

Virtually all of the terrestrial weeds targeted for biological control worldwide are non-indigenous perennial rangeland weeds (Julien 1992). Most noxious weeds infesting western rangelands in North America were introduced from Europe. The negative impacts of non-indigenous weeds are exacerbated by a complementary problem: the majority of these plants arrived without the complex of insects, mites, pathogens and other natural enemies that coexisted with them in their native countries. In their native ranges, these natural enemies helped to keep the plant populations at low, relatively stable densities by mediating the host's ability to compete in its environment. However, in the absence of these natural enemies, and in conjunction with favorable environmental and climatic conditions, disturbed rangeland habitats, and declining biodiversity, these non-indigenous plants became aggressive invaders of North American rangelands.

The aim of biological control is not to eradicate the target weed, but rather to exert enough pressure on the weed to reduce its dominance to a more acceptable level. The advantages of biological weed control are numerous. Biological control is cost-effective, environmentally safe, self-perpetuating, and is well-suited to integration in an overall weed management program. The limitations of biological control are that it is a long-term undertaking, and its effects are neither immediate nor always adequate. Following introduction, several years are generally required for a natural enemy population to establish in the new environment, reproduce, and increase its density on the host weed to produce a discernible impact on the weed population. In addition, several introductions may be required to enable natural enemy populations to establish over a large area.

When natural enemies are imported from a weed's native country and released into the new environment, the process is called "classical biological control." Biological control of weeds generally uses the classical approach, since most rangeland weeds and their natural enemies are not indigenous to North America. The distinguishing features of classical biological control are: 1) the importation of natural enemies for use as biological control agents; 2) the reliance on the natural enemies' ability to reproduce in the field and build large populations; and 3) the stability of self-sustaining populations of the agent(s) in the field (McClay 1989). Other approaches to biological control exist. These depend largely on mass-rearing programs and periodic release of biological control agents (known as "augmentation"), and steps taken to protect or conserve the agent population (known as "conservation"). While these latter approaches are used primarily in the biological control of crop pests such as insects and mites, redistribution of weed natural enemies and adjustments in land management practices to benefit weed biological control also fall into these latter categories.

Ecological Basis of Biological Control

Progress toward the effective use of biological control of weeds depends on an understanding of the factors influencing the interaction between the natural enemy and its weed host. Biological weed control is founded on two broad ecological principles: that one organism can be used to control another, and that some control organisms have a limited host range (Kok 1974, McEvoy 1996). In natural ecosystems, plant-eating (phytophagous) organisms are largely responsible for regulating plant populations (Strong et al. 1984), usually in conjunction with competition (Luck et al. 1995, Sheppard 1996). Figure 1 illustrates the desirable outcome of biological control of weeds. Biological control is generally initiated when the weed population reaches a density above the economic threshold, the level at which eradication and containment is considered economically acceptable. Following the release of one or more natural enemies, the weed population density declines over time, ideally to a stable, non-damaging level.

Host-specific insects, those that exhibit a limited host range, are adapted to their host plant (Strong et al. 1984). Similarly, plants have evolved mechanisms to counteract the action of their closely co-evolved herbivores (Trumble et al. 1993). Some researchers have proposed that these highly evolved associations are a hindrance to biological control, because plants can tolerate the effect of insects (Myers 1985), or they can compensate for the damaging influence of the herbivores (Maschinski and Whitham 1989). Hokkanen and Pimental (1984) proposed the concept of "new associations," which suggests that new insect-plant associations might have a greater likelihood of successful weed control. However, this hypothesis has not been widely accepted because of concerns for host-specificity and safety (Goeden and Kok 1986).

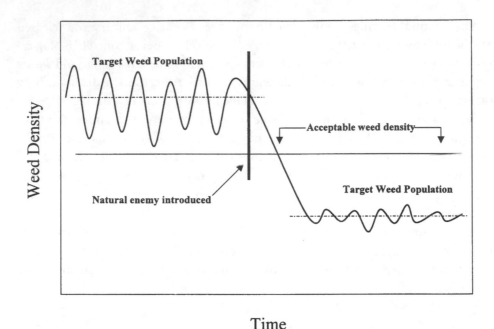

Figure 1. Generalized model of biological control of weeds (adapted from Luck et al. 1995).

Natural Enemies

Natural enemies used as biological control agents for rangeland weeds include phytophagous insects, mites, nematodes, pathogens, and vertebrates (e.g., sheep and goats). Of the 165 species of natural enemies imported for biological control of terrestrial weeds worldwide, 114 are arthropods (insects and mites) (Julien 1989), with the majority of biological weed control organisms being insects. Julien (1992) published a world catalog of invertebrate natural enemies that have been imported into various countries to control non-indigenous plants. Since 1920, a total of 69 invertebrate species (insects, mites, and nematodes) have been introduced to control 23 rangeland plant species in the United States (Julien 1992). Table 1 provides a partial list of the major groups of biological control agents established on noxious rangeland weeds in the western United States.

Natural enemies damage weeds either directly or indirectly. Direct feeding damage can severely injure or kill the plant. For example, insects suck out plant fluids, defoliate, eat seeds, and bore into roots, shoots and stems. These actions directly impact the plant, since the loss of valuable nutrients, tissues, and structures can kill or weaken the plant and render it less competitive in its environment. Indirect damage, resulting from secondary pathogen infections, can also weaken the plant and make it less competitive with other vegetation.

Table 1. Rangeland weeds and biological control agents in the western United States. For a more complete list of organisms and for more information, refer to Rees et al. (1996) and Coombs et al. (1998).

Target weed	Natural enemy	Type
Diffuse and spotted knapweed (*Centaurea diffusa* and *C.maculosa*)	*Agapeta zoegana*	root moth
	Cyphocleonus achates	root weevil
	Larinus minutus	seedhead weevil
	L. obtusus	seedhead weevil
	Metzneria paucipunctella	seedhead moth
	Sphenoptera jugoslavica	root gall beetle
	Urophora affinis	seedhead gall fly
	U. quadrifasciata	seedhead gall fly
Yellow starthistle (*Centaurea solstitialis*)	*Bangasternus orientalis*	seedhead weevil
	Chaetorellia australis	seedhead fly
	Eustenopus villosus	seedhead weevil
	Larinus curtus	seedhead weevil
	Urophora sirunaseva	seedhead gall fly
Leafy spurge (*Euphorbia esula*)	*Aphthona abdominalis*	root beetle
	A. cyparissiae	root beetle
	A. czwalinae	root beetle
	A. flava	root beetle
	A. lacertosa	root beetle
	A. nigriscutis	root beetle
	Chamaesphecia hungarica	root moth
	Hyles euphorbiae	defoliating moth
	Oberea erythrocephala	stem and root beetle
	Spurgia esulae	leaf bud gall midge
Musk thistle (*Carduus nutans*)	*Rhinocyllus conicus*	seedhead weevil
	Trichosirocalus horridus	bud and leaf weevil
Rush skeletonweed (*Chondrilla juncea*)	*Cystiphora schmidti*	gall midge
	Eriophyes chondrillae	gall mite
	Puccinia chondrillina	rust fungus
Tansy ragwort (*Senecio jacobae*)	*Longitarsus jacobaeae*	root beetle
	Pegohylemyia seneciella	seedhead fly
	Tyria jacobaeae	defoliating moth
St. Johnswort (*Hypericum perforatum*)	*Chrysolina hyperici*	leaf beetle
	C. quadrigemina	leaf beetle
	Agrilus hyperici	root-boring beetle
Mediterranean sage (*Salvia aethiopis*)	*Phrydiuchus tau*	root weevil

• *Insects*

Insects are prominent in biological control of weeds programs worldwide. Insects are often selected as control agents because many phytophagous species exhibit a high degree of plant host specificity, feeding on only a few plant species. Also, insects are relatively easy to study and rear. In addition, a number of insect species may occur on a single plant or within the same part of the plant. For example, there may be several insects such as moths, beetles, wasps, and flies feeding on or within the plant's roots. A number of species present in the native habitat increases the likelihood of at least one species being a candidate agent for biological control. Moreover, insects can be selected for particular ecological situations, i.e., moist versus dry conditions (Frick 1974). Generally, the insect larva or nymph (juvenile stage) is the life stage damaging to the host, but in some cases adult feeding severely damages the plant, as in the case of the hairy weevil (Eustenopus villosus) on yellow starthistle (*Centaurea solstitialis*).

Beetles (Order: Coleoptera) have been the most widely used and successful insects in weed biological control. Families of beetles used in biological control have been weevils (Family: Curculionidae and Apionidae), flea beetles (Family: Chrysomelidae), long-horn beetles (Family: Cerambycidae), stem-boring beetles (Family: Buprestidae), and seed beetles (Family: Bruchidae). Beetles may feed on seedheads, as in the case of the four yellow starthistle weevils (Table 1); bore in stems and roots, as in the case of *Sphenoptera jugoslavica* on knapweed (*Centaurea* spp.) and *Agrilus hyperici* on St. Johnswort (*Hypericum perforatum*); or defoliate plants, as in the case of the *Aphthona* spp. flea beetles on leafy spurge (*Euphorbia esula*).

Flies (Order: Diptera) are well represented in weed biological control programs in the western United States. Most are gall midges (Family: Cecidomyiidae) or fruit flies (Family: Tephritidae). Examples of gall midges include the stem/leaf gall midge (*Cystiphora schmidti*) on rush skeletonweed (*Chrondrilla juncea*) and the shoot tip gall midge (*Spurgia esulae*) introduced for the control of leafy spurge. Most flies occur on knapweeds and starthistles, and include *Urophora* spp. and *Chaetorellia* spp.

Wasps and sawflies (Order: Hymenoptera), true bugs (Order: Heteroptera), and thrips (Order: Thysanoptera) also have been used in biological control of weeds, but to a lesser extent than the aforementioned insects. Insects from these groups have been used for biological control in other areas of the United States; none occur on western rangelands.

• *Mites*

Mites are the other group of arthropods used in biological weed control. Though similar to insects, mites have only two body regions, four pairs of legs, and are microscopic in size. Both gall mites and spider mites damage the plant by sucking valuable fluids from the stems and leaves; gall mites also cause plants to develop galls. Mites have been used in three weed biological control programs in the United

States and Canada. A gall mite, *Eriophyes chondrillae*, was introduced for the control of rush skeletonweed. A spider mite, *Tetranychus lintearius*, was introduced for control of gorse (*Ulex europaeus*), and two gall mites (*Aceria malherbae* and *A. convolvuli*) were introduced for the control of field bindweed (*Convolvulus arvensis*).

• Nematodes

Nematodes are microscopic, unsegmented roundworms. Those used in biological control live in the soil. Like mites, nematodes damage their host plant primarily by sucking fluids from the roots and stems. Secondary damage results when pathogenic organisms (fungi, bacteria, or viruses) infect the plant at the site injured by the nematode. Only one species of nematode, *Subanguina picridis*, used for the control of Russian knapweed (*Acroptilon repens*), has been introduced into the United States as of 1998.

• Pathogens

Plant pathogens are microscopic organisms that include fungi, bacteria, and viruses. One of the greatest problems limiting the use of pathogens in biological control of weeds is finding a strain sufficiently host-specific. Factors that influence the effectiveness of plant pathogens are host specificity, virulence (toxicity), type and amount of damage, and longevity in the field, mediated by the need for specific environmental conditions. As of 1997, all of the plant pathogens used in biological control of weeds programs are fungi. The rust fungus *Puccinia chondrillina* is widely established on rush skeletonweed on rangelands throughout the West (Hasan et al. 1995). The fungus *Puccinia jaceae* is being evaluated for control of knapweeds and starthistles, and the *Sclerotinia* spp. fungi are being screened for control of thistles in the genera *Carduus* and *Cirsium*.

• Other Natural Enemies

In addition to the natural enemies listed above, other types of organisms have been used in weed biological control. For example, vertebrates (primarily fish) have been used to control aquatic weeds (Center et al. 1989), and goats, sheep, and cattle for rangeland weed control (Van Dreische and Bellows 1996).

For a more complete list of organisms, and for more information on specific biological control programs, refer to the relevant chapters in this book, or to *Biological Control of Weeds in the West* (Rees et al. 1996). *The Pacific Northwest Weed Control Handbook* (Coombs et al. 1998) provides a complete list of current biological weed control organisms.

Considerations in Developing and Implementing a Biological Control Program

Classical biological control programs are developed according to the following sequence of protocols: 1) determine the extent of the weed problem and its suitability for biological control; 2) survey the naturalized and native ranges (typically Europe and Asia) of the weed for natural enemies and select candidate biological control agents; 3) determine the feeding range of the potential organisms and their general suitability as biological control agents; 4) following a period of limited importation and quarantine, release approved agents into the field; and 5) document the impact of the agents or their failure (Harris 1989). The costs associated with introducing a weed's natural enemies in classical biological control make it imperative that careful consideration be given to selecting weeds for biological control, assessing their suitability as biological control targets, and evaluating the feasibility and probability of successful establishment of biological control.

• *Suitability of the Target Weed for Biological Control*

It is necessary, before expensive surveys and testing are undertaken, to assess whether a weed is suitable for biological control. Suitability of a weed for biological control depends on the type of plant (forb, shrub, or grass), its life cycle (annual, biennial, or perennial) (Crawley 1989), its population genetic structure (Burdon and Marshall 1981), its mode of reproduction (Chaboudez and Sheppard 1995), the habitat(s) it infests, its relatedness to economic plants or crops, its relatedness to native, threatened or endangered plants, and other conflicts of interest (McClay 1989, Peschken and McClay 1995). Weeds that are closely related to native or economically important plants are less likely to be selected for classical biological control because of the difficulty in finding agents that will not damage the economic plant species. For example, knapweed and thistles are related to native thistles and economic plants like safflower (*Carthamus tinctorius*) and artichoke (*Cynara scolymus*). Non-indigenous hawkweeds (*Hieracium* spp.) are closely related to native hawkweed species, some of which are rare. Problems with biological control of weedy grasses arise because it is difficult to find natural enemies that are specific to only the weedy grasses, and that do not attack native or non-weedy grasses desirable in the plant community, or cereal crops.

Burdon and Marshall (1981) examined the relationship of weed mode of reproduction with biological control success. They argued that the genetic structure of the target weed species might have important implications to bioagent selection. They also suggested that asexually reproducing plant species were more effectively controlled than sexually producing species, probably because of the genetic diversity and structure of the weed populations.

• Suitability of the Natural Enemy as a Biological Control Agent

Considerations in selecting an organism for use as a biological control agent include its phenology (seasonal life cycle), its synchrony with the target weed (how the life cycle of the natural enemy relates to that of the weed), its expected rate of population growth (reproductive rate, number of generations per year), and its impact on the weed (Harris 1973, Goeden 1983, Crawley 1989).

Host specificity is the set of plant species that enable an insect (or other organism) to feed, develop, and complete its life cycle in nature. Host specificity is one of the most important aspects of biological control, particularly when considering the public concerns for biodiversity, native flora, threatened and endangered plant species, and other threats to non-target organisms, including food crops. Insects that are "generalist" feeders (i.e., grasshoppers) and that have a broad host range are not suitable for biological control of weeds. Insects selected for use in biological control have a very narrow host range; some insect species can survive only on the weed. Insects that feed inside the plant are more likely to have a smaller host range than externally feeding insects (Strong et al. 1984).

• Taxonomy and Systematics

Taxonomic considerations are vitally important in biological control. It is essential that both the target weed and the candidate biological control agents be properly identified to ensure accurate insect-host associations. A number of programs have failed because of taxonomic uncertainties. For example, in the biological control program of rush skeletonweed, natural enemies tended to show a preference for a particular biotype of the weed. It was later determined that there are actually three biotypes of rush skeletonweed (Hasan et al. 1995). Careful selection of better adapted species or biotypes can increase chances of successfully establishing beneficial organisms on rangeland weeds. New applications of chemical ecology and molecular genetics in insect and plant systematics are particularly useful to compare species and to describe the relationships between plants and their natural enemies (Rowe et al. 1997).

• Surveys and Exploration

Exploration for natural enemies begins with identifying the native range of the weed. Most of the invasive rangeland weeds in the western United States are native to Europe, North Africa, the Middle East, and western Asia. Surveys and exploration throughout the weed's native range are conducted to collect most of its natural enemies. Following collection and identification of organisms associated with the weed, a number are selected for further study. The selection of natural enemies will depend on: 1) the degree of specificity with the host; 2) ease of collecting and rearing the natural enemy; and 3) habitat similarities between the native range in Europe or Asia and the invaded range in North America.

In the United States, surveys of the weeds are conducted to determine whether natural enemies are already present on the weeds. In some instances, it was found that the insect was already present on the weeds. For example, the fly *Urophora quadrifasciata* was already found on diffuse knapweed (*Centaurea diffusa*) in the United States before it was intentionally introduced; presumably it had spread from release sites in Canada.

• *Release and Redistribution*

After the natural enemies pass host specificity testing, a limited number of individuals are imported into quarantine facilities in the United States for further testing. Testing in the United States enables scientists to confirm that the imported biological control agents are free from parasitoids and pathogens, and to screen the agents against certain plants that were not evaluated overseas.

Natural enemy releases are made at selected locations. The locations may vary slightly in microclimatic conditions, and are carefully chosen to maximize the chance that the newly introduced biological control agents will establish, reproduce, and persist in the environment. It is often useful to release the insects in several different habitats to assess the optimal environment for the agent, since different biotypes of the same insect species may have unique habitat preferences. Failure of an agent to establish at a particular site may simply be due to some environmental factor, such as temperature or moisture regime.

• *Evaluation*

The success of biological control of weeds programs historically has been based on the degree to which the weed population declines following the introduction of the biological control agent(s). Unfortunately, objective assessments in weed biological control programs are generally lacking; many programs have been judged solely on basis of "before-and-after" photographs. During the 1980s and 1990s, a number of authors proposed varying models or guidelines to evaluate the success of biological control, including manipulative experiments (McEvoy et al. 1993, McClay 1995), economic cost-benefit analyses (Tisdell and Auld 1989), or categorical analysis (Harris 1991). Harris (1991) proposed a sequential, four-step process for evaluating the success of a biological control release: "establishment," "biological success," "host impact," and "control success." These categories of success provide evaluation of biological control within a practical context. Similarly, Cullen (1995) proposed that the ultimate success of an agent is a result of a combination of three major factors: 1) the damage an individual agent can do to a plant; 2) the ecology of the agent in determining its density and, therefore, the total damage produced; and 3) the ecology of the weed in determining whether the damage is significant for reducing its population. In his review of the debate on how to evaluate biological control programs, Cullen (1995) concluded that biological control programs are inherently unique and must be evaluated on a case-by-case basis, and not according to a list of standard protocols.

Campbell and McCaffrey (1991) studied the biology and impact of leaf- and root-feeding beetles (*Chrysolina quadrigemina, C. hyperici,* and *Agrilus hyperici*) some 35 years after the release of five biological control agents that led to successful control of St. Johnswort in Idaho. Local legend attributed the successful control of St. Johnswort to the *Chrysolina* beetles. However, Campbell and McCaffrey showed that the root borer, *A. hyperici,* damaged St. Johnswort plants as much or more than the *Chrysolina* beetles. Furthermore, Tisdale (1976) suggested a contributing factor in the successful biological control of St. Johnswort in Idaho was its low seedling survivorship (about 3%). Population densities of another rangeland weed in Idaho, Mediterranean sage (*Salvia aethiopis*), have fluctuated since the root-crown weevil *Phyrdiuchus tau* was introduced in 1979. While it is tempting to attribute the current population dynamics of Mediterranean sage to effects of the insect, it is undoubtedly a combination of plant competition, drought, severe winters, and other habitat factors that are mediating population fluctuations of this weed. More critical studies are needed to reveal the key factors, both biotic and abiotic, mediating populations of this weed in Idaho and elsewhere.

Although biological control of rangeland weeds has made advances in the area of population ecology and evolutionary ecology (Luck et al. 1995), a likely explanation for the historic lack of evidence and evaluation of biological control of weeds on the basis of continuous change in a weed population is that biological control of weeds programs have traditionally been the purview of the entomologist. As more plant population ecologists enter the biological control arena, empirical evaluation and critical analysis of biological control of weeds will likely become more common.

Safety Issues in Biological Control of Weeds

Biological control has a good record of environmental safety (Turner 1985), but it is not without certain risks. Risks include threats to non-target organisms and dangers to the environment. There is a great concern today for the safety of native plant species, particularly those threatened or endangered. Careful selection and monitoring of candidate biological control agents is imperative, to minimize risks to native and other non-target plants. Agents selected for biological control of weeds are imported into the United States only after very rigorous testing for host specificity to ensure that the potential biological control organism attacks only the target weed. The federal and state governments provide rigid guidelines for testing, importation, and quarantine of selected natural enemies, and biological data are required by state and federal agencies for all non-indigenous arthropod introductions released from quarantine (Antognini et al. 1995). Foreign exploration, quarantine, rearing, release and host specificity testing all follow a specific set of guidelines and protocols, established and monitored by the Technical Advisory Group on the Introduction of Biological Control Agents of Weeds (TAG) of the USDA Animal and Plant Health Inspection Service (APHIS).

• Pre-release Studies

Pre-release studies are designed to determine some important relationship between the planned insect release and target weed biology that will increase or expand the opportunity for early success, establishment, and impact of the selected biological control agents. Pre-release studies can also aid habitat selection based on: the known requirements or preferences of the planned biological control agent (DeClerck-Floate 1996); more detailed studies of plant-insect interactions before the biological control agent is introduced (Reeder et al. 1996); or studies that involve target site assessment to identify the weed's "Achilles' heel." Pre-release studies to determine the most vulnerable parts of the plant are useful to limit the selection of natural enemies that attack the most susceptible parts of the weeds (Wilson and McCaffrey, 1997-1998, unpublished data). Biological studies such as these give scientists some ability to predict the outcome of natural enemy-weed interactions. A better understanding of the population dynamics of natural enemies in relation to the weed population dynamics will lead to more accurate predictions of the outcome of deliberate natural enemy introductions. This moves us away from the "lottery" approach, where a number of species are introduced with the hope that one will be successful (Myers 1985).

• Conflicts of Interest

Conflicting interests in biological control are critical issues that continue to spawn considerable debate. In his review of conflicts of interests in weed biological control, Turner (1985) proposed two issues fundamental to the problem of conflicting interests in classical biological control of weeds: uncertain impacts and conflicting perspectives. While host specificity testing validates the limited host range of the candidate organism, these tests are necessarily conducted in artificial, controlled laboratory situations. It is generally difficult to predict how the natural enemy will behave after it is released into the field. Host range expansion, or the temporal and spatial distribution of natural enemies once introduced, may induce conflicts between different interests (Andres 1981, Tisdell et al. 1984). For example, several cases of natural enemies feeding on non-target plants, particularly native species, have been reported. The introduced St. Johnswort beetles (*Chrysolina* spp.) reportedly feed on native *Hypericum* species (Andres 1985). Several introduced leafy spurge insects are able to develop on native spurge (*Euphorbia* spp.) species (Pemberton 1985). The seedhead weevil *Rhinocyllus conicus*, imported for the control of musk and plumeless thistle (*Carduus nutans* and *C. acanthoides*), is known to attack native and rare thistles (Turner and Herr 1996, Louda et al. 1997). Conflicts of interest are not only restricted to introduced insects, but are also reported for pathogens (Freeman and Charudattan 1985) and vertebrates (van Zon 1985).

Socioeconomic issues contribute to the conflicting interest in weed biological control as well. For example, the emergence of St. Johnswort during the late 1990s as a pharmacological plant (Miller 1998) has created conflicts between biological weed control and those who want to cultivate the weed. Honey-producers in California have been opposed to the biological control of yellow starthistle (C.E. Turner 1994, personal communication), a plant highly preferred by honeybees in that state (Vansell 1931).

Biological Weed Control: Implications for Integrated Pest Management and Long-Term Rangeland Rehabilitation

The aim of biological control is to shift the competitive balance away from the weed to desirable grasses and forbs, the intended outcome being to return weed-infested rangelands to more diverse and productive plant communities. In some ecosystems in the western United States, biological control agents, together with competition from perennial grasses, combine to achieve successful weed suppression and restoration of desirable, perennial vegetation (McEvoy et al. 1989). In a series of manipulative studies conducted in western Oregon, McEvoy et al. (1993) described the mechanisms enabling a disturbed plant community infested with tansy ragwort (*Senecio jacobaea*) to return to a desirable pasture, following the introduction of two biological control agents. The favorable outcome of these experiments, and tansy ragwort biological control in general, can be attributed, in part, to the presence of an established community of perennial grasses that filled in the space opened by the removal of the weed (Figure 2A). Figure 2 represents a comparison of the outcomes of biological control of weeds in perennial grass (A) and non-indigenous annual grass (B) systems. Non-indigenous annual grass communities are less stable, and are susceptible to repeated disturbances and subsequent invasion by herbaceous weeds.

In other parts of the western United States, particularly in the arid rangelands of the intermountain West, the absence of desirable vegetation is common. These former perennial grass-dominated ecosystems have largely been replaced with non-indigenous annual grasses such as downy brome (*Bromus tectorum*), medusahead (*Elymus caput-medusae*), and ventenata (*Ventenata dubia*) (Northam and Callihan 1990, Young 1992). The space opened by the suppression of one or more weed species is usually filled by other non-indigenous weedy forbs (Figure 2B). The herbaceous weedy plant communities persist because of constant recruitment into areas not closed by stable, perennial vegetation. Even if a new weed is not recruited, the plant community reverts to the weedy annual grass understory that prevailed prior to infestation by the weedy forb (Tisdale 1976, Müller-Schärer and Schroeder 1993). For example, Campbell and McCaffrey (1991) found that following successful suppression of St. Johnswort in Idaho, the study sites reverted to the non-indigenous annual grass community that predated introduction of the weed. In 1997, these and similar sites were dominated by weedy forbs such as common crupina (*Crupina vulgaris*), Mediterranean sage, knapweed, yellow starthistle, hawkweed, or sulfur cinquefoil (*Potentilla recta*) (Wilson and McCaffrey, unpublished data). This phenomenon has been referred to as the "biological control treadmill" (McCaffrey

109

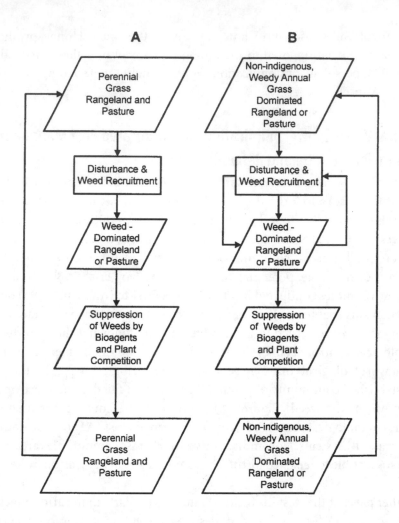

Figure 2. Conceptual models of weed biological control dynamics in perennial pasture (A) and annual grass-dominated rangelands (B). The perennial model is based on the work of McEvoy et al. (1993) and the annual grassland model is based on studies conducted by Campbell and McCaffrey (1991).

and Wilson 1994). It describes how suppression of one weedy forb leads to subsequent invasion and community dominance by another weedy forb. The need for new biological control efforts brings additional costs associated with the need for additional programs (McEvoy 1996), providing the new target alien weed is amenable to biological control (Harshman 1956, McCaffrey and Wilson 1994).

Campbell and McCaffrey (1991) found the presence of a native perennial, bluebunch wheatgrass (*Agropyron spicatum*), and the introduced perennial Kentucky bluegrass (*Poa pratensis*) at one of their St. Johnswort study sites, following biological suppression. While the grasses were not dominant, they could be conserved and managed with appropriate inputs to increase their presence and role in the plant community. This is an important consideration. In rangeland habitats, management inputs beyond release of the biological control agents are often necessary to stabilize (ecologically) the habitat and to make it more productive (Müller-Schärer and Schroeder 1993). Successful weed management that incorporates biological weed control must address more than the response of plants to herbivory by single or multiple herbivores. It is the characteristics of the weed, its relationship to the environment, and its role in the plant community that largely determine the success of a biological control program, not just the intrinsic characteristics of the biological control agent(s).

Biological weed control researchers have long recognized the importance of competitive grass species in the management of rangeland and pasture weeds (Kennett et al. 1992, Müller-Schärer and Schroeder 1993), as well as the need to integrate careful use of herbicides (Messersmith and Adkins 1995), enhanced plant biology studies (Navas 1991), and other tools (Kok 1974, Andres 1982). Such approaches can greatly improve our land management capabilities for any target weed before and after biological control stress is exerted.

It is widely accepted that sustainable rangeland management depends not only on the understanding of the ecological principles that regulate plant community structure, but more specifically, the successional dynamics of weed infested habitats. Sheley et al. (1996) identified three general causes of succession that should be addressed in land management: 1) disturbance, a common factor in most rangeland environments; 2) colonization, or the establishment of species within a habitat; and 3) species performance, or the response of individual species to the environment. They proposed that these processes should be managed by way of designed disturbance, controlled colonization, and controlled species performance. Biological weed control affects colonization and performance of weed species; therefore, it is an important component of the successional process (Brown and Gange 1992). This is the theoretical context within which future biological control undertakings should be developed and assessed. Ultimately, biological control programs treated as large, ecological field experiments (Myers 1978) will enable land managers to move away from the biological control treadmill, particularly in western rangeland ecosystems.

111

Literature Cited

Andres, L.A. 1981. Conflicting interests in the biological control of weeds. In: Delfosse, E.S. (ed.), Proc. V Int. Symp. Biol. Contr. Weeds, July 22-27, 1980, Brisbane, Australia. CSIRO, Melbourne, 11-20.

Andres, L.A. 1982. Integrating weed biological agents into a pest-management program. Weed Sci. 30:25-30.

Andres, L.A. 1985. Interaction of *Chrysolina quadrigemina* and *Hypericum* spp. in California. *In*: Delfosse, E.S. (ed.), Proc. VI Int. Symp. Biol. Contr. Weeds, 19-25 August 1984, Vancouver, Canada. Agric. Canada, 235-39.

Antognini, J., P.C. Quimby, Jr., C.E. Turner, and J.A. Young. 1995. Implementing effective noxious range weed control on rangelands. Rangelands 17(5):158-63.

Brown, V.K., and A.C. Gange. 1992. Secondary plant succession: how is it modified by insect herbivory? Vegetatio 101:3-13.

Burdon, J.J., and D.R. Marshall. 1981. Biological control and the reproductive mode of weeds. J. Appl. Ecol. 18:649-58.

Campbell, C.L., and J.P. McCaffrey. 1991. Population trends, seasonal phenology, and impact of *Chrysolina quadrigemina*, *C. hyperici* (Coleoptera: Chrysomelidae), and *Agrilus hyperici* (Coleoptera: Buprestidae) associated with *Hypericum perforatum* in northern Idaho. Environ. Entomol. 20:303-15.

Center, T.D., A.F. Cofrancesco, and J.K. Balciunas. 1989. Biological control of aquatic and wetland weeds in the southeastern U.S. *In*: Delfosse, E.S. (ed.), Proc. VII Int. Symp. Biol. Contr. Weeds, 6-11 March 1988. Rome, Italy. Ist. Sper. Patol. Veg. (MAF), 239-62.

Chaboudez, P., and A.W. Sheppard. 1995. Are particular weeds more amenable to biological control? A reanalysis of mode of reproduction and life history. *In*: Delfosse, E.S., and R.R. Scott (eds.), Proc. VIII Int. Symp. Biol. Contr. Weeds, 2-7 February 1992, Lincoln Univ., Canterbury, New Zealand. DSIR/CSIRO, Melbourne.

Charudattan, R., and H L. Walker. 1982. Biological Control of Weeds with Plant Pathogens. John Wiley & Sons, New York.

Coombs, E.M., G L. Piper, and J.P. McCaffrey. 1998. Biological control. *In*: R.D. William, D. Ball, T.L. Miller, R. Parker, J.P. Yenish, R.H. Callihan, C. Eberlein, G.A. Lee and D.W. Morishita (eds.), Pacific Northwest Weed Control Handbook. Oregon St. Univ., Coop. Ext. Serv., Corvallis, OR, 3-5.

Crawley, M.J. 1986. Population biology of invaders. Phil. Trans. R. Soc. Lond. 314:711-31.

Crawley, M. J. 1989. Plant life-history and the success of weed biological control projects, *In*: Delfosse, E.S. (ed.), Proc. VII Int. Symp. Biol. Contr. Weeds, 6-11 March 1988, Rome, Italy. Ist. Sper. Patol. Veg. (MAF).

Cullen, J.M. 1995. Predicting effectiveness: Fact and fantasy. *In*: Delfosse, E.S. and R.R. Scott (eds.), Proc. VIII Int. Symp. Biol. Contr. Weeds, 2-7 February 1992, Lincoln Univ., Canterbury, New Zealand, 103-109. DSIR/CSIRO, Melbourne.

DeBach, P. 1964. Biological Control of Insect Pests and Weeds. Chapman and Hall, London.

DeClerck-Floate, R. 1996. The role of pre-release studies in developing a biocontrol strategy for hound's-tongue in Canada. *In*: V.C. Moran and J.H. Hoffman (eds.), Proc. IX Int. Symp. Biol. Contr. Weeds, 19-26 January 1996, Stellenbosch, South Africa, Univ. Cape Town, 19-26.

Freeman, T.E., and R. Charudattan. 1985. Conflicts in the use of plant pathogens as biocontrol agents for weeds. *In*: Delfosse, E.S. (ed.), Proc. VI Int. Symp. Biol. Contr. Weeds, 19-25 August 1984, Vancouver, Canada. Agric. Canada, 351-57.

Frick, K.E. 1974. Biological control of weeds: Introduction, history, theoretical and practical implications. *In*: Maxwell, F.G. and F.A. Harris (eds.), Proc. Summer Inst. Biol. Contr. of Plants, Insects and Diseases. Univ. Press of Mississippi, Jackson, MS, 204-23.

Goeden, R.D. 1983. A critique of Harris' system for selection of insect agents in biological control of weeds. Protect. Ecol. 5:287-301.

Goeden, R.D. 1988. A capsule history of biological control of weeds. Biocontr. News Info. 9:55-61.

Goeden, R.D. 1993. Arthropods for suppression of terrestrial weeds. *In*: Lumsden, R.D. and J.L. Vaughn (eds.), Pest Management: Biologically based technologies. Amer. Chem. Soc., Washington, 231-37.

Goeden, R.D., and L.T. Kok 1986. Comments on a proposed "new" approach for selecting agents for the biological control of weeds. Can. Entomol. 118:51-58.

Harris, P. 1973. The selection of effective agents for the biological control of weeds. Can Entomol. 105:1495-1503.

Harris, P. 1989. Practical considerations in a classical biocontrol of weeds program. Proc. Int. Symp. Biol. Cont. Implementation, McAllen, Texas, 4-6 April 1989.

Harris, P. 1991. Classical biological control of weeds: its definition, selection of effective agents, and administrative-political problems. Can. Ent. 123:827-49.

Harshman, E.P. 1956. The changes of vegetation following biological control of *Hypericum perforatum*. MS Thesis, Univ. Idaho, Moscow, ID.

Hasan, S., P. Chaboudez, and C. Espiau. 1995. Isozyme patterns and susceptibility of North American forms of *Chrondrilla juncea* to European strains of the rust fungus *Puccinia chrondrillina*. *In*: Delfosse, E.S. and R.R. Scott (eds.), Proc. VIII Symp. Biol. Contr. Weeds, 2-7 February, 1992, Lincoln University, Canterbury, New Zealand. DSIR/CSIRO, Melbourne, 367-73.

Hokannen, H., and D. Pimental. 1984. New approach for selecting biological control agents. Can. Ent. 116:1109-21.

Huffaker, C.B., and P.S. Messenger (eds.). 1976. Theory and Practice of Biological Control. Academic Press, New York.

Julien, M.H. 1989. Biological control of weeds worldwide: trends, rates of success and the future. Biocontr. News Info. 10(4):299-306.

Julien, M.H. (ed.). 1992. Biological Control of Weeds: A World Catalogue of Agents and Their Target Weeds, 3rd ed., CAB International, Wallingford, U.K.

Kennett, G.A, J.R. Lacey, C.A. Butt, K.M. Olson-Rutz, and M.R. Haferkamp. 1992. Effects of defoliation, shading and competition on spotted knapweed and bluebunch wheatgrass. J. Range Manage. 45:363-69.

Kok, L.T. 1974. Principles and methodology of biological weed control. FAO Plant Protect. Bull. 22:77-81.

Louda, S.M., D. Kendall, J. Conner, and D. Simberloff. 1997. Ecological effects of an insect introduced for the biological control of weeds. Science 277:1088-90.

Luck, R.F., M.J. Tauber, and C.A. Tauber. 1995. The contributions of biological control to population ecology and evolutionary ecology. *In*: Nechols, J.R., L.A. Andres, J.W. Beardsley, R.D. Goeden, and C.G. Jackson (eds.), Biological Control in the Western United States: Accomplishments and Benefits of Regional Research Project W-84, 1964-1989. Univ. Calif, Div. Agric. Res. Pub. 3361.

McCaffrey. J.P., and L.M. Wilson. 1994. Assessment of biological control of exotic broadleaf weeds in intermountain rangelands. *In*: Monsen, S.B and S.G. Kitchen (eds.), Proc. Symp. on Ecology, Management of Annual Rangelands, Boise, ID, May 18-22, 1992. USDA-ARS, Gen. Tech. Report INT-GTR-313, 101-2.

McClay, A.S. 1989. Selection of suitable target weeds for classical biological control in Alberta. Alberta Environmental Centre, Vegreville, AB.

McClay, A.S. 1995. Beyond "Before-and-after": Experimental design and evaluation on classical weed biological control. *In*: Delfosse, E.S. and R.R. Scott (eds.), Proc. VIII Int. Symp. Biol. Cont. Weeds, 2-7 February, 1992, Lincoln University, Canterbury, New Zealand. DSIR/CSIRO, Melbourne, 213-9.

McEvoy, P.B. 1996. Host specificity and biological pest control. Bioscience 46(6):401-5.

McEvoy, P.B., C.S. Cox, R.R. Jameds, and N.T. Rudd. 1989. Ecological mechanisms underlying successful biological weed control: Field experiments with ragwort *Senecio jacobaea*. *In*: Delfosse, E.S. (ed.), Proc. VII Int. Symp. Biol. Contr. Weeds, 6-11 March 1988, Rome, Italy. 1st. Sper. Patol. Veg. (MAF), 55-66.

McEvoy, P.B., N.T. Rudd, C.S. Cox, and M. Huso. 1993. Disturbance, competition, and herbivory effects on ragwort *Senecio jacobaea* populations. Ecol. Monogr. 63(1): 55-75.

McFadyen, R.E.C. 1998. Biological control of weeds. Ann. Rev. Entomol. 43:369-93.

Maschinski, J., and T.J. Whitham. 1989. The continuum of plant responses to herbivory: the influence of plant association, nutrient availability, and timing. Amer. Nat. 134(1):1-19.

Messersmith, C.G., and S.W. Adkins. 1995. Integrating weed-feeding insects and herbicides for weed control. Weed Technol. 9:199-208.

Miller, A.L. 1998. St. John's Wort (*Hypericum perforatum*): clinical effects on depression and other conditions. Altern. Med. Rev. 3:18-26.

Moran, V.C., and H.G. Zimmerman. 1984. The biological control of cactus weeds: achievements and prospects. Biocontr. News Info. 5:297-320.

Müller-Schärer, H., and D. Schroeder. 1993. The biological control of *Centaurea* spp. in North America: do insects solve the problem? Pestic. Sci. 37:343-53.

Myers, J.H. 1978. Biological control introductions as grandiose field experiments: Adaptations of the cinnabar moth to new surroundings. *In*: Freeman, T.E. (ed.), Proc. IV Int. Symp. Biol. Contr. Weeds, 30 August-2 September 1976, Univ. Florida, Gainesville, FL, 181-88.

Myers, J.H. 1985. How many insect species are necessary for successful biocontrol of weeds ?, in Delfosse, E.S. (ed.), Proc. VI Int. Symp. Biol. Contr. Weeds, 19-25 August 1984, Vancouver, Canada. Agric. Canada.

Navas, M.L. 1991. Using plant population biology in weed research: a strategy to improve weed management. Weed Res. 31:171-79.

Northam, F.E., and R.H. Callihan. 1990. Grass adaptation to semi-arid, yellow starthistle-infested canyonland. Proc. West. Soc. Weed Sci, 13-15 March 1990, Reno, NV.

Pemberton, R.W. 1985. Native plant considerations in the biological control of leafy spurge. *In*: Delfosse, E.S. (ed.), Proc. VI Int. Symp. Biol. Contr. Weeds, 19-25 August 1984, Vancouver, Canada. Agric. Canada, 365-90.

Peschken, D.P., and A.S. McClay. 1995. Picking the target: a revision of McClay's scoring system to determine the suitability of a weed for classical biological control. *In*: Delfosse, E.S. and R.R. Scott (eds.), Proc. VIII Int. Symp. Biol. Contr. Weeds, 2-7 February 1992, Lincoln Univ., Canterbury, New Zealand. DSIR/CSIRO, Melbourne, 137-43.

Reeder, R.H., C.A. Ellison, and M.B. Thomas. 1996. Population dynamic aspects of the interaction between the weed *Rottboellia cochinchinensis* (itch grass) and the potential biological control agent *Sporisorium ophiuri* (head smut). *In*: V.C. Moran and J.H. Hoffman (eds.), Proc. IX Int. Symp. Biol. Contr. Weeds, 19-26 January 1996, Stellenbosch, South Africa, Univ. Cape Town, 205-11.

Rees, N.E., P.C. Quimby, Jr., G.L. Piper, E.M. Coombs, C.E. Turner, N.R. Spencer, and L.V. Knudson (eds.). 1996. Biological Control of Weeds in the West. West. Soc. Weed Sci., USDA-ARS, Montana St. Univ., Bozeman, MT.

Rowe, M.L., D.J. Lee, S.J. Nissen, B.M. Bowditch, and R.A. Masters. 1997. Genetic variation in North American leafy spurge (*Euphorbia esula*) determined by DNA markers. Weed Sci. 45:446-54.

Sheley, R.L., T.J. Svejcar, and B.D. Maxwell. 1996. A theoretical framework for developing successional weed management strategies on rangeland. Weed Technol. 10:766-73.

Sheppard, A.W. 1996. The interaction between natural enemies and interspecific plant competition in the control of invasive pasture weeds. *In*: V.C. Moran and J.H. Hoffmann (eds.), Proc. IX Int. Symp. Biol. Contr. Weeds, 19-26 January 1996, Stellenbosch, South Africa, Univ. Cape Town, 47-53.

Strong, D.R., J.H. Lawton, and R. Southwood. 1984. Insects on Plants: Community Patterns and Mechanisms. Harvard Univ. Press, Cambridge, MA.

Tisdale, E.W. 1976. Vegetational responses following biological control of *Hypericum perforatum* in Idaho. Northwest Sci. 50(2):61-75.

Tisdell, C.A., and B.A. Auld. 1989. Evaluation of biological control projects. *In*: Delfosse, E.S. (ed.), Proc. VII Int. Symp. Biol. Contr. Weeds, 6-11 March 1988, Rome, Italy. 1st. Sper. Patol. Veg (MAF), 93-100.

Tisdell, C.A., B.A. Auld, and K.M. Menz. 1984. On assessing the value of biological control of weeds. Protect. Ecol. 6:169-79.

Trumble, J.T., D.M. Kolodny-Hirsch, and I.P. Ting. 1993. Plant compensation for arthropod herbivory. Annu. Rev. Entomol. 38:93-119.

Turner, C.E. 1985. Conflicting interests in biological control of weeds. *In*: Delfosse, E.S. (ed.), Proc. VI Int. Symp. Biol. Contr. Weeds, 19-25 August 1984, Vancouver, Canada. Agric. Canada, 203-25.

Turner, C.E., and J.C. Herr. 1996. Impact of *Rhinocyllus conicus* on a non-target, rare, native thistle (*Cirsium fontinale*) in California. *In*: V.C. Moran and J.H. Hoffmann (eds.), Proc. IX Int. Symp. Biol. Contr. Weeds, 19-26 January 1996, Stellenbosch, South Africa, Univ. Cape Town, 103.

Van Driesche, R.G., and T.S. Bellows, Jr. 1996. Biological Control. Chapman & Hall, New York.

Vansell, G.H. 1931. Nectar and pollen plants in California. Univ. Calif. College Agric., Agric. Exp. Sta., Berkeley, Bull. 517.

van Zon, J.C.J. 1985. Conflicts of interest in the use of the grass carp. *In*: Delfosse, E.S. (ed.), Proc. VI Int. Symp. Biol. Contr. Weeds, 19-25 August 1984, Vancouver, Canada. Agric. Canada, 399-403.

Wilson, F., and C.B. Huffaker. 1976. The philosophy, scope and importance of biological control. *In*: Huffaker, C B. and P.S. Messenger (eds.), Theory and Practice of Biological Control. Academic Press, New York, 3-15.

Wilson, J.M., and J.P. McCaffrey. 1997-1998. Unpublished data: Target site assessment for biological control of meadow hawkweed. University of Idaho, Moscow, ID.

Young, J.A. 1992. History and use of semiarid plant communties—changes in vegetation. *In*: Monsen, S.B. and S.G. Kitchen (eds.), Proc. Symp. on Ecology and Management of Annual Rangelands, Boise, ID, 18-22 May 1992. USDA-ARS, Gen. Tech. Rep. INT-GTR-313, 5-8.

Herbicides and Rangeland

Alvin J. Bussan and William E. Dyer

A key objective of weed management systems is to maintain the productivity of rangeland while minimizing the occurrence and impact of weeds. Strategies may include increasing the amount of desirable forage, establishing stable plant communities that prevent erosion or weed invasion, and providing habitat for wildlife. Herbicides have been an integral part of management systems for rangeland weeds since the late 1940s, and have reduced the presence of noxious and common weeds that interfere with grazing, decrease productivity, and lower the aesthetic value of rangeland. Of the 988 million acres of rangeland in the United States, about 25% were treated with herbicides in 1997.

Herbicides decrease growth, seed production, and competitiveness of susceptible weeds. Weeds treated with herbicides use less soil moisture and nutrients, which then become available for more desirable members of the plant community. In addition, decreasing weed growth with herbicides can reduce their canopy cover. Neighboring plants can take advantage of increased light penetration through the canopy, and better compete with weeds.

A key to the long-term success of rangeland weed management systems is to decrease the population of undesired plant species. Even if target weeds are not killed by herbicide applications, their vigor, seed production, and vegetative reproduction (for perennial plants) will usually be reduced. Decreased photosynthesis in leaves and stems results in less energy available for replenishment of root, stolon, and rhizome reserves. Lower energy reserves in vegetative reproductive parts may also cause stress and increase the potential for winter-kill or dessication. Translocated herbicides can also effectively prevent some of the vegetative reproductive parts from sprouting, or perhaps kill them completely.

Herbicides seldom provide long-term control of weeds when used alone and outside the context of an integrated weed management system. Often, weeds dominate in certain areas within a field, especially perennial noxious weeds such as knapweed (*Centaurea* spp.) and leafy spurge (*Euphorbia esula*). In other situations, the stand of desirable grasses and forbs may be stressed by overgrazing or other factors. As a result, annual weeds like cheatgrass (*Bromus tectorum*) are able to become established and further stress the desirable plant community. In these latter cases, herbicides may control existing weeds, but the lack of a healthy community of desirable plants usually allows that weed or another species to become established after the residual effects of the herbicide have dissipated.

This book focuses on managing rangeland weeds using integrated approaches, including the use of herbicides. Successful weed management requires not only the development of a strategy for killing existing weeds, but long-term plans for preventing their reestablishment or the invasion of other weedy species through careful land management. Once a weed management plan is developed, herbicides can be selected that will help meet the initial objective of reducing weed populations to reasonable numbers. To select the proper herbicide for a specific weed management plan, an understanding of how herbicides kill plants, how they are applied, and their fate in the environment is required. Therefore, the objectives of this chapter are: 1) to describe how to apply herbicides; 2) to define the modes of action of rangeland herbicides; and 3) to describe the fate of selected herbicides in the environment.

Herbicide Application

One of the keys to a successful herbicide application is a uniform spread of the product across the desired area. To achieve a uniform herbicide application, one must be aware of the various types of herbicide formulations and the many possible methods of application.

• Formulations

The term "formulation" refers to how herbicides are packaged. Herbicide formulations include the active and inert ingredients. The active ingredient is the toxicant or chemical that inhibits growth or kills the plant. Many herbicides in their pure form may not be water-soluble or able to enter the plant, making them essentially useless in the field. Therefore, herbicides are prepackaged or formulated with solvents, diluents, and/or various adjuvants to make them water soluble and able to penetrate leaf tissues. Most rangeland herbicides are formulated so they can be diluted in water and sprayed, although some are applied as dry granular material.

Sprayable formulations are diluted with water, fertilizer, or oil-based carriers, and sprayed on soil or vegetation. Sprayable formulations include water-soluble liquids, water-soluble powders, emulsifiable concentrates, wettable powders, water-dispersible liquids, and water-dispersible granules.

Dry formulations of herbicides are packaged as granules or pellets, and can be applied directly to the field without dilution in water. Dry formulations usually have lower concentrations of herbicide active ingredient. Dry formulations are usually less hazardous to the applicator because of the decreased concentration of the herbicide active ingredient.

117

Table 1. Herbicide formulations and their description.

Formulation	Formulation Type	Abbreviation	Solubility and Description
Water Soluble Liquid	Sprayable	S, SL	Liquid formulation that dissolves completely in water requiring little agitation. May require wetting agents to allow the spray to spread and stick on the leaf surface.
Water Soluble Powder	Sprayable	SP	Dry formulations that dissolve completely in water. Once in solution requires little agitation. Not widely used today because of fine dusts that become airborne during mixing.
Emulsifiable Concentrates	Sprayable	E, EC	An emulsifier is used to suspend nonpolar herbicides in water. Upon mixing, an emulsion forms and the spray solution has a cloudy appearance. Mild agitation is required.
Wettable Powders	Sprayable	W, WP	Dry herbicide that uses a dispersing agent to suspend insoluble particles in water. Care must be taken to ensure WPs are thoroughly mixed in water. Requires vigorous agitation.
Water Dispersable Liquids	Sprayable	WDL, L, F	Powders suspended in liquid. Like WPs, the herbicide is insoluble in water. Shake herbicide well before dispersing into spray tank. Moderate agitation is required.
Water Dispersable Granules	Sprayable	WDG, DF	Insoluble herbicides ground into a powder mixed with dispersing agents, then formed into granules. Safer than WPs or WDL because they produce less dust when mixing. Agitation is required.
Granules	Dry application	G	Dry herbicide formulated into granules less than 10 mm^3 that is applied directly to the soil. Rainfall is required to leach the herbicide into the soil and activate it.
Pellets	Dry application	P	Pellets are similar to granules, but are larger than 10 mm^3.

• *Carriers*

Carriers are gases, solids, or liquids used to dilute or suspend herbicides during application and allow for proper placement of the herbicide, whether it be to the soil or on foliage. Gas carriers are used for fumigation or soil sterilization, and are seldom used in rangelands.

Liquid carriers include water, liquid fertilizers, diesel, and other similar low-viscosity oils. Water is by far the most widely used carrier because it is universally available, cheap, and is generally effective with a wide range of herbicides. However, problems can be encountered when using hard or dirty water. Hard water (water with high levels of dissolved calcium and magnesium salts) can react or bind with herbicides having an ionic charge, and decrease their activity or cause them to precipitate in the spray tank. Glyphosate and 2,4-D salt or amine formulations are particularly susceptible to problems when applied in hard water. Similarly, these and other herbicides are adsorbed to suspended clay and silt in dirty water, reducing their effectiveness. Dirty water that contains silt, sand, or algae can clog screens, wear out nozzles, and abrade other sprayer parts, resulting in uneven spray patterns and poor weed control.

Liquid fertilizers are rarely used as herbicide spray carriers in rangeland, although diesel or other mineral oils may be used for dormant applications to woody species. Diesel or other mineral oils increase the activity of herbicides because they are phytotoxic, and thus cause injury to the plants directly. The added effect of the carrier may be important for difficult weeds or woody or brushy species.

Dry carriers are used to apply herbicides without further dilution, and are the major components of granules and pellets. In the case of granules or pellets, the formulation also serves as the carrier. The inert ingredients of granules and pellets include attapulgite, kaolinite, or vermiculite clays, dry fertilizers, polymers, starch, and other solid substances.

Dry fertilizer is rarely used as a dry carrier in rangelands. Some herbicide manufacturers have used dry fertilizer in the formulation of lawn and turf herbicides. Dry fertilizer can be impregnated with herbicides. During impregnation, the dry fertilizer is placed in a rolling drum while liquid herbicide formulations are applied and mixed thoroughly with the fertilizer. Uniform applications of the herbicide/fertilizer mixture to the soil should provide control equivalent to applications with liquid carriers. Dry fertilizer impregnated with herbicide may be useful in rangeland revegetation by improving soil fertility and controlling weeds in a single-pass application.

• *Spray Additives*

Spray additives can be included in formulated herbicides, or can be added to the spray mixture to improve the effectiveness of the spray solution. Adjuvants are classified by their uses rather than their chemistry, although chemical properties determine their suitability for use with different herbicides. Adjuvants include surfactants, antifoaming agents, compatibility agents, crop oil or crop oil concentrates, activators, and drift control agents.

• Application Considerations

The total volume of spray solution including the actual product, spray additives, and carrier applied per acre is called "spray gallonage." Correct spray gallonage is crucial for ensuring uniform coverage of leaf foliage with foliar-active herbicides. In general, higher spray gallonage results in more and larger spray droplets. As a result, higher spray gallonage results in more spray solution contacting the leaf surface, leading to increased uptake of the herbicide. Minimum spray gallonage required for acceptable herbicide performance is indicated on the herbicide label. In general, systemic herbicides such as 2,4-D or glyphosate require lower spray gallonage than contact herbicides such as paraquat. Soil applied herbicides only require enough spray gallonage to ensure uniform coverage of the soil surface.

Managing weeds successfully involves controlling plants at the time of application and preventing future weed populations. Annual weeds should be treated before they flower to prevent seed set, the source of future generations. In general, the smaller the annual weed, the easier it is to manage with herbicides. More importantly in rangelands, perennial weed management depends on preventing production of seeds and vegetative reproductive parts. To prevent seed set, herbicides must be applied prior to flowering. In addition, herbicides must be applied when they will be translocated to roots or rhizomes. Because phloem-mobile herbicides follow the flow of sugars within plants, they are most effective on perennial weeds when applied to plants storing carbohydrates in the vegetative reproductive tissues. The best application time for perennials is in the late fall prior to frost, and to a lesser extent in late spring prior to flowering.

Timing of herbicides that have soil residual activity (such as picloram) is less critical than herbicides with no residual activity because weeds that emerge and begin to grow within the treated soil zone are still exposed to herbicides through the roots. The central factor for successful weed control with soil-applied herbicides is activation via rainfall or snow melt. Some herbicides are subject to photodegradation or volatilization, so the best application times are in the spring or fall, but not during the heat of summer.

The timing of non-residual systemic herbicide application is much more critical. Because these herbicides enter only through foliar tissue, the weeds must be actively growing to maximize absorption. Herbicide efficacy decreases when plants are stressed. Translocation from the shoots to the roots is essential for perennial weed control, so non-residual systemic herbicides should be applied when plants are actively moving carbohydrates to the root.

• The Herbicide Label

The herbicide label is the legal guide for proper use and application of herbicides. Herbicide labels are approved by the U.S. Environmental Protection Agency to ensure

Table 2. Spray additives and their use.

Additive	Purpose	Examples	When to Use
Surfactants	Surfactants improve the emulsifying, dispersing, spreading, and wetting of solutions. In general, they decrease surface tension of water.	Non-ionic surfactant (NIS), Emulsifiers, Dispersing agents, Wetting Agents	Most sulfonylurea herbicides require surfactant for foliar applications. Consult the label.
Antifoaming agents	Antifoaming agents reduce the amount of foam in spray solutions, allowing pumps and nozzles to work better and improving spray uniformity.		Use with surfactants and certain herbicides that produce foam when mixed.
Drift control agents	Drift control agents prevent fine spray droplets and movement of herbicide to non-target sites.		Use when conditions during application are conducive to drift.
Compatibility agents	Compatibility agents aid in the suspension or mixing of herbicides with other herbicides, other pesticides, or liquid fertilizer.		Use when mixing pesticides or fertilizers.
Crop oil	Usually non-phytotoxic and contains NIS to help form emulsions. Activate herbicides by increasing absorption of the herbicide.	Crop oil, Crop oil concentrate (COC), Methylated seed oil (usually from soybean)	Use when weeds are stressed. Consult the label.
Activators	Activators enhance herbicide activity. Herbicides with multiple ionic states benefit most from activators		Improves performance of herbicide salts (i.e., 2,4-D amine and glyphosate).
Fertilizers	Fertilizers used in small doses buffer the pH of spray solutions and bind free ions. Fertilizers also increase absorption of some herbicides.	Ammonium sulfate (AMS), liquid N (28-0-0 or 32-0-0)	Use with glyphosate and herbicide salts.
Dyes	Dyes are used to temporarily color the foliage or ground to aid in application.		Do not use if not approved on the herbicide label.

safety of the applicator and to protect desirable vegetation, non-target areas, wildlife, and the environment. The herbicide label also contains information on the product's mechanism of toxicity, susceptible weed species, grazing restrictions, and proper application techniques. The applicator should read and become familiar with the herbicide label before application, because improper application is against the law.

Herbicide Classification

As of 1998, more than 150 herbicide active ingredients are used for managing weeds. Many of these products are formulated, packaged, or pre-mixed with other herbicides, resulting in several hundred commercial products for sale. Herbicides are classified or grouped by several different methods to distinguish among their uses and activity. Herbicides are often grouped by their chemical structure, how they are used, or how they kill plants. The latter scheme is the most logical from a management standpoint, because modes of action can be targeted to particular weeds, and users can keep track of how often a particular mode of action is used.

• *Herbicide Mode of Action*

A herbicide's mode of action is the mechanism by which it kills plants, and includes all the plant processes affected from the time it is applied until the weed dies. Some of these plant processes are herbicide uptake or absorption, translocation, metabolism, and interference at the site of action. A herbicide must first be taken up or absorbed into the interior of the plant; this can occur through the roots, through the foliage, or both, depending on how the herbicide is applied. Once inside the plant, most herbicides are translocated or moved throughout the plant via the vascular system. Plant metabolism may alter the herbicide by removing or adding essential functional groups, which usually inactivate (but sometimes activate) the molecule. Herbicides eventually kill plants by inhibiting or affecting a single enzyme or enzyme pathway, termed the site of action, which is essential for plant growth.

• *Uptake or Absorption*

Because herbicides are applied to bare soil or to existing plant vegetation, they may enter the plant through roots or shoots. Most rangeland herbicides that persist in soil are absorbed by both roots and shoots.

• *Root Uptake*

Herbicides that persist in the soil and continue to affect newly emerging plants or sprouting perennial shoots are considered residual herbicides. The length of soil residual period varies widely across herbicide families, but plant control can be obtained as long as the herbicide is present in the soil. Most herbicides must be dissolved in the soil solution (water phase) to be taken up by plant roots, although a few enter from the soil vapor phase.

All plant roots are covered with an outer layer of cells called the epidermis. Root hairs are single-celled extensions of epidermal cells which greatly increase the surface area of a root system and are responsible for absorption of water and dissolved compounds (including herbicides) from the soil. To be effective, herbicides must not only be taken up by the roots, but must enter the vascular system of the plant.

Plant roots have a specialized layer of cells inside the root that acts as a barrier between outer cells and the vascular tissue. This barrier ensures that all materials entering the root must pass through a living cell, instead of just moving through intercellular spaces. This layer can exclude herbicides if they are particularly oily or not formulated correctly. Once past this barrier, herbicides can enter the xylem, or water-transporting vascular tissue. Herbicides mobile in the xylem usually travel only upward in the plant, and tend to accumulate in leaf margins.

The other part of the plant vascular system is the phloem, which transports sugars and other nutrients both upward and downward. Herbicides that can enter the phloem have a better chance of accumulating in growing points, or meristems, of the plant, and thus are usually more effective on perennial plants.

• Foliar Uptake

To be effective, herbicides applied to vegetation must be applied to actively growing, green vegetation, and not to plants that are drought-stressed or dormant. Foliar absorption occurs mostly through leaves and stems, although small amounts of some herbicides can enter through small openings in the leaf called stomata. In order for herbicides to be effectively absorbed, they must remain in contact with the leaf or stem surface, penetrate into the leaf tissue, and enter living cells.

Herbicides must pass through tissues before they can enter living cells. The first and most effective barrier to herbicide absorption is called the cuticle. This is a waxy layer covering the entire aboveground portion of the plant. The natural function of the cuticle is to protect the plant from excessive water loss through leaf surfaces. Because it is made of several waxy layers, the cuticle can also be very effective at preventing entry of herbicides into the plant. Cuticle thickness varies by plant species and, most importantly, is strongly affected by environmental conditions. Any kind of drought stress, as well as low humidity, high light, and high temperatures will trigger plants to develop thicker cuticles. Therefore, herbicides are usually much less effective on drought-stressed plants, and applications during stressful conditions should be avoided.

Foliar absorption of herbicides is affected by many factors, the most important of which are the type of herbicide, spray solution components, and the species of weed and condition of its leaf surfaces. The key factor for the herbicide is that it remain in solution once the spray droplet is on the leaf surface. If the herbicide crystalizes, it is highly unlikely it will return to solution and penetrate the leaf. For example, amine forms of phenoxy herbicides can react with dust and crystalize on the leaf surface, whereas phenoxy ester formulations are much less likely to do so.

The most important factor of the spray solution is its overall surface tension. Decreasing the surface tension by including spray adjuvants will enhance the amount of spray solution contacting the leaf surface, and can help prevent the spray droplets from drying too quickly. In addition, herbicide absorption is affected by the amount of pubescence (small hairs), cuticle structure, and cuticle composition. A rough cuticle surface and pubescence can decrease the contact between spray droplets and the leaf surface. Cuticle structure and composition refer to a leaf's waxiness and other components, and again vary widely across weed species. The type and amount of waxes in the cuticle can affect herbicide absorption even more than cuticle thickness.

• *Translocation*

Herbicide translocation refers to the movement of a herbicide, once inside the plant. Some herbicides are translocated throughout the plant (systemic), whereas other herbicides have very limited movement (contact herbicides). Complete spray coverage is the key to good management when applying contact herbicides. Contact herbicides effectively burn off the vegetation of perennial weeds, such as spotted knapweed (*Centaurea maculosa*) or leafy spurge, but have no effect on root systems because they are not translocated. In contrast, systemic herbicides are translocated throughout the plant, including root buds and other reproductive structures, and therefore are more effective for controlling perennial weeds. Glyphosate, 2,4-D, triasulfuron, and picloram are systemic herbicides commonly used on rangelands.

Only a few rangeland herbicides (most notably, tebuthiuron) are xylem-mobile. Such herbicides move in the direction of water from the roots to the top of the plant. If a xylem-mobile herbicide is applied only to the lower leaves, new vegetation emerging from the growing point will not show injury. However, xylem-mobile herbicides applied to soil are absorbed through roots and the whole plant will show injury.

Most rangeland herbicides are phloem-mobile and are translocated throughout the plant. Herbicides translocated through the phloem tend to accumulate in meristematic regions, flowers, new leaves, roots, and vegetative propagules. As a result, such herbicides tend to be most effective at managing perennial plants with various mechanisms of vegetative propagation.

• *Metabolism*

Plants have the ability to metabolize many foreign compounds with which they come in contact, including herbicides. However, not all plant species contain the same metabolic enzymes, and thus different species are sensitive to different herbicides.

124

Plant metabolism can affect a herbicide in one of two ways. Usually metabolism of the herbicide deactivates it to nontoxic products. However, a few cropland herbicides must be metabolized by the plant to become toxic to the plant.

• *Site of Action*
A herbicide's site of action is the specific enzyme or enzyme pathway that is inactivated and causes plant death. By inhibiting an enzyme or pathway, the herbicide prevents the plant from synthesizing critical compounds necessary for life. In other cases, inhibition causes the accumulation of toxic intermediates or highly reactive molecules which destroy cells. The site of action is known precisely for many herbicides, but several are still unknown.

Mode of Action of Rangeland Herbicides
Of the numerous herbicides commercially available, only a handful are widely used in rangeland weed management. Table 3 shows some of the commonly used rangeland herbicides and summarizes their characteristics.

• *Growth-Regulating Herbicides*
Growth-regulating herbicides are chemical mimics of natural plant hormones called auxins. Plants synthesize minute quantities of auxin to regulate their own growth and development, after which they are immediately inactivated by metabolism. However, growth-regulating herbicides, which stimulate the same growth processes, cannot be quickly metabolized by sensitive species, and thus the plants essentially grow themselves to death.

Growth-regulating herbicides are the most widely used herbicides in rangeland. 2,4-D alone is applied on an estimated 4.9 million acres of rangeland per year. Most growth-regulating herbicides are applied directly to the foliage, although several groups within this family have residual activity and are root absorbed. Plant growth-regulating herbicides used on rangeland include phenoxies, benzoic acid, and picolinic acid.

Phenoxies. Phenoxy herbicides include 2,4-D, 2,4-DB (rarely if ever used in rangeland), MCPA, and 2,4,5-T (no longer available). Phenoxy herbicides have numerous trade names, but are usually referred by their common names. They are commonly formulated as amines or esters, and rarely as water-soluble salts. The characteristic chemical structure of phenoxy herbicides is a chlorinated benzene ring.

Phenoxy herbicides are generally applied as foliar sprays and are absorbed through leaves. Phenoxy herbicides can also be absorbed through roots, but are rarely applied to soil because of their very short residual period. Once inside the plant, phenoxy herbicides are translocated in the phloem throughout the plant, and are thus effective on perennial broadleaf weeds. Grass species can metabolize phenoxy herbicides, which is the primary mechanism of selectivity. The site of action of phenoxy herbicides is unclear, since they appear to affect a multitude of enzymes and physiological processes in sensitive plants. Because phenoxy herbicides have relatively low toxicity and are rapidly degraded in soil, they are commonly used for rangeland

125

Table 3. Commonly used rangeland herbicides.

Common Name	Trade Name	Mode of Action	Weed Spectrum	Soil Residual
glyphosate	Roundup®	amino acid synthesis inhibitor	non-selective	no
picloram	Tordon®	growth regulator	broadleaf species	yes
2,4-D	many	growth regulator	broadleaf species	no
dicamba	Banvel®	growth regulator	broadleaf species	no
triasulfuron	Escort®	amino acid synthesis inhibitor	broadleaf species	yes
tebuthiuron	Spike®	photosynthesis inhibitor	trees and shrubs	yes
clopyralid	Stinger®	growth regulator	broadleaf species	yes
imazapyr	Arsenal®	amino acid synthesis inhibitor	non-selective	yes

weed control in ditches, along streambanks and lakeshores, and for aquatic weed control.

Benzoic acids. Dicamba, the only benzoic acid herbicide used in rangelands, is marketed under the trade names Banvel® or Clarity®. Dicamba is available in liquid or granular formulations. The characteristic structure of benzoic acids is a benzene ring with a carboxylic acid group.

Dicamba is usually applied to foliage as a spray, but can be soil applied as a granule. Dicamba is absorbed through leaves and roots, and is translocated throughout the plant via the phloem. Dicamba has longer soil activity than 2,4-D, with a half-life of about 14 days under most conditions. Monocots are tolerant to dicamba because of rapid metabolism. Dicamba's specific site of action is unknown.

Dicamba is rarely used alone, but instead is applied in a tankmix with other herbicides. Dicamba may be tankmixed with 2,4-D or other phenoxy herbicides to increase its activity and spectrum of weed control, but cannot be applied near open water. Dicamba's relatively short soil residual activity limits its use in rangeland, but it may be mixed with picloram or other longer residual herbicides.

Picolinic acid. Picolinic acid herbicides include picloram, clopyralid, and triclopyr, which are marketed as Tordon®, Curtail® or Transline®, and Garlon® or Remedy®, respectively. All three members are commonly formulated as liquid herbicides. The characteristic structure of picolinic acids is a benzene ring substituted with a nitrogen and a carboxylic acid side chain.

Picolinic acid herbicides can be applied to green vegetation and to soil. These herbicides are absorbed through roots and leaves and translocated via the phloem, similarly with the phenoxy and benzoic acid herbicides. However, picolinic acids are much more persistent in the soil than phenoxy or benzoic acid herbicides. Depending on soil type and environmental conditions, picloram may have soil activity for two to three years or longer, while clopyralid and triclopyr remain active in the soil for two to four months. Picolinic acids are metabolized by grasses and certain broadleaf weeds, which serves as the major mechanism of plant selectivity. The specific enzyme or pathway influenced by picolinic acids is unknown.

Picolinic acid herbicides can be up to 10 times more active than 2,4-D on certain broadleaf weeds, such as knapweed species, leafy spurge, and other perennial broadleaves. For example, clopyralid is very effective for controlling Canada thistle (*Cirsium arvense*), while triclopyr is commonly mixed with 2,4-D or dicamba and used as a brush killer. Increased efficacy combined with long soil residual activity make picolinic acid herbicides a common choice for weed control in rangelands. However, because picolinic acid herbicides are highly mobile in soil and persistent, they can have a high potential for leaching. Care should be taken when picolinic acids are applied to areas with shallow water tables, high rainfall, or near open water.

• Amino Acid Synthesis Inhibitors

Amino acid synthesis inhibitor herbicides kill plants by preventing the plant from synthesizing key amino acids. Amino acids are the building blocks of proteins, which are themselves absolutely essential for all life processes. Therefore, depriving the plant of even one or a few amino acids is lethal.

Amino acid synthesis inhibitors are generally used as foliar sprays for killing existing vegetation, and can be used as nonselective sprays or as selective herbicides for killing specific weeds. These herbicides are characterized by very slow action, with plant stunting and yellowing as the first symptoms. Complete plant death may take several weeks.

Amino acid derivatives. Amino acid derivative herbicides include glyphosate and sulfosate, which are marketed as Roundup® or Rodeo® and Touchdown®, respectively. Amino acid derivatives are formulated only as liquid herbicides.

Glyphosate is a nonselective herbicide used only as a foliar spray. Glyphosate has no soil residual properties, because it binds very tightly to soil particles, rendering it inactive. Since it is a systemic herbicide, glyphosate is translocated throughout the plant in the phloem, and accumulates in growing points. As a result, glyphosate is very effective at killing vegetative reproductive organs of certain perennial weeds, particularly grasses. Plants cannot metabolize glyphosate to any appreciable extent, although some species like horsetail (*Equisetum arvense*) are naturally tolerant. Glyphosate can be used selectively using rope-wick applicators to treat weeds that are growing above the desirable plant canopy, or by spraying when weeds are growing but desirable vegetation is still dormant.

Glyphosate inhibits an enzyme that is responsible for synthesis of the aromatic amino acids. By blocking this pathway, glyphosate not only deprives the plant of

127

these essential building blocks, but also a whole host of secondary compounds such as lignin, alkaloids, and flavonoids. The very slow injury symptoms and plant death after glyphosate treatment are a result of starvation for these compounds.

Glyphosate is most often used for total vegetation control in rangeland situations. However, glyphosate will selectively control winter annual weeds such as downy brome if applied early in the spring before desirable grasses and forbs break dormancy. For perennial weeds, glyphosate is most effective if applied in late summer or early fall after seed set. Since glyphosate has no soil residual activity, reseeding or other revegetation practices can be carried out immediately after application.

Sulfonylureas. Sulfonylurea herbicides include metsulfuron, triasulfuron, and a number of other products that are used primarily to control weeds in cropland. Metsulfuron is marketed as Escort® or Ally®, while triasulfuron is marketed as Amber®. Triasulfuron is labeled for some uses in rangeland. Sulfonylureas are commonly formulated as water dispersible granules.

Sulfonylurea herbicides are normally applied as foliar sprays, although they are readily absorbed through roots and leaves. Soil residual periods for sulfonylurea herbicides vary widely among individual members, and range from only a few weeks to three years or more. Sulfonylureas are translocated throughout the plant via the phloem and accumulate in growing points. These herbicides inhibit an enzyme in the pathway responsible for synthesis of the branch chained amino acids leucine, isoleucine, and valine. As with glyphosate, loss of these essential compounds causes sensitive plants slowly to starve to death.

Unlike glyphosate, sulfonylurea herbicides are metabolized by different grass and broadleaf plant species, and the degree of metabolism varies widely. Thus, depending on the individual herbicide, different sulfonylureas can control a wide range of plant species. They can be used selectively to control undesirable rangeland species, particularly perennial broadleaf weeds. Because of the highly variable weed control spectra for individual members, sulfonylureas may have advantages over growth-regulating herbicides such as picloram in certain situations. For example, some desirable broadleaf plants and forbs may be tolerant to certain sulfonylureas, while weeds like spotted knapweed could be controlled.

Imidazolinones. Imidazolinones are a large family of herbicides used for cropland and rangeland weed control. For rangeland, imazapic marketed as Plateau® and imazapyr marketed as Arsenal® are the most commonly used. Imidazolinones are sold as liquid and granular formulations. These herbicides have the same site of action as sulfonylureas, and their translocation patterns are similar.

Plateau® has moderate soil residual activity and will provide one to two years of residual control of winter annual grasses such as downy brome, as well as some perennial broadleaf weeds. Imazapyr has extremely long soil residual activity (sometimes longer than five years) and is typically used as a soil sterilant to control weeds along railroads and in industrial settings.

• *Photosynthesis Inhibitors*

Photosynthetic inhibitors have long been used to selectively control grass, broadleaf and brushy weeds in crops, rangelands, and rights of way. Photosynthetic inhibitors are absorbed through roots and shoots, and usually have long soil residual activity. These herbicides are effective on certain perennial weeds, because shoot death prevents replenishing of food supplies for underground vegetative buds. Photosynthesis inhibitors used in rangelands are mobile only in the xylem, so leaf edges and oldest tissues show injury symptoms first, including leaf yellowing and browning. Unlike some other residual herbicides, photosynthesis inhibitors leave skeletons of dead weeds in the treated area.

Photosynthesis inhibitors block the energy flow in a photosynthetic pathway. The released energy causes the creation of high energy free radical molecules which rapidly destroy cell membranes.

Substituted urea. The only substituted urea used in rangelands is tebuthiuron, marketed under the trade name Spike®. Tebuthiuron is formulated as a granular herbicide and has several years of residual activity. It is primarily used for brush management in rangeland, especially sagebrush. Because of its long soil residual activity, tebuthiuron causes repeated defoliation and eventual death of sensitive brush species. Tebuthiuron is also highly susceptible to leaching via water movement.

Herbicide-Resistant Weeds

Herbicide-resistant weeds are thought to start from one or a few plants present in a field before spraying, so they are not noticed during the first few years a herbicide is used. However, by repeatedly using the same herbicide over several years, the applicator "selects" for those few resistant individuals by killing all the susceptible plants. Each additional year that the applicator uses the same herbicide or herbicides with the same mode of action, the more prevalent resistant weeds become until they become the most predominant form.

The most important factors controlling the appearance of resistant weeds are:

1) Selection Intensity. This term refers to how strongly resistant biotypes will be chosen from a population, and is related to the efficacy of the herbicide and how often the herbicide is applied. If the herbicide is very effective, is applied often, has long soil residual activity, and/or is the only practice for controlling a particular weed, then the selection intensity for resistance is very high. Under these conditions, selection for resistant weeds will be rapid.

2) Weed Biology. Some weed species have substantial amounts of genetic diversity, meaning that a single species consists of many different varieties or hybrids. Generally, weeds like spotted knapweed that outcross (pollen is spread from one plant to another by insects or wind) have more diversity than those that pollinate themselves, like quackgrass (*Agropyron repens*). Weeds with more genetic diversity have a higher potential to develop resistance to herbicides, since the initial incidence of resistant individuals should be higher.

129

3) Herbicide Mode of Action. Herbicides that kill weeds the same way belong to the same mode of action family. For example, sulfonylurea and imidazolinone herbicides target the same plant enzyme, and so weeds that are resistant to one herbicide are often also resistant to other members of the family. The target enzyme for this herbicide family can exist in many forms, some of which are already herbicide-resistant. Therefore, there is a relatively high proportion of weeds in an unsprayed field that are already resistant (maybe 1 in 10,000). In contrast, resistant forms of the enzyme that glyphosate targets appear to be much more rare (maybe 1 in 100 million).

The most important practice to prevent herbicide resistance is to integrate management methods and rotate among different herbicide modes of action. Applications of the same herbicide (or herbicides with the same mode of action) should not be made in successive years. It is important to remember that residual herbicides are still imposing selection pressure for resistant weeds for however long they persist in the soil. If available, another mode of action should be used, or if not possible, other means of weed control should be substituted in non-application years. Such measures could include burning, mechanical control, timed grazing, or high intensity, short-term grazing.

Herbicide Fate

Herbicides applied in field settings have several fates, including uptake by plants, degradation in plants or soil, off-target movement, and unintended deposition in the environment. In a well-planned management system, herbicides should effectively control weeds with little or no adverse environmental effects. The important characteristics that determine the environmental fate of a particular herbicide include the chemical and physical properties of the herbicide, soil characteristics, climatic conditions, and members of the target plant community.

• Plant Uptake

One obvious fate of herbicides is entry into the plant. In general, only a small percentage of applied herbicide enters the plant. Once the herbicide molecule enters the plant it may remain in its active form or can be metabolized into inactive forms. If it is not metabolized, the herbicide may be leached from dead and decaying plant material into the soil.

• Soil-Herbicide Interactions

Most herbicides eventually end up in the soil, either through direct application or indirectly through a plant. Once on the soil surface, some herbicides react with sunlight and are photodegraded to nontoxic products, while others are unstable and spontaneously degrade into inactive forms. If they enter the soil, all herbicides are broken down into inactive forms, through chemical or biological processes, or both. However, the rates at which individual herbicides are degraded can vary widely, depending on the chemical properties of the herbicide and the microbial populations in the soil. Herbicide half-lives in soil can range from a few minutes to many years. In general, soils that are fertile and well-watered support substantial microbial populations, and are therefore more likely to degrade herbicides more rapidly. Little microbial activity occurs in dry and nonfertile soils, especially under extreme environments.

Once in the soil, the extent of movement of a particular herbicide depends on a number of factors, most importantly its affinity for organic matter and clay particles and its water solubility. Herbicides like glyphosate bind very tightly to soil components immediately upon contact, thus preventing their further movement and allowing soil microorganisms to begin degradation processes. However, other herbicides have very little affinity for soil, and can be leached extensively if applications are followed by heavy rainfall or surface water movement. Even if bound to soil particles, herbicides can be moved off-site or into surface waters via erosion of treated soil. Excessive water movement can even cause herbicides with reasonable soil affinity to move through subsoil and into groundwater.

Herbicides should always be used so that the likelihood of moving into surface waters or leaching into groundwater is minimized. Some scenarios that may lead to water contamination by herbicides include: 1) using herbicides on soils with very low organic matter content and/or high sand content; 2) using highly persistent herbicides in leachable soils or in high rainfall areas; 3) using herbicides on soils over a shallow water table; 4) using herbicides with low affinity for soil under leachable conditions.

• Off-Target Movement

Herbicide volatilization and drift are the primary mechanisms of off-target movement. Off-target movement can result in unintended injury to plant species, contamination of surface waters, and contamination of ecologically sensitive areas.

Volatilization occurs when herbicides become suspended in the atmosphere after application, and can happen within a few minutes of application, or several hours to days later. Under certain environmental conditions, herbicide droplets evaporate from the soil or leaf surfaces and are moved into the air. Again, depending on barometric pressure and wind conditions, volatilized herbicides can affect non-target areas long distances from the application site. Volatilization can be minimized by choosing nonvolatile herbicide formulations (2,4-D amine is much less volatile than 2,4-D ester, for example), using dry carriers when possible, and avoiding applying herbicides during hot summer days.

Herbicide drift occurs when herbicide droplets land on non-targeted areas. Herbicide drift occurs at the time of application and is usually limited to 100 to 200 yards (90 to 180 m). Application conditions that minimize herbicide drift include using low sprayer pressures and nozzles with large orifices that produce larger spray droplets, spraying in calm conditions, adjusting boom height as low as possible, and using windshields on the boom.

131

Conclusion

Herbicides can be a critical component of weed management systems. The choice of which particular herbicide to use depends on a number of factors, including target weed species, density of weeds, presence of desirable grasses and forbs, soil attributes, proximity to water, and environmental conditions. Because of these interrelated factors, herbicide choices are best made in consultation with experienced Extension specialists, county weed supervisors, field scouts, custom applicators, and others. There are seldom blanket recommendations for individual weed control situations.

Herbicides are likely to be the most expensive component of a rangeland weed management program. Therefore, when using herbicides it is crucial to ensure optimum performance of the product. To improve herbicide performance, use appropriate spray additives and spray gallonage, and apply at the correct time. Finally, choose the right herbicide, know what it will and will not do, use it carefully and according to the label, and use it in rotation with other practices.

In general, herbicides are most effectively used to gain initial control of a new weed invasion or a severe infestation. Herbicides are rarely, if ever, a complete solution to weed problems. Instead, they must be incorporated into long-term management plans that include replacement of weeds with desirable species, proper land use by grazing animals, and prevention of new infestations.

Other Sources

Anderson, W.P. 1996. Weed Science, Principles and Applications, 3rd Edition. West Publishing Company, St. Paul, MN.

Devine, M.D., S.O. Duke, and C. Fedtke. 1993. Physiology of Herbicide Action. P T R Prentice Hall, Inc.

Esau, K. 1953. Plant Anatomy, 2nd Edition. John Wiley and Sons, Inc.

Fuerst, E.P., and M.A. Norman. 1991. Interactions of herbicides with photosynthetic electron transport. Weed Sci. 39:458-64.

Lehninger, A.L., D.L. Nelson, and M.M. Cox. 1993. Principles of Biochemistry, 2nd Edition. Worth Publishers.

Powles, S., and J. Holtum. 1994. Herbicide Resistance in Plants: Biology and Biochemistry. CRC Press. Boca Raton, FL.

Ray, T.B. 1984. Site of action of chlorsulfuron. Plant Physiol. 75:827-31.

Ross, M.A., and C.A. Lembi. 1985. Applied Weed Science. Macmillan Publishing Company. New York, NY.

Shimabukuro, R.H. 1985. Detoxification of Herbicides. In Weed Physiology Vol. II. S.O. Duke, (ed.). CRC Press. Boca Raton, FL, 215-40.

Stidham, M.A. 1991. Herbicides that inhibit acetohydroxyacid synthase. Weed Sci. 39:428-34.

Revegetating Noxious Weed-Infested Rangeland

James S. Jacobs, Michael F. Carpinelli, and Roger L. Sheley

Weed management efforts often focus on simply controlling weeds, with limited regard to the existing or resulting plant community. Because of environmental, ecological, and economic concerns, the appropriateness and effectiveness of rangeland weed management practices are being questioned. It has become clear that weed management decisions must consider these concerns. The development of future weed management practices must be based on our understanding of the biology and ecology of rangeland ecosystems.

Land-use objectives must be developed before rangeland weed management plans can be designed. Strictly killing weeds is an inadequate objective, especially for large-scale infestations. However, a generalized objective might be to develop a healthy plant community that is relatively weed-resistant, while meeting other land-use objectives such as forage production, wildlife habitat development, or recreational land maintenance.

Highly degraded rangeland dominated by noxious weeds is often devoid of competitive desirable plants. On these sites, rangeland weed control is often short-lived because desirable species are not available to occupy niches opened by weed control procedures (James 1992, Sheley et al. 1996). Introducing and establishing competitive plants is essential for successful management of weed infestations and the restoration of desirable plant communities (Hubbard 1975, Larson and McInnis 1989, Borman et al. 1991). However, revegetation often is not included in a weed management plan because it is costly and has a high risk of failure. The process of revegetation must first identify the desirable plant community that meets management objectives, and then determine the seeding method, herbicide treatment, species to be seeded, and follow-up treatments to best achieve the desirable plant community.

The decision to revegetate must consider direct costs (seedbed preparation, seeds and seeding, follow-up management), indirect costs (risk of failure, non-use during establishment period), and benefits (increased forage, improved ecosystem function, soil conservation). Revegetation efforts should focus on sites and methods with the greatest potential for increasing net benefits in the shortest amount of time.

Rangeland revegetation is costly because current methods use agronomic practices. Typically, revegetation of weed-infested rangeland requires multiple entries. First, the site is disked in late fall to loosen the soil surface and encourage the germination of weed seeds in the seedbank. A few weeks later, a non-selective herbicide such as glyphosate is applied to kill the newly establishing weeds. The combination of disking and herbicide application reduces the number of weed seeds in the seedbank and reduces weed competition the following spring. Soon after the herbicide is applied, fall-dormant grasses are seeded. The following spring, some of the remaining weed seeds in the seedbank and seeded grasses germinate and emerge. With adequate spring precipitation, both grass and weed seedlings survive. If grass seedlings survive until mid-summer, a reduced rate of 2,4-D or mowing is usually applied to weaken weeds and retard them from going to seed. Although revegetation with aggressive species has been shown to inhibit weed reinvasion (Hubbard 1975, Huston et al. 1984, Larson and McInnis 1989, Borman et al. 1991), managers are reluctant to attempt it, because of the high probability of failure and expense associated with this multiple-entry approach. Effective, single-entry methods must be developed for revegetation to be affordable and applicable to remote areas.

Failures in revegetation of weed-infested rangeland are usually caused by a combination of factors. The most important are insufficient soil moisture and intense weed competition. While some factors, such as climate, are beyond our control, we can use what we know about ecology to improve our success rate.

Seedling establishment is the most critical phase of revegetation (James 1992). Seedling establishment appears associated with the availability of safe sites (Harper et al. 1965, Wright et al. 1978) and the availability of seeds (Pickett et al. 1987). Rehabilitating weed-infested rangeland with desirable grasses typically fails, however, because of competition with weeds for safe sites during the initial stages of establishment (Borman et al. 1991, James 1992). In addition, density-dependent factors (such as competition) and density-independent factors (such as climate) interact to determine seedling survival during grass establishment in weed-infested rangeland (Velagala et al. 1997). Revegetation methods must address both of these factors to improve revegetation success.

Seeding Methods

Seedbed preparation and seeding depth affect seedling survival. The ideal seedbed for range seeding consists of pulverized surface soil with moderate amounts of mulch or plant residue, firm soil below seeding depth, and no residual competitive plants. Ideal seeding depths are about $1/4$ inch (6 mm) for small seeds, and $1/2$ inch (12 mm) for large seeds. Site conditions and cost will determine which seedbed preparation method is most appropriate.

• Plowing and Drilling

Plowing is the most effective method for preparing an ideal seedbed. However, it is costly and only practical on sites that are accessible to machinery and have fertile, deep, and rock-free soils. Plowing not only removes competitive vegetation, it increases the establishment of seeded species. Seed placement is enhanced by plowing because the roughening of the soil surface increases the number of safe sites. Plowing also loosens the upper layer of soil, thus facilitating root extension of establishing seedlings. In areas that can be farmed and intensively managed, plowing and drill-seeding has the best chance of insuring revegetation success. A typical three-year intensive revegetation plan includes plowing and seeding an annual hay crop in the first and second years, and replowing and seeding to a grass-legume mixture in the third year. The hay-cropping allows for the germination and removal of weeds from the seedbank while providing immediate financial return.

• No-Till Drill

The no-till drill is a tractor-pulled machine that opens a furrow in untilled ground, drops seeds in the furrow at a specified rate and depth, and then rolls the furrow closed. This method is the most practical and commonly used method on rangeland that is accessible to machinery, because it is less expensive than tillage and reduces the risk of erosion associated with tillage. This is especially important in arid and semi-arid areas where tillage may exacerbate wind and water erosion. Because this method does not remove competitive plants, no-till-drill seeding is commonly preceded by a non-selective herbicide application.

• Broadcast Seeding

On sites that are inaccessible to machinery, site preparation is limited to removal of competitive plants via herbicide or fire. On small-scale projects, seeds are typically broadcast with a hand-held seed-scattering device. Large-scale projects are usually helicopter-seeded. Untilled soil usually lacks safe sites, which may be countered by using a higher seeding rate than would be used on plowed ground. Covering seeds with soil or mulch improves germination and establishment. Other disadvantages to broadcast seeding are poor seed distribution, loss of seeds to rodents and birds, and slower establishment.

• Alternative Methods

Recent studies show that increasing the seeding rate above agronomic levels improves grass seedling establishment on weed-infested rangeland (Jacobs et al. 1996, Velagala 1996, Velagala et al. 1997). Alternative revegetation methods may be developed that use high seeding rates without increasing cost. For example, a strip-tilling/seeding method could seed desirable species at high densities on evenly spaced tilled strips. Successful seedling establishment would be increased because of the high seeding rate, and under proper management the established species within the strips would spread naturally to the interstrip area. One way to facilitate the expansion of seeded

135

species to the interstrips would be first to apply a herbicide to remove existing vegetation from the entire area prior to strip-tilling/seeding. Another way would be to use selective grazing of broadleaf species by sheep or goats to control the establishment of weeds in the interstrips. Similarly, islands of desirable plants could be established using high seeding densities, so that the plant cover is dominated by desirable species. From these islands, desired species could spread naturally.

Herbicides

Herbicides provide an alternative to tillage for the removal of unwanted vegetation. Herbicides are often used where accessability or erosion is a concern or tillage is cost-prohibitive. Herbicide selectivity, persistence, and timing of application are important considerations when deciding how to use herbicides for weed control in revegetation.

• Non-Selective Herbicides

Most revegetation seedings are preceded by an application of a non-selective herbicide such as glyphosate. On rangeland where fall-dormant seedings are most practical, a late-season application of glyphosate can be used to eliminate fall-germinating annual weeds like cheatgrass (*Bromus tectorum*) and yellow starthistle (*Centaurea solstitialis*) that have emerged prior to the application. This may substantially reduce weed competition for early season moisture the following spring.

• Broadleaf Herbicides

Broadleaf herbicides are often used in rangeland revegetation, because most weeds are broadleaved species and most seeded species are grasses. An advantage to some broadleaf herbicides is that their soil residual will control broadleaf seedlings for up to 18 months. Picloram applied at a rate of 0.28 kg active ingredient per hectare (0.25 lb/ac) provides control of many weeds for two or three years (Davis 1990). The combination of glyphosate and a residual broadleaf herbicide, such as picloram, may provide the best control of weeds for a one-pass revegetation procedure where only grasses are seeded. Broadleaved desirable species can be incorporated once the grasses are established and the residual effect of the herbicide is gone. Of course, where broadleaved desirable species are seeded in a one-pass operation, herbicide selection will be limited to a non-residual broadleaf herbicide such as 2,4-D or a non-selective, non-residual herbicide such as glyphosate.

Species Selection

Selection of desirable plant species is determined by intended use, soils, precipitation, temperature, and establishment characteristics. Another important consideration is the ability of the desired species to withstand reinvasion. For example, many aggressive non-indigenous species and some native species have been shown to reduce leafy spurge infestations.

• Intended Use

The intended use of a revegetation site is important in determining what species to plant. If livestock grazing is the intended use, a perennial with high forage production is an obvious choice. One such species, crested wheatgrass (*Agropyron cristatum*), is a good spring pasture bunchgrass that can withstand a 60% defoliation without affecting its biomass production (Sheley and Larson 1997). Intermediate wheatgrass (*Elytriga intermedia*), a late-maturing grass, may be more appropriate if the intended use is summer pasture. Some areas such as state and federal parks mandate the use of native plants. Here, the plant community composition prior to weed invasion should be used as a guide to determine which species to seed.

• Soils

Soil texture affects the establishment success of seeded species. While medium- to fine-textured soils are optimal for most species, some species do best in either sandy or clay soil. For example, Indian ricegrass (*Oryzopsis hymenoides*) and pubescent wheatgrass (*Thinopyrum intermedium* subsp. *barulatum*) are well adapted to sandy soils, western wheatgrass (*Agropyron smithii*) does well on clay soils, and most other species commonly used in revegetation do well on medium- to fine-textured soils (Table 1).

• Precipitation

Seeded species need to be adapted to the precipitation level of the site. Crested wheatgrass and Russian wildrye (*Psathyrostachys juncea*) (non-indigenous species), and Indian ricegrass and bluebunch wheatgrass (*Pseudoroegneria spicata*) (native species), are adapted to rangeland sites receiving 10 to 12 inches (25 to 30 cm) of annual precipitation. Pubescent wheatgrass and intermediate wheatgrass (non-indigenous species), and green needlegrass (*Stipa viridula*) and western wheatgrass (native species), are adapted to 13 to 15 inches (33 to 38 cm) of annual precipitation. Requirements for some other common revegetation species are summarized in Table 1.

• Temperature

Temperature zones should be considered when designing seed mixes for revegetation. Warm season, C4 grasses, for example big (*Andropogon gerardii*) and little bluestem (*A. scoparius*), are well-adapted to the midwestern prairie. Cool season, C3 grasses, for example Idaho fescue (*Festuca idahoensis*) and bluebunch wheatgrass, are well-adapted to mountain meadows of the Great Basin. Seed supply companies are good sources of information on the environmental requirements of revegetation species.

• Establishment

Species differ in how fast and how well they establish. Crested and pubescent wheatgrasses are some of the easiest species to establish. Natives are generally slower and more difficult to establish. Seed size presents a trade-off between quick

137

Table 1. Characteristics of rangeland revegetation grass species.

Grass Species	Growth form	Soil type	Precipitation range	Establishment
Russian wildrye	Bunchgrass	Silty-clay	Less than 13 inches	Difficult
Crested wheatgrass	Bunchgrass	Silty-clay	Less than 13 inches	Easy
Siberian wheatgrass	Bunchgrass	Silty-clay	Less than 13 inches	Easy
Orchardgrass	Bunchgrass	Silty-clay	Less than 13 inches	Easy
Bluebunch wheatgrass	Bunchgrass	Silty-clay	Less than 13 inches	Fair
Streambank wheatgrass	Rhizomatous	Silty	Less than 13 inches	Fair
Thickspike wheatgrass	Rhizomatous	Sandy-silt	Less than 13 inches	Fair
Slender wheatgrass	Bunchgrass	Silty-clay	Less than 13 inches	Quick
Bluegrass	Rhizomatous	Shallow	Less than 13 inches	Easy
Indian ricegrass	Bunchgrass	Sandy	Less than 13 inches	Easy
Sheep fescue	Bunchgrass	Silty	Less than 13 inches	Easy
Basin wildrye	Rhizomatous	Silty-clay	Less than 13 inches	Slow
Pubescent wheatgrass	Rhizomatous	Sandy-silt	13 to 15 inches	Easy
Beardless wheatgrass	Bunchgrass	Silty-clay	13 to 15 inches	Fair
Western wheatgrass	Rhizomatous	Clay	13 to 15 inches	Fair
Green needlegrass	Bunchgrass	Silty-clay	13 to 15 inches	Slow
Prairie sandreed	Rhizomatous	Sandy	13 to 15 inches	Fair
Intermediate wheatgrass	Rhizomatous	Silty-clay	14 to 15 inches	Fair
Smooth bromegrass	Rhizomatous	Silty-clay	14 to 15 inches	Easy
Altai wildrye	Rhizomatous	Saline	14 to 15 inches	Slow
Meadow bromegrass	Rhizomatous	Silty-clay	15 to 18 inches	Quick
Tall wheatgrass	Bunchgrass	Saline	15 to 18 inches	Fair
Timothy	Bunchgrass	Silty-clay	15 to 18 inches	Easy
Idaho fescue	Bunchgrass	Sandy-silt	15 to 18 inches	Slow

establishment and seed placement. Generally, larger seeds establish quickly, but do best if drill-seeded or sown in tilled ground. Smaller seeds are more likely to find a safe site in untilled ground, but their limited amount of stored carbohydrates may inhibit establishment.

• Designing Seed Mixes

Though grasses dominate rangeland and are the most commonly used species in revegetation, it is advantageous to use a combination of species with differing growth forms when designing seed mixes. For example, seed mixtures of grasses with legumes improve the rate of microbial and soil structure recovery on Conservation Reserve Program (CRP) land, compared to grasses alone. In addition, a species mix of contrasting root growth forms (e.g., fibrous-rooted grasses and taprooted forbs) more efficiently fills niches and uses resources (Jacobs and Sheley 1999). In turn, where resource use is maximized, productivity is maximized. A diverse plant community is likely to be weed-resistant because few resources are available to a potential invader. Niche occupation can also be maximized by combining species that grow at different times of the year.

Seeding Rate

Increased seeding rates can alter the competitive interaction between desired species and weeds, and can aid stand establishment. Increasing densities of intermediate wheatgrass from less than 1,000 seeds/m^2 to more than 1,000 seeds/m^2 removed the effect of spotted knapweed on intermediate wheatgrass where interspecific interference occurred (Velagala et al. 1997). In a field study, Velagala (1996) found the greatest seedling establishment at the highest seeding rates, especially when combined with tillage. In that study, intermediate wheatgrass did not establish at a seeding rate of 500/m^2, which is the standard recommended seeding rate. The wide range of seeding rates studied by Velagala (1996) showed the potential for using seeding rates to enhance establishment.

Seed Treatments

Seed treatments may enhance the establishment phase of revegetation. Seed priming (Callan et al. 1990) is a treatment that initiates the germination process in a seed, allows it to continue to a certain point, and then suspends it. The primed seed is then ready to continue germination in the field when conditions are favorable. The idea behind seed priming is that the first seedling to capture resources has a competitive advantage (Harper 1980). Another advantage to priming is that dormancy is broken and germination is assured. This treatment has been shown to aid establishment under agricultural conditions, and it holds promise in rangeland revegetation.

Soil pathogens may accelerate the death of seeds and seedlings. Their role is best understood in crop species, though Tadros (1957) found fungal decomposition of seeds or very young seedlings in the soil could account, in part, for the different floristic composition of the natural vegetation. Seed fungicide treatments confer numerous advantages, including protection against diseases and pests and enhancement of growth (Powell and Mathrews 1988). A variety of these fungicides are commercially available for grass seeds, and are intended to improve seedling establishment in reclamation projects (Taylor and Harmon 1990). Perennial grass seeds and seedlings can be protected from soil-borne organisms, including *Pythium* and *Rhizoctonia* spp., and many of the systemic fungicides provide the crowns and roots with a longer protection against *Fusarium* spp. and common root rots (Sprague 1950).

Putting It All Together

For revegetation of weed-infested rangeland to become more widely applicable to the various rangeland conditions, cost-effective and reliable methods need to be developed. Developing strategies that enhance our ability cost-effectively to establish desired plant communities may provide ranchers and land managers with a sustainable method for managing noxious weed-infested rangeland. Reducing the

139

number of entries onto the land will reduce the cost of revegetation. Combining the factors discussed above to improve the success of establishing desirable species, then applying them in a single pass, will be the most cost-effective and reliable way to revegetate rangeland. New equipment has made possible the simultaneous application of herbicides, tillage, and seeding, as well as seeding using no-till methods.

Revegetation will be most successful if it works with successional processes. The three processes that influence the direction of succession are site availability, species availability, and species performance (Sheley et al. 1997). Combining these three processes in a revegetation application will provide the highest probability for long-term success. Tilling, herbicides, and intensive grazing create available safe sites. Species availability is accomplished by successfully selecting and distributing seeds. Herbicides, fertilizers, biological controls, and selective grazing can be used to enhance desirable species performance.

Follow-up Management

Money and effort spent on revegetation will be wasted unless management practices are changed to favor the desirable species that were seeded. Rangelands are dynamic plant communities that are constantly being shaped by the process of succession. Successful revegetation requires that managers continuously monitor the land and adjust management practices to direct succession in a way that maintains a desirable plant community. For example, timing and frequency of cattle grazing can be adjusted to minimize the impact on grasses. Sheep or goats can be used to target broadleaved weed species. Biological control can be used to reduce the performance and seed production of weeds. Livestock given feed containing seeds of desirable species can be used as a tool to spread seeds. There are many possibilities. Success is dependent on the creativity and vigilance of the land manager.

Literature Cited

Borman, M.M., W.C. Krueger, and D.E. Johnson. 1991. Effects of established perennial grasses on yields of associated annual weeds. J. Range Manage. 44:318-26.

Callan, N.W., D.E. Mathre, and J.B. Miller. 1990. Bio-priming seed treatment for biological control of Pythium ultimum preemergence damping-off in sh2 sweet corn. Plant Disease 74:368-72.

Davis, E.S. 1990. Spotted knapweed (Centaurea maculosa L.) Seed longevity, chemical control and seed morphology. M.S. Thesis. Montana State Univ, Bozeman, MT.

Harper, J. 1980. Population biology of plants. Academic Press. New York.

Harper, J.L., J.T. Williams, and G.R. Sagar. 1965. The behavior of seeds in soil. 1. The heterogeneity of soil surfaces and its role in determining the establishment of plants from seed. J. Ecology. 53:273-86.

Hubbard, W.A. 1975. Increased range forage production by reseeding and the chemical control of knapweed. J. Range Manage. 28:406-407.

Huston, C.H., R.H. Callihan, and R.L. Sheley. 1984. Reseeding intermediate wheatgrass in yellow starthistle-infested rangeland. *In*: Proc. Knapweed Symp., Mont. St. Univ. Coop. Ext. Bull. 1315, 42-44.

Jacobs, J.J., R.L. Sheley, and B.D. Maxwell. 1996. Effect of *Sclerotinia sclerotiorum* on the interference between bluebunch wheatgrass (*Agropyron spicatum*) and spotted knapweed (*Centaurea maculosa*). Weed Technol. 10:13-21.

Jacobs, J.S. and R.L Sheley. 1999. Competition and resource partitioning among *Pseudoroneria spicatum, Hedysarum boreale,* and *Centaurea maculosa.* Great Basin Nat. (*In press.*)

James, D. 1992. Some principles and practices of desert revegetation seeding. Arid Lands Newsletter. 32:22-27.

Larson, L.L., and M.L. McInnis. 1989. Impact of grass seedlings on establishment and density of diffuse knapweed and yellow starthistle. Northwest Sci. 63:162-166.

Pickett, S.T.A., S.L. Collins, and J.J. Armesto. 1987. Models, mechanisms and pathways of succession. Bot. Rev. 53:335-71.

Powell, A.A., and S. Mathews. 1988. Seed treatments: developments and prospects. Outlook on Agriculture 17:97-103.

Sheley, R.L., and L.L. Larson. 1997. Cheatgrass and yellow starthistle growth at 3 soil depths. J. Range Manage. 50:146-50.

Sheley, R.L., T.J. Svejcar, and B.D. Maxwell. 1996. A theoretical framework for developing successional weed management strategies on rangeland. Weed Technol. 10:712-20.

Sheley, R.L., B.E. Olson, and L.L. Larson. 1997. Effect of weed seed rate and grass defoliation level on diffuse knapweed. J. Range Manage. 50:39-40.

Sprague, R. 1950. Disease of cereals and grasses in North America (fungi, except smuts and rusts). The Ronald Press Co. N.Y.

Tadros, T.M. 1957. Evidence of the presence of an edaphobiotic factor in the problem of serpentine tolerance. Ecology 38:14-23.

Taylor, A.G., and G.E. Harman. 1990. Concepts and technologies of selected seed treatments. All. Rev. Phytopathol. 28:321-39.

Velagala, R.P. 1996. Using seed rate and plant densities to enhance intermediate wheatgrass establishment in spotted knapweed dominated rangeland. M.S. Thesis. Montana State Univ, Bozeman, MT.

Velagala, R.P., R.L. Sheley, and J.S. Jacobs. 1997. Influence of density on intermediate wheatgrass and spotted knapweed interference. J. Range Manage. 50:523-29.

Wright, D.L., H.D. Perry, and R.E. Blaser. 1978. Persistent low maintenance vegetation for erosion control and aesthetics in highway corridors. *In*: F.W. Schaller and P. Sutton (eds.), Reclamation of Drastically Disturbed Lands. Amer. Soc. Agron., Crop Sci. Soc. of America, and Soil Sci. Soc. Amer., Madison, WI., 553-83.

The Weeds

Biennial Thistles

K. George Beck

Biennial plants live for two growing seasons. They germinate and grow into rosettes in their first year and spend the winter in this growth stage. In their second year, rosettes resume growth in early spring, then bolt in mid- to late spring. Bolted plants then flower, set seed, and die by summer's end. Bull thistle (*Cirsium vulgare*), musk thistle (*Carduus nutans*), and Scotch thistle (*Onopordum acanthium*) typically are biennials. They are members of the sunflower family (Asteraceae) and thistle tribe. They are found frequently in disturbed or degraded pastures and rangeland where they decrease livestock carrying capacity and weight gains. These biennial thistles also occur in disturbed areas such as along trails, roadsides, and on abandoned crop and pasture lands. Bull, musk, and Scotch thistles are listed as noxious in four, sixteen, and seven states, respectively.

Identification
• *Bull Thistle*
The accepted common name is bull thistle, but it has been called spear thistle, Fuller's thistle, and lance-leafed thistle. Bull thistle has a short, fleshy taproot and grows 2 to 5 feet (0.6 to 1.5 m) tall, with many spreading branches (Whitson 1991). It is green or brownish, shoots have spiny wings, and it is sparsely hairy. Bull thistle is the only *Cirsium* species with spiny-winged stems. Leaves are more or less lance-shaped, pinnately lobed, and 3 to 6 inches (7.6 to 15.2 cm) long. They are prickly hairy on the topside and very hairy (woolly) underneath. Triangular to lance-shaped lobes on leaves are tipped with stout, needle-like spines.

Gumdrop-shaped flowers are $1^1/_2$ to 2 inches (3.8 to 5 cm) in diameter, 1 to 2 inches (2.5 to 5 cm) long, usually solitary, and more or less clustered at the ends of shoots and branches. Flowers are bright purple, sometimes white, fragrant, and subtended by narrow involucre bracts that are spine-tipped. Seeds are light-colored, with dark brown to black longitudinal stripes, $1/_{16}$ inch (0.15 cm) long, oblong, somewhat flattened, sometimes curved, with a long, white, hairy plume that is easily detached.

• Musk Thistle

The musk thistle group is comprised of *Carduus nutans*, *C. macrocephalus*, and *C. thoermeri*. They have been referred to as true species (McCarty et al. 1980) and as subspecies differing only in size of the flowerhead and foliage pubescence and texture (Great Plains Flora Assoc. 1986). Regardless, the plants are strikingly similar, all are commonly referred to as musk thistle, and their management is the same. The accepted common name is musk thistle, but it also is called nodding thistle.

Musk thistle, a biennial, germinates and grows into a rosette the first year. Seedling leaves are not deeply lobed, but each lobe has three to five points tipped with a spine. As the rosettes age the first year, lobes become deeply cut to twice pinnately lobed, each with three to five points and tipped with a stiff white or yellow spine. Rosettes can be 2 feet (0.6 m) or more in diameter (Lym and Christianson 1996). The plant develops a large, fleshy, corky taproot that is hollow near the soil surface. In its second year, musk thistle bolts, and shoots grow from 2 to 6 feet (0.6 to 1.8 m) tall. Flowering shoots are very branched. Leaves are dark green with a light green mid-rib and mostly white margins. Leaves on shoots are 3 to 6 inches (7.6 to 15.2 cm) long, alternate, clasped down the shoot, and are deeply lobed. Each lobe has three to five points, tipped with a spine. Shoots have spiny wings their full lengths, except for a few inches below flowerheads, where shoots are smooth.

Flowerheads are solitary and terminal on shoots. Flowers may bend or nod approximately 90 degrees on primary shoots. They are $1^1/2$ to 3 inches (3.8 to 7.6 cm) in diameter, bright purple, or rarely white. Flowers are subtended by numerous large, lance-shaped, spine-tipped, green to brown bracts that resemble a pinecone. Seeds are $1/8$ to $3/16$ inch (0.3 to 0.5 cm) long, shiny, striated, yellow-brown, with a white hairlike plume.

• Scotch Thistle

The primary species of Scotch thistle in the United States is *Onopordum acanthium*. A second species may be found in some locations. In Colorado, for example, *O. tauricum* is found in the southern section of the state along the foothills of the Sangre De Cristo mountains from Pueblo to Walsenburg. Both Scotch thistle species are biennials and should be managed similarly to bull and musk thistles. Scotch thistle is the accepted common name, but it also has been called cotton thistle, woolly thistle, downy thistle, silver thistle, winged thistle, Queen Mary's thistle, heraldic thistle, and asses' thistle.

Onopordum acanthium leaves are large, green, spiny, and covered with fine dense hairs on both sides, giving the leaf a grayish-green, cottony appearance (Whitson 1991). *Onopordum tauricum* leaves are similar except that they lack hairs, and are smooth and bright green. First-year rosettes are 10 to 12 inches (25 to 30 cm) or more in diameter. Leaves may be 2 feet (0.6 m) long and have a distinct white mid-rib. Leaves of young plants are oblong, while leaves of older plants are more rectangular. Scotch thistle has a fleshy taproot.

Flowering shoots may grow 8 feet (2.4 m) tall or more. Shoots are very hairy, or cottony (*O. tauricum* lacks hairs), and have a distinct winged appearance. Prominent triangular lobes occur on leaf margins and winged margins of shoots. Lobes end with a prominent, sharp, green to white spine. Flowerheads are numerous and terminal on primary and axillary shoots. Flowers are 1 to 2 inches (2.5 to 5 cm) in diameter, pale purple to red, flat on top, and subtended by a series of overlapping bracts, each tipped with a spine. Seeds are about $3/16$ inch (0.5 cm) long, oblong to obovate, four-angled, deep brown to black, and distinctly wrinkled. Seeds are tipped with a pappus that is bristle-like but not feathery.

Origin, History, and Distribution

The thistles have a long association with humans (Mitich 1988). They are mentioned in Greek and Norse mythology (Lym and Christianson 1996). Blessed thistle (*Silybum marianum*) was named for the Virgin Mary (Mitich 1988). *Cirsium* comes from the Greek *kirsos*, which means swollen vein, perhaps referring to the effect of being pricked by the spines (Plowden 1970). *Carduus* is from the Greek *kardos* or thistle (Jaeger 1947).

Thistles have been used medicinally as well as for food. However, they are notoriously associated with unkempt agricultural land. While this may be true of the invasive thistles that are not native to North America, the vast majority of native thistles fill specific ecological niches and have traits useful to humans.

• *Bull Thistle*

The distribution of bull thistle is limited primarily to the world's north and south temperate zones (Holm et al. 1997). It is not native to North America, and is thought to have entered the northwest United States via overseas shipping through Portland, Oregon, in the late nineteenth century (Forcella and Harvey 1988). Over the following 30 years, bull thistle moved east to Montana, then south to Idaho, and finally further east and west from these points. In western North America, it is the most common and widespread of the pasture and rangeland thistles. Bull thistle's typical habitat is disturbed or degraded land such as roadsides, fence rows, overgrazed pastures and rangeland, eroded gullies, ditch banks, and vacant lots. Although bull thistle has been reported in all of the lower 48 states, it usually is not considered as problematic as musk or Scotch thistles (Dewey 1991).

• *Musk Thistle*

As with bull thistle, the worldwide distribution of musk thistle is almost entirely limited to the temperate zones of the northern and southern hemispheres, although occasional sightings of the weed in the tropics at high elevations have been reported (Holm et al. 1997). In the intermountain region of western North America, it occupies habitats ranging from saline soils in low altitude valleys to acidic soils at 8,000 feet (2,438 m). In Colorado, it also can be found at elevations between 9,000 and 10,000 feet (2,743 to 3,048 m). In Canada, musk thistle infestations can be found in soils with pH ranging from 6 to 9, which represents most of the soils in the southern part of that country (Holm et al. 1997). Musk thistle will germinate and grow under a wide range of environmental conditions. It infests arid areas in Nevada to relatively high moisture areas of Virginia and the East Coast.

Musk thistle was introduced into the United States from Europe. The earliest records of occurrence were in central Pennsylvania in 1852 (Stuckey and Forsyth 1971). There are many records of musk thistle's occurrence along the east coast in the late 1800s, apparently associated with ships' ballast. The weed began to appear in the Midwest about the turn of the twentieth century (McCarty et al. 1980). By the early 1940s, musk thistle was being regarded as a potential problem. Musk thistle is found in at least 40 states (Dunn 1976).

• *Scotch Thistle*

Scotch thistle is native to Europe and Asia, where it is common. When Malcolm I ruled Scotland, Scotch thistle saved the Scots from invading Norsemen (Lym and Christianson 1996). The invaders tried to capture Staines Castle by wading across the moat, but they found it dry and infested with thistles. Their cries of pain from stepping on the thistles with their bare feet were heard by guards, and the Norsemen were defeated. This event made Scotch thistle the emblem of Scotland.

Scotch thistle was introduced into the eastern United States in the late 1800s (Gray 1889, Bentham and Hooker 1904). Scotch thistle is found in most western states and occupies sites characterized by high soil moisture, especially in dry climates. For example, it is a particular problem in the bottoms and sides of draws in the Snake River breaks in central and northern Idaho. However, it also will occupy dry sites.

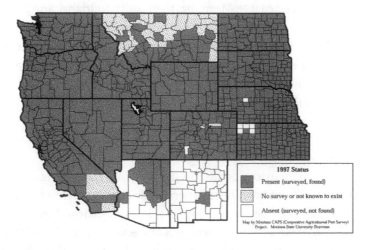

Distribution of bull thistle by county in the western United States. Data were collected by surveying various weed authorities in each state.

Potential Invasion

Disturbance favors bull, musk, and Scotch thistles. Bull thistle proliferates and thrives in pastures that are heavily grazed and subject to nitrogen fertilization (Doing et al. 1969, Michael 1970), but is rare in ungrazed pastures (Forcella and Wood 1986). Seed production and seedling establishment are enhanced under disturbed conditions, which create open, habitable sites for weeds. Over three times as many bull thistle seeds were produced per plant in heavily grazed pastures than in those that were ungrazed, and 24% more seedlings established (Forcella and Wood 1986). These researchers concluded that heavily grazed pastures were at the greatest risk of bull thistle invasion. Other researchers found that bull thistle invasion was enhanced in pastures with decreased vegetative cover (George et al. 1970).

Adequate moisture and sunlight favor musk thistle establishment, so degraded pastures and rangeland with bare patches are subject to invasion (Feldman et al. 1968, Doing et al. 1969). Musk thistle is common in areas covered by snowdrifts, such as gullies, fence lines, and brush patches (Holm et al. 1997). Abundant soil nitrogen also may favor musk thistle (Medd and Lovett 1978a).

Scotch thistle seems to favor habitats with high soil moisture. It is frequently associated with waterways in the western United States, as well as abandoned or degraded land (Dewey 1991). Often it is associated with plant communities dominated by the annual weedy grass downy brome (*Bromus tectorum*) where it has displaced native bunchgrasses or sod-forming grasses. Especially in these dry environments, Scotch thistle seems most abundant in (if not restricted to) areas of high soil moisture, such as the bottoms of swales, gullies, draws, and roadside borrow pits. This may be partially due to the presence of a water-soluble germination inhibitor in Scotch thistle seeds (Young and Evans 1972).

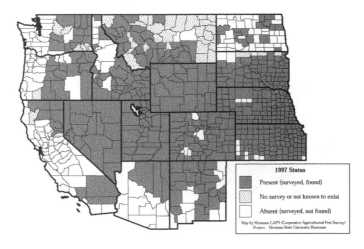

Distribution of musk thistle by county in the western United States. Data were collected by surveying various weed authorities in each state.

Impacts

Bull, musk, and Scotch thistles are problems in pastures because they compete with and decrease desirable forage. Sharp spines on all three species deter livestock, and presumably wildlife, from grazing. One adult bull thistle plant per square yard decreased spring or summer liveweight gains of sheep by about 3.8 lb per animal (Hartley 1983). Musk thistle reduced the growth of desirable pasture species in an area 50% to 100% greater in diameter than that of the thistles themselves (Thompson et al. 1987). This effect may be caused by musk thistle allelopathy (Wardle et al. 1993). The robust nature of musk or Scotch thistle, in particular, can act as a living barbed wire fence, preventing access to areas they border.

Biology and Ecology

• Bull Thistle

Bull thistle is normally a biennial. Germination occurs in spring or fall in response to adequate soil moisture (Forcella and Wood 1986). Plants grow the first year as a rosette, and develop a fleshy taproot that does not creep or spread like Canada thistle. In spring of the second year, plants resume growth and bolt (shoots elongate) to 2 to 5 feet (0.6 to 1.5 m) tall. By mid-summer, shoots bear from 10 to 200 flowerheads (Forcella and Wood 1986), although the lower end of this range is most common. Bull thistle reproduces and spreads solely from seeds. Its growth, development, and fecundity in pastures is influenced by the level of management.

Bull thistle averages 100 seeds per flowerhead, although it may produce up to 350 under favorable conditions (Klinkhamer et al. 1988, Michaux 1989). A healthy plant may produce 5,000 seeds, whereas very robust individuals have produced up to 50,000 seeds. From 60% to 75% of bull thistle seeds may be dormant at maturity (Williams

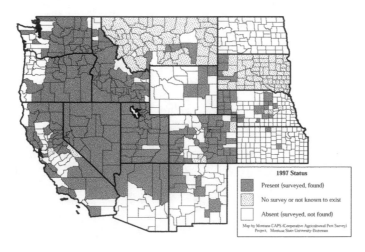

Distribution of Scotch thistle by county in the western United States. Data were collected by surveying various weed authorities in each state.

Biennial thistles
top left: Bull thistle
above: Musk thistle
lower left: Scotch thistle

(Photos by K. George Beck, Colorado State University)

Canada thistle
(Photo courtesy of Don Morishita, University of Idaho)

Cheatgrass
(Photo courtesy of Roger Sheley, Montana State University)

Common crupina
(Photo by D.W. Thill, University of Idaho)

left: Dalmatian toadflax, right: Yellow toadflax
(Photos by Reeves Petroff, Montana State University)

Diffuse knapweed

(Photo by R.D. Richard, USDA-APHIS, Bozeman, MT)

Dyer's woad

(Photo by Steve Dewey, Utah State University)

left: Meadow hawkweed, right: Orange hawkweed
(Photos by Linda M. Wilson, University of Idaho)

above: Leafy spurge
(Photo courtesy of Roger Sheley, Montana State University)

left: Mediterranean sage
(Photo by Joseph McCaffrey, University of Idaho)

top: Medusahead
(Photo courtesy of Michael Borman, Oregon State University)

below: Oxeye daisy
(Photo by Bret Olson, Montana State University)

Poison-hemlock
(Photo courtesy of Joe DiTomaso, University of California-Davis)

Purple loosestrife
(Photo by R.D. Richard, USDA-APHIS, Bozeman, MT)

left: Rush skeletonweed, right: Russian knapweed

(Photos courtesy of Roger Sheley, Montana State University)

Snakeweeds

(Photo courtesy of Tracy Sterling, New Mexico State University)

Sowthistles

(Photo courtesy of Richard Zollinger, North Dakota State University)

Spotted knapweed
(Photo by Cindy Talbott Roché, Asotin, WA)

left: Squarrose knapweed
(Photo by Cindy Talbott Roché, Asotin, WA)

right: St.Johnswort
(Photo by Gary Piper, Washington State University)

left: Sulfur cinquefoil
(Photo by Peter Rice, University of Montana)

right: Tansy ragwort
(Photo by Eric Coombs, Oregon Dept. of Agriculture)

left: Whitetop
(Photo courtesy of Roger Sheley, Montana State University)

right: Yellow starthistle
(Photo by Cindy Talbott Roché, Asotin, WA)

1966), but up to 90% may germinate within a year, and less than 1% may remain viable in soil up to five years (Roberts and Chancellor 1979).

Water, animals, and human activities disperse seeds, and wind to a lesser extent. The movement of bull thistle seeds in agricultural products such as hay may be the most common way to transport large quantities great distances (Holm et al. 1997). The feathery pappus readily detaches at maturity. The majority of seeds fall into an area about 1.5 times the height of the producing plant (Michaux 1989). Half of the seeds produced by a bull thistle plant were found within 1 yard (0.9 m) of the parent plant (Klinkhamer et al. 1988).

Vegetative cover influences bull thistle population dynamics. Seed production per plant averaged 12,200 in heavily grazed pastures, compared to 4,125 in ungrazed pastures (Forcella and Wood 1986). Rosette populations in grazed pastures (2.6 per square yard) were more than twice that in ungrazed pastures (1.2 per square yard), and there were almost twice as many flowering plants in grazed pastures. Forcella and Wood (1986) also found that the transition from seedlings to rosettes is when the greatest attrition in bull thistle populations occurred. About 15% and 10% of seeds from grazed and ungrazed pastures, respectively, produced seedlings. The average survival of seedlings in grazed and ungrazed pastures was 1.0% and 0.2%, respectively. However, 50% of rosettes in both pasture types survived and grew into adults.

• Musk Thistle

Musk thistle typically is a biennial, but it may complete its life cycle as a winter annual or occasionally as an annual (Feldman et al. 1968). Musk thistle spends approximately 90% of its life cycle as a rosette, then bolts, flowers, produces seed, and dies (Roeth 1979). Seed production typically occurs within 45 to 55 days after bolting.

Musk thistle depends upon seed production for reproduction and spread (McCarty 1982). Flowering begins with the terminal (primary) bud and proceeds downward toward the base of the plant. The terminal flowerhead is solitary, and the topmost branch usually develops a solitary flowerhead approximately the same size as the terminal one. These first flowerheads are the largest and produce the most seeds (McCarty 1964). Lower branches often develop secondary and sometimes tertiary flowerheads. It may take nine weeks for all flowerheads on a plant to mature, and seed may be shed from seven days (McCarty 1982) to two months (Doing et al. 1969) after first bloom. Musk thistle is an outcrossing species, but may self-pollinate (McCarty et al. 1980). Thus, an isolated plant can easily expand into a large infestation.

Seed production by musk thistle is quite variable and dependent upon habitat conditions. Seed production per plant has been reported from a few thousand to more than 100,000 (Holm et al. 1997). An average plant may produce 10,000 to 11,000 seeds, 33% of which are capable of germination and seedling establishment

(McCarty 1982). Potentially, about 3,600 seedlings could be produced from the average musk thistle plant.

Wind, water, wildlife, livestock, and human activities disperse seed. *Carduus* seeds have been found in the crops of birds and in their nests (Holm et al. 1997). The majority of seeds produced fall within 50 inches (1.3 m) of the parent plant (Delahunty 1962). Although the seed has a pappus, very few seeds are windborne. Under controlled windy conditions, fewer than 1% of musk thistle seeds moved 100 yards from the parent plant (Smith and Kok 1984). However, it takes only one seed to start a new infestation.

Musk thistle germination is favored on poorly vegetated sites; seedlings establish only on bare soils (Feldman et al. 1968, Doing et al. 1969). Abundant red light reaches bare ground on poorly vegetated sites and stimulates musk thistle germination, although red light is not necessary (Medd and Lovett 1978a). Also, potassium nitrate solutions enhanced musk thistle germination, and excess nitrogen in soils may cause seeds to germinate and create a flush of seedlings.

Field and laboratory studies showed that musk thistle requires vernalization for floral initiation (Medd and Lovett 1978b). Short days (decreasing day length, as in fall) before a vernalization period reduced the length of the vernalization period necessary to initiate floral development and the need for subsequent long days (increasing day length, as in spring) after vernalization. The short-day substitution for the vernalization requirement may explain why musk thistle displays a biennial nature for those plants germinating in spring, and a winter annual nature for those that germinate in fall.

Musk thistle seeds may survive in the soil a decade or more, and it may take 15 years to decrease germination of buried musk thistle seeds to 1% (Burnside et al. 1981). Therefore, land managers must exercise a diligent effort to control musk thistle for 15 years or more to eradicate this weed.

Inadequate soil moisture may hinder musk thistle germination and stand establishment. Musk thistle germination under dry (-10 bars) (Medd and Lovett 1978a) to very dry (-16 bars) (McCarty et al. 1969) experimental conditions was decreased 99% and 50%, respectively. Seedling growth was reduced 50% at -6 bars, but seedlings still grew at -20 bars, although growth was decreased 91%. Even though germination and growth of musk thistle will be decreased in very dry soil conditions, the data clearly indicate that at least some musk thistle may germinate and establish.

Nebraska researchers (Feldman et al. 1968) found more musk thistle in pastures with larger amounts of litter, which they believed created better soil moisture conditions for establishment and competition with desirable grasses. Their observations seem to contradict the finding that musk thistle establishes best on bare soil in poorly vegetated sites. However, improved establishment under litter may be partially due to allelopathy. Wardle et al. (1993) found musk thistle to be most allelopathic when it is bolting and rosette leaves are decomposing and releasing soluble inhibitors, and when mature plants are dying and releasing soluble inhibitors.

Musk thistle seedlings were stimulated by the addition of thistle tissue to soil. They proposed that as musk thistle invades an area, it stimulates recruitment of its own species.

Musk thistle seedlings may be sensitive to the competition of neighboring plants for light. When musk thistle seedlings received 2%, 8%, or 14% of full sunlight, growth was decreased 97%, 68%, and 35%, respectively (Medd and Lovett 1978a). Management that allows grasses to grow taller in spring to shade musk thistle seedlings may decrease the seedlings' establishment and growth.

• Scotch Thistle

Scotch thistle typically grows as a biennial, although historical literature indicates that it may grow as an annual as well. Young and Evans (1969) found Scotch thistle to grow as an annual, biennial, or short-lived, simple perennial depending upon the environmental conditions in which it was growing. They believed this variation gave it a competitive advantage. Its life cycle was not bound by strict photoperiod (daylength) or temperature requirements. They also found seed production to be independent of plant density. Over two years, flowering plant density ranged from 0.1 to 2.1 plants per square foot, and the number of flowering heads per plant ranged from 70 to 310. Seed production per flowering head ranged from 110 to 140. Eight to 14% of Scotch thistle seeds were non-dormant when freshly harvested.

Scotch thistle seeds contain a water soluble germination inhibitor (Young and Evans 1972). Light quality and photoperiod influenced germination of freshly harvested seeds, but not as much as the germination inhibitor. Light rich in red wavelengths (such as that reflecting on bare soil) decreased germination. This may suggest that seed must be buried in soil or under litter for germination to be stimulated, and that optimum soil moisture is more critical for Scotch thistle germination than musk thistle. The presence of a water-soluble inhibitor also supports this hypothesis. Young and Evans (1972) concluded that two systems regulated Scotch thistle dormancy and germination: phytochrome (light quality), and the presence of a water soluble inhibitor. Approximately 85% to 90% of Scotch thistle seeds display innate dormancy upon maturity, which assures a soil seed reserve and perpetuation of the population.

Water, wind, wildlife, livestock, and human activities disperse Scotch thistle seed. As with bull and musk thistles, it is most probable that while long-distance transport of Scotch thistle seed can and does occur, the overwhelming majority of seeds fall nearby the parent plant.

Management

Bull, musk, and Scotch thistles reproduce only from seeds, so the key to managing them successfully is to prevent seed production. Cultural, chemical, and mechanical methods to control these species are quite similar, and there is some overlap for biological control as well. Combining control methods into an integrated management system will result in the best long-term biennial thistle population decreases. Always monitor and evaluate weed management programs to determine whether and when control treatments must be invoked again or adapted.

• *Prevention*

Prevention of weed infestations always should be the first line of defense. Preventing or drastically decreasing seed production will help decrease the spread of these weeds. Cleaning mowers, vehicles, and tillage equipment after operating in a stand of biennial thistle will prevent spread. When seeding is necessary, use clean, certified seed and mulch to ensure that these or other weeds are not being sown. Develop monitoring programs to locate infestations, and place a priority on controlling small infestations so they do not expand.

• *Cultural Control*

Bull and musk thistle germination and establishment is favored in open areas, and all three species are favored by disturbance. Therefore, re-establishment of desirable vegetation will usually be necessary for successful management of biennial thistles. The choice of perennial grass species to sow will depend upon climate and location. The Natural Resource Conservation Service and land grant universities are good sources of information about appropriate perennial grass species for a particular locale.

Perennial grasses are most competitive with broadleaf weeds in the western United States. In Australia, ryegrass (*Lolium rigidum*) was more competitive than subclover (*Trifolium subterraneum*). Researchers have recommended conservation of ryegrass in pastures infested with bull thistle (Forcella and Wood 1986). Feldman et al. (1968) compared musk thistle seedling establishment and development into rosettes under three grazing management regimes and three pasture grass types. In all grass pasture types and grazing management systems, musk thistle declined over time. The authors concluded that musk thistle will invade pastures that are in good to excellent condition for grazing, but grazed pastures that are managed carefully may enhance grass competition and deter musk thistle survival from seedlings to rosettes. Their results indicate that cool- or warm-season grass competition is an essential component of any effective musk thistle management system. This is most likely true for all weed species invading rangeland.

• Chemical Control

Clopyralid, dicamba, MCPA, picloram, 2,4-D, metsulfuron, and chlorsulfuron will control bull, musk, and Scotch thistles. Herbicide choice and rates are influenced by growth stage, stand density, and environmental conditions, for example, drought or cold temperatures. Herbicide recommendations to control biennial thistles are in Table 1. Check with state or county weed specialists for appropriate local use rates.

When biennial thistles are in the rosette growth stage, clopyralid, dicamba, MCPA, picloram, or 2,4-D are the best choices, but after bolting begins, metsulfuron or chlorsulfuron should be used. Musk thistle susceptibility to dicamba, picloram, and 2,4-D decreases after the weed begins to bolt (Feldman et al. 1968, Roeth 1979). When these herbicides were applied to bolting musk thistle, 43% to 65% were controlled, but when applied to rosettes, 90% to 100% were controlled. Chlorsulfuron at 0.75 oz active ingredient per acre or metsulfuron at 0.3 oz active ingredient per acre eliminated viable seed production when applied during bolting or bud growth stages, but did not when applied in the rosette stage (Beck et al. 1990). Clopyralid, dicamba, dicamba plus 2,4-D, or picloram did not eliminate viable seed production when applied at bolting, bud, or bloom growth stages.

Fall is a good time to control biennial thistles with herbicides because all live plants will be seedlings or rosettes. Cool or dry weather conditions commonly associated with autumn may decrease biennial thistle control from 2,4-D or dicamba, but not from picloram (Roeth 1979).

Table 1. Herbicides and rates to control bull, musk, and Scotch thistles in pastures, rangeland, and non-crop areas.

Herbicide	Rate (lb ai/ac)	Timing/remarks
Clopyralid	0.13 to 0.5	Apply to rosettes in spring or fall
Clopyralid + 2,4-D	0.2 + 1.0 to 0.3 + 1.5	Apply to rosettes in spring or fall
Dicamba	0.5 to 1.0	Apply to rosettes in spring or in fall if good growing conditions exist
2,4-D	1.5 to 2.0	Apply to rosettes in spring
2,4-D + dicamba	1.0 + 0.5	Apply to rosettes in spring
Picloram	0.13 to 0.25	Apply to rosettes in spring or fall
Chlorsulfuron	0.047 (0.75 oz ai)	Spring from bolting to bud stages; add a non-ionic surfactant
Metsulfuron	0.019 (0.3 oz ai)	Spring from bolting to bud stages; add a non-ionic surfactant

• Mechanical Control

Because biennial thistles do not reproduce from their roots, any mechanical or physical method that severs the root below the soil surface will kill the weed. However, it is essential to revegetate the site with desirable plants, particularly competitive grasses, to compete with biennial thistles that invariably will re-invade from seeds left in the soil.

Most mechanical methods are not practical to use on rangeland, but could be functional in improved pastures or roadsides. Tillage, hoeing, or even hand-pulling should be successful, providing these operations are done before the reproductive growth stages to prevent seed production. When musk thistle was mowed two days after terminal flowerheads opened to full bloom, viable seed production was eliminated (McCarty and Hatting 1975). Some plants recovered after mowing at each growth stage (late bud to late bloom) and seeds were produced. Although seed set was decreased 99% when mowed in late bloom, data suggest that seven seedlings could be produced from each mowed musk thistle plant the following year.

A single mowing will not satisfactorily control musk or other biennial thistles, because of growth stage variability in natural populations. Mowing alone is not a viable control measure for biennial thistles, because invariably seed will still be produced.

• Biological Control

A combination of insects that target different growth stages of the biennial thistles will be more effective at decreasing their populations than using one insect. As of 1998, there are no plant pathogens developed as biological control agents for the biennial thistles. Always carefully evaluate the need to integrate biological control with seeding competitive grasses (cultural control).

The seedhead weevil *Rhinocyllus conicus* is native to central and eastern Europe, western Asia, and the Mediterranean (Mellini 1951). It lives in a variety of climates including those that are extremely cold. The seedhead weevil was introduced into the United States from Europe in 1968 and has been released in several western states to control musk thistle. In Colorado, for example, the seedhead weevil was released in 1974 and has spread throughout the state and can be found at elevations from 4,500 to 10,000 feet (1,350 to 3,048 m) (Colorado Dept. of Agriculture 1993).

The seedhead weevil limits seed production of musk thistle (Hodgson and Rees 1976). Females deposit eggs on flower bracts, and eggs hatch in six to eight days. Young larvae burrow into the flower receptacle and consume developing seeds. The weevil may have one or two generations per year (Mellini 1951). The seedhead weevil attacks terminal and early developing lateral flowerheads much more than later developing flowerheads (Kok and Surles 1975).

Decreases in viable seed production are highly variable. Several researchers have assessed the effects of *R. conicus* on musk thistle seed production in different areas of the United States and found decreases from 10% to 78% (Surles and Kok 1978,

McCarty and Lamp 1982). The weevil does not destroy 100% of musk thistle seeds. Viable seeds will be produced from plants infested with *Rhinocyllus conicus,* and infestations will perpetuate.

R. conicus will use *Carduus, Cirsium, Silybum,* and *Onopordum* genera as hosts, but prefers the *Carduus nutans* group (Rees 1991). It is not host specific to the non-native, invasive, biennial thistles. It will use numerous species of native thistles, some of which are endangered (Louda et al. 1997).

Trichosirocalus horridus is a European weevil introduced into the United States in 1974. Although this weevil was imported to control musk thistle, it will use bull thistle also. *T. horridus* has one generation per year, and larvae feed on the growing points of musk thistle rosettes and developing shoots, which decreases plant vigor and flowering (Rees 1991). Surviving plants produce fewer and smaller flowerheads and fewer seeds. *T. horridus* is established in Virginia, Kansas, Missouri, and Wyoming, although it was released in other states (Colorado Dept. of Agriculture 1993). In Colorado, for example, *T. horridus* was received in 1983 and is well established. During the summer of 1993, approximately 31,000 weevils were collected and redistributed to 65 locations in 31 counties. As with the musk thistle seedhead weevil, *T. horridus* appears well adapted to Colorado, and is spreading throughout the state.

Cheilosa corydon was released in the United States in 1990 to control musk thistle. Eggs are deposited in young leaves and shoots near the center of the plant. Larvae burrow into shoots and move up and down, causing shoots to break or dry prematurely (Rees 1991). By summer, the third instar larvae burrow into roots where they remain until fall precipitation begins. Plant water and nutrient transport are impaired, flowering and seed production are decreased, and secondary invasion by soil microbes occurs through lesions in roots caused by feeding larvae.

A seedhead fly, *Urophora stylata,* was found in Germany and Switzerland (Julien 1992). Larvae feed on developing seeds in bull thistle flowerheads and decrease seed production up to 65%. *Urophora stylata* was released in Canada in 1973 and established in British Columbia, Nova Scotia, and Quebec, but not in Ontario. It was reported to die in sparse bull thistle stands, and weed populations have not decreased. It was released in the United States in 1983. *Urophora stylata* established in Colorado, Maryland, and Oregon. Galls in bull thistle flowerheads were observed in Colorado in 1993 (Colorado Dept. of Agriculture 1993).

• *Grazing*

Good grazing management will stimulate grass growth and keep pastures and rangeland in a healthy state. Healthy pastures and rangeland may be more resistant to biennial thistle invasion. Bare spots caused by overgrazing are prime habitable sites for biennial thistles. In many instances, grazing lands will have to be rested from grazing for grasses to recover. This should be coupled to precipitation cycles, so adequate soil moisture will be available to stimulate grass growth.

157

• Integrated Management Systems

Thistles in general invade disturbed or degraded areas where competition with desirable plants has been decreased or eliminated. Improving the desirable plant community by seeding with competitive grasses may be necessary to control biennial thistles and return the site to a productive state. Seeding is usually done after other control methods have been invoked, and the weed population has been decreased effectively.

When herbicide applications were followed by dormant seeding of competitive, perennial grasses in the western United States, leafy spurge (*Euphorbia esula*) (Whitson et al. 1989) and Russian knapweed (*Acroptilon repens*) (Benz et al. 1999) were effectively managed. Using herbicides in spring to control biennial thistles followed by dormant seeding (sowing in late fall) of competitive perennial grasses is an example of an effective management system for biennial thistles in the western United States.

Integrating the use of herbicides with insect biological control agents is likely to decrease biennial thistle populations more effectively than using insects alone. Few research projects have been conducted to assess the influence of herbicides on insect biological control agents. Apparently, applications of 2,4-D do not increase *Rhinocyllus conicus* mortality (Lee 1980). Until more such research is conducted, it may be safest to create a spatial or temporal separation of herbicide use and biological control. Release insects in the center of a biennial thistle infestation, then spray the borders to prevent further weed spread. Or apply the herbicide at a time when direct exposure of the insect to the herbicide will be minimized, such as when the insect is pupating.

• Long-Term Management Strategies

Chemical, mechanical, and biological control data suggest that the threshold for viable seed production by biennial thistles is zero to achieve long-term population decreases. Zero seed production may not be a realistic goal. This underscores the importance of desirable plant competition in any biennial thistle management strategy to deter the establishment of thistle seedlings and the transition to the rosette growth stage. As with bull thistle, the transition from seedling to the rosette growth stage in musk and Scotch thistles may be the most precarious stage in their life cycles. Seedling and rosette biennial thistles are the most logical growth stages to target with control efforts. Recovery of infested areas should not be considered complete until a diverse population of desirable plants has replaced invasive biennial thistles and they are a minor to non-existent component of the plant community.

Literature Cited

Beck, K.G., R.G. Wilson, and M.A. Henson. 1990. The effects of selected herbicides on musk thistle (Carduus nutans) viable achene production. Weed Technol. 4:482-86.

Bentham, G., and J.D. Hooker. 1904. Handbook of the British Flora. L. Reeve and Co., London.

Benz, L.J., K.G. Beck, T.D. Whitson, and D.W. Koch. 1999. Reclaiming Russian knapweed infested rangeland. J. Range Manage. (*In press*).

Burnside, O.C., C.R. Fenster, L.L. Evetts, and R.F. Mumm. 1981. Germination of exhumed weed seeds in Nebraska. Weed Sci. 29:577-86.

Colorado Department of Agriculture. 1993. Division of Plant Industry Annual Report 1992-1993. Colorado Department of Agriculture, Lakewood, CO.

Delahunty, E. 1962. Nodding thistle. Proc. 15th New Zealand Weed and Pest Control Conf., 15:24-28.

Dewey, S.A. 1991. Weedy thistles of the western U.S. In: Noxious Range Weeds. L.F. James, J.O. Evans, M.H. Ralphs, and R.D. Child (eds.), Westview Press, Boulder, CO, 247-53.

Doing, H., E.F. Biddiscombe, and S. Knedlans. 1969. Ecology and distribution of the *Carduus nutans* group (nodding thistles) in Australia. Vegetatio 17:313-51.

Dunn, P.H. 1976. Distribution of *Carduus nutans*, *C. acanthoides*, *C. pycnocephalus*, and *C. crispus* in the United States. Weed Sci. 24:518-24.

Feldman, I., M.K. McCarty, and C.J. Scifres. 1968. Ecological and control studies of musk thistle. Weed Sci. 116:1-4.

Forcella, F., and S. Harvey. 1988. Patterns of weed migration in northwestern United States. Weed Sci. 36:194-201.

Forcella, F., and H. Wood. 1986. Demography and control of *Cirsium vulgare* (Savi.) Ten. in relation to grazing. Weed Res. 26:199-206.

George, J., K. Hutchinson, and B. Mottershead. 1970. Spear thistle *Cirsium vulgare* invasion of grazed pastures. Proc. 11th Internat. Grassland Conf., Australia 11:685-88.

Gray, A. 1889. Manual of the Botany of the Northern United States. S. Watson and J.M. Coulter (eds.), American Book Co., NY.

Great Plains Flora Association. 1986. The flora of the Great Plains. University Press of Kansas. Lawrence, KS.

Hartley, M.J. 1983. Effect of Scotch thistles on sheep growth rates. Proc. 36th New Zealand Weed and Pest Cont. Conf., 86-89.

Hodgson, J.M., and N.E. Rees. 1976. Dispersal of *Rhinocyllus conicus* for biocontrol of musk thistle. Weed Sci. 24:59-62.

Holm, L., J. Doll, E. Holm, J. Pancho, and J. Herberger. 1997. World Weeds. Natural Histories and Distributions. John Wiley and Sons, New York.

Jaeger, E.C. 1947. A Source-book of Biological Names and Terms. Charles C. Thomas, Springfield, IL.

Julien, M.H. 1992. Biological Control of Weeds. A world catalogue of agents and their target weeds. CAB International, Wallingford, UK.

Klinkhamer, P., T. DeJong, and E. van der Meijden. 1988. Production, dispersal, and predation of seeds of biennial *Cirsium vulgare*. J. Ecol. 76:403-14.

Kok, L.T., and W.W. Surles. 1975. Successful biocontrol of musk thistle by an introduced weevil, *Rhinocyllus conicus*. Environ. Entomol. 4:1025-27.

Lee, R. 1980. The influence of selected herbicides on the development of *Rhinocyllus conicus* Froelich. An insect used for biocontrol of musk thistle. M.S. Thesis, Utah State University, Logan, UT.

Louda, S.M., D. Kendall, J. Connor, and D. Simberloff. 1997. Ecological effects of an insect introduced for the biological control of weeds. Science 277:1088-90.

Lym, R.G., and K.M. Christianson. 1996. The thistles of North Dakota. North Dakota St. Univ. Ext. Svc. Pub. No. W-1120, Fargo, ND.

McCarty, M.K. 1964. New and problem weeds: musk thistle. Proc. 20th N. Central Weed Control Conf., 20:62-63.

McCarty, M.K. 1982. Musk thistle (*Carduus thoermeri*) seed production. Weed Sci. 30:441-45.

McCarty, M.K., and J.L. Hatting. 1975. Effects of herbicides or mowing on musk thistle seed production. Weed Rcs. 15:363-67.

McCarty, M.K., and W.O. Lamp. 1982. Effect of a weevil, *Rhinocyllus conicus*, on musk thistle (*Carduus thoermeri*) seed production. Weed Sci. 30:136-40.

McCarty, M.K., C.J. Scifres, A.L. Smith, and G.L. Horst. 1969. Germination and early seedling development of musk and plumeless thistle. Univ. of Nebraska Res. Bull. 229.

McCarty, M.K., H.J. Gorz, and F.A. Haskins. 1980. Inheritance of flower color in musk thistle (*Carduus thoermeri*). Weed Sci. 28:347-51.

Medd, R.W., and J.V. Lovett. 1978a. Biological studies of *Carduus nutans* (L.) spp. *nutans*. I. Germination and light requirements of seedlings. Weed Res. 18:363-67.

Medd, R.W., and J.V. Lovett. 1978b. Biological studies of *Carduus nutans* (L.) spp. *nutans*. II. Vernalization and phenological development. Weed Res. 18:369-72.

Mellini, E. 1951. Insetti del *Carduus nutans* L. II. *Rhinocyllus conicus* Frol. e *Larinus jaceae* F. (Col. Curculionidae). Boll. Ent. Bologna 18:319-49.

Michael, P.W. 1970. Weeds of grasslands. *In:* Australian Grasslands. R.M. Moore (ed.), Australian National University Press, Canberra, 349-360.

Michaux, B. 1989. Reproductive and vegetative biology of *Cirsium vulgare*. New Zealand J. Bot. 27:401-14.

Mitich, L.W. 1988. Intriguing world of weeds. Thistles I: *Cirsium* and *Carduus*. Weed Technol. 2:228-29.

Plowden, C.C. 1970. A Manual of Plant Names (2nd ed.). Philosophical Library, NY.

Rees, N.E. 1991. Biological control of thistles. *In:* Noxious Range Weeds, L.F. James, J.O. Evans, M.H. Ralphs, and R.D. Child (eds.), Westview Press, Boulder, CO, 264-73.

Roberts, H., and R. Chancellor. 1979. Periodicity of seedling emergence and achene survival of some species of *Carduus, Cirsium,* and *Onopordum*. J. Appl. Ecol. 16:641-47.

Roeth, F.W. 1979. Comparisons of dicamba, picloram, and 2,4-D for musk thistle (*Carduus nutans*) control. Weed Sci. 27:651-55.

Smith, L., and L. Kok. 1984. Dispersal of musk thistle *Carduus nutans* seed. Weed Sci. 32:120-25.

Stuckey, R.L., and J.L. Forsyth. 1971. Distribution of naturalized *Carduus nutans* (Compositae) mapped in relation to geology in northwestern Ohio. Ohio J. Sci. 71:1-15.

Surles, W.W., and L.T. Kok. 1978. *Carduus* thistle seed destruction by *Rhinocyllus conicus*. Weed Sci. 26:264-69.

Thompson, A., A.E. Saunders, and P. Martin. 1987. The effect of nodding thistle (*Carduus nutans*) on pasture production. Proc. 40th New Zealand Weed and Pest Control Conf., 222-25.

Wardle, D.A., K.S. Nicholson, and A. Rahman. 1993. Influence of plant age on the allelopathic potential of nodding thistle (*Carduus nutans* L.) against pasture grasses and legumes. Weed Res. 33:69-78.

Whitson, T.D., D.W. Koch, A.E. Gade, and M.A. Ferrell. 1989. The control of leafy spurge (*Euphorbia esula* L.) by the integration of herbicides and perennial grasses. Univ. Wyo. Rangeland Research and Extension Demonstrations, 42-47. Coop. Ext. Serv., Agric. Exp. Sta., Laramie, WY.

Whitson, T.D. (ed). 1991. Weeds of the West. Western Society of Weed Science.

Williams, J. 1966. Variation in germination of several *Cirsium* species. Trop. Ecol. 7:1-7.

Young, J.A., and R.A. Evans. 1969. Control and ecological studies of Scotch thistle. Weed Sci. 17:60-63.

Young, J.A., and R.A. Evans. 1972. Germination and persistence of achenes of Scotch thistle. Weed Sci. 20:98-101.

Canada Thistle

Don W. Morishita

Canada thistle (*Cirsium arvense*) is an aggressive perennial weed that infests arable and nonarable lands. Also sometimes called California thistle, creeping thistle, and field thistle, it is a member of the Asteraceae family, and is listed as a noxious weed in at least 35 states (Dewey 1991). Canada thistle is just one of many thistles found in the United States. Utah alone recognizes 30 species or varieties of thistles from the genera *Cirsium, Conicus, Carduus,* and *Onopordum.* Twenty-four of these species are native to the region (Dewey 1991). *The Flora of the Pacific Northwest* lists 17 *Cirsium* species (Hitchcock and Cronquist 1994). Canada thistle and bull thistle (*C. vulgare*) are the only *Cirsium* species that are nonindigenous. Canada thistle is probably the most widespread of all thistle species, and thus is considered by many to be the most difficult thistle to control.

Excellent reviews of Canada thistle biology and management have been published (Moore 1975, Donald 1990, Donald 1994). These reviews present updated information on Canada thistle control and management, as well as physiological ecology and plant biology.

Identification

Canada thistle is an erect perennial dicot with extensive creeping horizontal roots up to 16 feet (5 m) long and 2 to 22 feet (0.6 to 6.75 m) deep. It reproduces vegetatively from horizontal creeping roots or from seeds (Holm et al. 1991). In the seedling stage, cotyledons are oblong and fleshy. The first true leaves are obovate and do not have a petiole. Leaf margins have regularly spaced, coarse, marginal hairs that resemble the spines of later-developing leaves.

As the plant develops, the leaf margins become serrate, then lobed, with spines becoming more prominent. By the time the plant has developed two true leaves the root system is branching, up to 6 inches (15 cm) long, and the main root has begun to thicken. Soon after this, root development includes vertical and horizontal growth.

At maturity, Canada thistle leaves are spiny with deeply lobed leaves and several branching flower stalks. The leaves are alternate, oblong or lanceolate, usually with crinkled edges and spiny-toothed margins, very irregularly lobed, 1 to 8 inches (2 to

20 cm) long, and terminate in a spine. Leaves may be hairy or smooth on the underside when mature. Upper leaves are much smaller than lower leaves, and have a narrow base that continues down the stem beyond the point of leaf attachment, giving the impression of a spiny stem. Hitchcock and Cronquist (1994) list two varieties or subspecies of Canada thistle. *C. arvense* var. *arvense* is described with leaves that are merely toothed or shallowly lobed, and not very spiny. *C. arvense* var. *horridum* has deeply pinnatifid leaves and is strongly spiny.

Canada thistle stems are erect, grooved, nearly smooth or slightly hairy when young, and increasingly hairy with age. Established Canada thistle plants arise as adventitious shoots from numerous underground buds and grow 1 to 4 feet (30 to 120 cm) tall.

Flowering heads form in clusters on the ends of branches. They are about $^1/_2$ inch (1.25 cm) in diameter and are normally pink, but can vary in color from white to purple. The heads are discoid with tubular flowers and have numerous bracts, but no spines.

Origin, History, and Distribution

Canada thistle is probably native to southeastern Europe and the eastern Mediterranean area (Moore 1975, Erickson 1983). It occurs in Europe from about 30° N latitude to north of the Arctic Circle at 68° N latitude. In addition to North America, it has spread to northern and southern Africa, the Middle East, Japan, India, New Zealand, Australia, and South America.

Canada thistle was first introduced to this continent from Europe, probably in the 1600s via contaminated grain seed, hay, and ship's ballast (Dewey 1901, Hansen 1918, Erickson 1983). The weed was recognized as a control problem before 1795, which led to noxious weed legislation enacted in Vermont in 1795 and New York in 1831. A University of Idaho bulletin published in 1898 reported that Canada thistle was known to exist in only two areas in Idaho: near Boise in southern Idaho and near Sandpoint in northern Idaho (Henderson 1898). Dewey (1901) reported it to be found in all states north of 37° N latitude by 1900. Canada thistle is now common throughout most of the United States north of 35° N latitude into Canada to 59° N latitude (Erickson 1983, Moore 1975).

Potential Invasion

Like most aggressive noxious weeds, Canada thistle has a wide habitat range and is fairly adaptable. It has been present in the United States long enough to have spread to the geographic areas in which it is currently adapted. Within these areas of current adaptation it has the ability to infest many more acres.

Canada thistle is usually found in open areas with moderate or medium moisture conditions (Moore 1975). It is found most frequently along roadsides and railroad rights-of-way, and on rangeland, forest land, lawns, gardens, cropland, and abandoned fields. It is also found on stream banks, lake shores, and other riparian areas, as well as occasionally inhabiting sand dunes and open sandy areas. In Colorado, American bulrush (*Scirpus americanus*), creeping bentgrass (*Agrostis palustris*), longstem spikerush (*Eleocharis macrostachys*), and saltgrass (*Distichlis stricta*) were most frequently associated with dense Canada thistle stands in undisturbed areas (Stachon and Zimdahl 1980). In moderately dense stands giant ragweed (*Ambrosia trifida*), horseweed (*Conyza canadensis*), common lambsquarters (*Chenopodium album*), redroot pigweed (*Amaranthus retroflexus*), prairie sunflower (*Helianthus petiolaris*), smooth dock (*Rumex altissimus*), and foxtail barley (*Hordeum jubatum*) were most often associated with Canada thistle. It also has been observed that annual broadleaf species diversity increases near the borders of Canada thistle patches. Thus, Canada thistle has the ability to invade many different habitat types.

Canada thistle grows best in a temperature range of 32° to 90° F (0° to 32° C) and precipitation of 16 to 30 inches (40 to 75 cm) per year (Moore 1975). Clay soils have been determined to be most favorable to Canada thistle growth (Detmers 1927), but Canada thistle is adaptable to many other soil types and conditions (Rogers 1928, Korsmo 1930, Reed and Hughes 1970). It is most productive in well-aerated soils, and can survive dry conditions better than wet.

The ability of Canada thistle to propagate sexually and asexually gives it a competitive advantage over many annual or biennial species. Asexual propagation represents a major form of Canada thistle reproduction. A Canada thistle patch can spread rapidly by vegetative means under favorable conditions. Horizontal root growth can extend more than 19 feet (6 m) in one season and may eventually penetrate into the soil as deep as 22 feet (6.75 m) (Rogers 1928, Hayden 1934). Most roots, however, will develop in the upper 2 feet (0.6 m) of soil. Left undisturbed, a single Canada thistle plant can produce 26 adventitious shoots, 154 adventitious

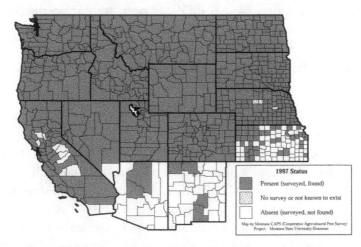

Distribution of Canada thistle by county in the western United States. Data were collected by surveying various weed authorities in each state.

root buds, and 364 feet (111 m) of roots after 18 weeks of growth (Nadeau and Vanden Born 1989).

Canada thistle seedlings are much less competitive than established thistle. They will survive only if competition is limited and the daytime light intensity remains above 20% of full sunlight. It does not take very long for seedlings to develop the ability to reproduce asexually. Lateral roots begin growing at the two-leaf stage when seedlings are four to five weeks old (Friesen 1968). At this time adventitious root buds will form on the tap root.

Impacts

If left unmanaged, Canada thistle has the potential to form dense infestations. An individual seedling can rapidly form a large patch through vegetative reproduction of the root system (Donald 1994). The spread of the clone may continue indefinitely, with groups of stems becoming independent as the root system breaks up. Canada thistle roots have a life span of about two years before being replaced by new roots that develop from the old roots (Rogers 1928).

Canada thistle may decrease or limit forage and livestock production on rangeland. Its spiny leaves make Canada thistle very unpalatable to livestock. Dense infestations decrease the forage productivity of pasture and rangeland by displacing desirable forbs and grasses. Canada thistle also limits the use of recreational areas when infestations become dense, because of the physical hazard presented by the spines on the leaves.

Land values for crop production, grazing, and recreation can decline with infestations of noxious weeds like Canada thistle. Besides reducing property value, Canada thistle increases land management costs, because control measures must be implemented to comply with state noxious weed laws.

The only known beneficial uses of this weed are as a nectar source for honeybees and occasional prevention of soil loss when growing in erosive locations. Canada thistle is not considered a poisonous plant, in part because of the spines on the leaves. The roots, however, are emetic when consumed (Lewis and Elvin-Lewis 1977).

Biology and Ecology

The creeping horizontal roots described earlier are key to the establishment and survival of Canada thistle. The above-ground shoots are killed by freezing temperatures in the winter, but the roots survive with accumulated carbohydrates as their food reserve. New shoots from established Canada thistle stands begin to emerge when average weekly temperature reaches 41° F (5° C) with optimum emergence after temperatures are at least 46° F (8° C) (Hodgson 1968).

Sexual reproduction of Canada thistle (by seed) may not be considered as important as asexual reproduction, but is nonetheless an important part of survival. Canada thistle is described as an imperfectly dioecious species which means that in most cases, female or seed-producing flowers are found on one plant and male or pollen-producing flowers are found on another plant. An individual plant can have many shoots (Hodgson 1968 and Moore 1975). Since Canada thistle typically grows in patches, it is not uncommon to find sterile heads of female flowers when a patch of pollen-producing flowers is not close enough for pollination. Flowers of Canada thistle are pollinated primarily by honeybees (Detmers 1927, Derscheid and Schultz 1960). Female flowers have a vanilla-like odor, while male flowers are not strongly scented (Holm et al. 1991).

A Canada thistle shoot can produce as many as 100 heads in a season (Detmers 1927), but 32 to 69 heads per shoot is more typical under favorable growing conditions (Bakker 1960). A female flower head may contain from 84 to 132 florets, and averages about 100 (Detmers 1927, Bakker 1960). Up to 93 seeds per head have been counted; the number of seeds produced depends on the level of pollination (Hayden 1934).

The most likely mechanism of seed dispersal today is through human activities. Canada thistle seeds are spread in contaminated crop seed, feed, manure, irrigation water, and by wind. Although Canada thistle seeds (achene is the correct taxonomic description) have a tuft of hairs or pappus attached to them, wind dispersal is not considered to be a major factor in its spread (Sheldon and Burrows 1973, Burrows 1975). The pappus does not remain firmly attached to the seed, which probably contributes to poor dispersal by wind. Canada thistle seeds have been shown to float and can be transported in irrigation water (Wilson 1980). This helps explain why Canada thistle commonly occupies ditchbanks and streambanks.

Canada thistle seeds have been shown to survive in soil up to 22 years (Madsen 1962). However, length of survival depends on depth of burial. Seeds that survive longest in the soil are buried more than 8 inches (20 cm) deep. In studies where Canada thistle seeds were buried 1 to 3 inches (2.5 to 7.5 cm) deep, less than 1% of the buried seeds remained viable after $2^1/_2$ to 5 years (Bakker et al. 1979). Canada thistle seeds can germinate on the soil surface and from at least as deep as $2^3/_8$ inches (6 cm), but the greatest percentage of seeds germinate $^3/_{16}$ to $^9/_{16}$ inches (0.5 to 1.5 cm) deep (Wilson 1979). In a pasture study no seedlings became established when Canada thistle seeds were spread on the soil surface (Amor and Harris 1975). However, when seeds were buried $^3/_{16}$ to $^7/_{16}$ inches (0.5 to 1.0 cm) deep, 6.8% to 12% of seeds emerged, and 78% and 93% of the emerged seedlings survived to maturity, respectively.

There are some reports that Canada thistle seeds have a dormancy period. However, most studies show fresh seeds will germinate (Moore 1975). In fact, some of the seeds formed during summer may germinate immediately, form rosettes before winter, and flower the following spring. Other seeds may remain dormant until spring or even later. Seed dormancy and longevity in the soil contribute to the long-term impact of weed infestations.

Canada thistle seedlings are very easy to control mechanically, chemically, and culturally. Before Canada thistle becomes perennial in nature, it is as susceptible to tillage and broadleaf herbicides as any other broadleaf weed seedling. Canada thistle seedlings are susceptible to low light intensity. If shaded by other plants, Canada thistle seedlings are not competitive, and are relatively easy to manage.

Canada thistle is a long-day plant, and requires a 16-hour day length to induce shoot elongation and flowering. Less than 16-hour photoperiods results in newly emerged shoots remaining as low-growing rosettes (Hunter and Smith 1972). This has been shown to be important for late summer control tactics.

Development of the root system is essential in the establishment and survival of Canada thistle. Lateral roots that develop from the primary root at the seedling stage will grow $2^7/_{16}$ to $4^{11}/_{16}$ inches (6 to 12 cm) horizontally before bending downward and growing vertically (Moore 1975). Aerial shoots can develop from the original vertical root or from buds on the lateral roots.

Studies in Montana showed that carbohydrate levels in the roots decreased from early spring to about the mid to latter part of June, then slowly increased until September before leveling off (Hodgson 1968). This information reveals the importance of applying translocated herbicides at the time of maximum carbohydrate translocation into the root system. This achieves maximum herbicide translocation into the root system. Another important characteristic of Canada thistle roots is their ability to regenerate from small root pieces. Root fragments as small as $^7/_{16}$ to $^1/_2$ inch (1.0 to 1.2 cm) long can grow into plants (Hayden 1934, Hamdoun 1972). A root fragment more than six weeks and less than two years old can regenerate an entire plant (Rogers 1928). Thus, disturbance of soil containing live Canada thistle roots can lead to the regeneration or establishment of new populations if no other management practice is utilized.

Management

Donald (1990) points out that nonchemical Canada thistle control methods were outlined in an 1846 publication (Stevens 1846). Farmers were advised to use the following practices for controlling Canada thistle: 1) deep plowing, 2) frequent plowing, 3) repeated summer fallow cultivation, 4) repeated mowing, 5) burning, 6) heavy planting of competitive forage legumes or grasses, 7) planting competitive row crops, 8) smothering with thick layers of straw or boards to exclude light, 9) applying salt solutions, and 10) using combinations of these practices. Many of these recommendations are still in use today, although some of the methods have changed.

• Prevention

As with the management of all weeds, prevention includes planting Canada thistle-free crop seeds, cleaning equipment before transporting from an infested area, using other sanitation practices, and controlling infestations when there are individual plants or small patches.

167

• Hand-Pulling

Hand-pulling or grubbing is not considered to be an economically effective means of controlling an established stand of Canada thistle. Because of its extensive root system and carbohydrate reserve, Canada thistle can survive grubbing once or twice a month for many years before its root reserves are depleted (Sheley et al. 1995).

• Mechanical Control

Cultivation can be an effective method of control if repeated regularly. Repeated cultivation at 21-day intervals over a 122-day period eradicated mild infestations in Idaho (Seely 1952). Six cultivations with duckfoot sweeps at 21-day intervals decreased Canada thistle shoots 98% after one season (Hodgson 1958). Earlier work (Dewey 1901, Hansen 1918) demonstrated that hoeing "as soon as shoots appear above ground" would eventually kill the plants. These studies all emphasized the importance of eliminating all shoots at each cultivation to prevent carbohydrate storage in the root system. Disking 3, 7, 10, 14, or 30 days after applying various herbicides did not improve Canada thistle control in uncropped, dryland fields compared to using herbicides without disking (Zimdahl and Foster 1993).

Repeated mowing is considered to be an effective control practice in alfalfa and other forage crops including pastures. Mowing alfalfa two times per year reduced Canada thistle populations 86% after one year and 100% after four years (Hodgson 1968). In other studies, mowing three or four times per year nearly eliminated Canada thistle in three years, and grazing for four days followed by mowing nearly eliminated Canada thistle after three years (Welton et al. 1929, Schreiber 1967). However, other research has indicated that mowing only kept Canada thistle stands in check, and did not control the weed (Willard and Lewis 1939). A more recent study combined the effects of mowing and herbicides for Canada thistle control (Beck and Sebastian 1993). Mowing two or three times consistently enhanced Canada thistle control following applications of picloram, picloram + 2,4-D, clopyralid + 2,4-D, and dicamba.

In other pasture studies, nitrogen was evaluated as a management tool. Ammonium nitrate fertilizer applied on a dryland pasture at 40 pounds per acre in Nebraska increased Canada thistle density and biomass over a two-year period. Canada thistle was thought to be able to utilize the nitrogen more efficiently than the forage grasses (Reece and Wilson 1983). With irrigation, the response to nitrogen was quite different (Thrasher et al. 1963). In a grass pasture, increasing nitrogen rate with irrigation reduced Canada thistle density. Without irrigation, density was not reduced consistently by nitrogen fertilization. With water as a limiting resource, addition of fertilizer did not help to reduce Canada thistle density.

• Chemical Control

Reviews of chemical control research by Donald (1994) and Moore (1975) show that single herbicide applications do not provide long-term control due to the difficulty in killing the root system, which can survive even though the shoots have been killed. Like other control methods, effective chemical control requires multiple applications. Several herbicides are registered for the control of Canada thistle in rangeland. Besides applying at the proper rate, application timing is critical for obtaining maximum control with these chemicals. Herbicides used for Canada thistle control in rangeland include 2,4-D, dicamba, clopyralid, metsulfuron, MCPA, glyphosate, and picloram. Before using these products or any other herbicide, it is important to read the label thoroughly for complete instructions.

2,4-D, a phenoxy acetic acid, is probably the most widely used herbicide for Canada thistle control. For best results this herbicide should be applied at a rate of 1.5 to 2.0 lb ae/acre (1.7 to 2.24 kg ae/ha) before Canada thistle reaches the bud stage. In one study where 2,4-D was applied at 1.5 lb ae/acre (1.7 kg ae/ha) for three years, Canada thistle control was better when applied at the bud stage than at the bloom stage (Hodgson 1968). At 0.5 lb ae/acre (0.6 kg ae/ha) or higher application rates applied at the late bud stage, 2,4-D has been shown to prevent flowering (Donald 1990). MCPA is another phenoxy acetic acid herbicide and is very similar to 2,4-D. Although not used as commonly as 2,4-D, it is registered for use in rangelands as a low volatile ester formulation at a rate of 1.0 to 1.4 lb ae/acre (1.12 to 1.57 kg ae/ha). The phenoxy herbicides have a relatively short residual, and are a good choice if grass is to be seeded after herbicide application.

Like 2,4-D and MCPA, dicamba (Banvel®) is a growth regulator-type herbicide. For broadcast applications dicamba should be applied at 2.0 lb ae/acre (2.24 kg ae/ha). Dicamba can be applied at any time during the growing season to actively growing plants. However, dicamba has been shown to be more effective in controlling Canada thistle when applied at the late vegetative to bud stage (Marriage 1981) or in the fall if Canada thistle has 8 to 12 inches (20 to 30 cm) of regrowth. Newly seeded areas may be severely injured with rates exceeding 0.5 lb ae/acre (0.6 kg ae/ha). Established grass growing under stress is susceptible to dicamba injury, especially if planted less than 45 days/pint of product used, excluding days when the ground is frozen.

Clopyralid is also a growth regulator-type herbicide. Curtail® is a preformulated mixture of clopyralid plus 2,4-D, and Stinger® contains only clopyralid. These herbicides can be applied when Canada thistle is actively growing. Recommended application rates for Curtail® are 1.2 to 1.8 lb ae/acre (1.3 to 2.0 kg ae/ha). The higher rate is recommended for dense infestations or applications under poor growing conditions such as drought. Stinger® can be used at a rate of 0.25 to 0.38 lb ae/acre (0.28 to 0.42 kg ae/ha) depending on the growth stage. The lower rate is recommended for applications from the rosette to prebud stage, and the higher rate is recommended for applications up to the bud stage or in fall. Field studies have shown Curtail® and Stinger® applied at 0.19 and 0.38 lb ae/acre (0.21 and 0.42 kg ae/ha) at the prebud stage to effectively control Canada thistle one year after application (Bixler et al. 1991). Bud stage applications one year later were not as effective. These herbicides should not be used on newly seeded areas until the grass is well established. Many desirable broadleaf forage plants are susceptible to both herbicides as well.

Sulfonylurea herbicides such as metsulfuron (Escort®) inhibit an enzyme in the synthesis of three branched-chain amino acids. Metsulfuron is not considered particularly effective for Canada thistle control, but can be tank-mixed with many other herbicides to broaden the spectrum of weed control. The herbicide can be applied at 0.6 oz ai/acre (42 gm ai/ha) with nonionic surfactant at 0.25 to 0.5% by volume when Canada thistle is in the bud to bloom stage or on fall rosettes. This herbicide suppresses growth.

A nonselective herbicide, glyphosate (Roundup®) inhibits an enzyme involved in the synthesis of three aromatic amino acids. For Canada thistle control glyphosate should be applied at a rate of 2 to 3 lb ai/acre (2.24 to 3.36 kg ai/ha). It is best to apply when the plants are at or beyond the bud growth stage. Since it is nonselective, care must be taken to avoid eliminating surrounding desirable vegetation.

Picloram (Tordon®) is in the same chemical family as Stinger®, but has been available longer. The recommended rate of application is 0.5 to 1.0 lb ai/acre (0.56 to 1.12 kg ai/ha). The lower use rate is recommended to be tank-mixed with 2,4-D. Several studies have demonstrated effective Canada thistle control for one year or longer with 1.0 lb ai/acre. However, picloram can persist in soil up to several years. Therefore, seeding grass or broadleaf cover can be affected.

• Grazing

Canada thistle is not a very palatable plant because of the spines on its leaves. However, intensive pasture grazing by sheep in Australia reduced the spread of Canada thistle compared to an ungrazed pasture (Amor and Harris 1974). Anecdotal evidence suggests that some livestock, such as goats, will prevent Canada thistle from flowering, but livestock feeding habits have not been evaluated in research conditions.

• Biological

Several biological agents have been evaluated for Canada thistle control. The organisms include insects, nematodes, and fungi (Moore 1975, Maw 1976, Donald 1994). Donald (1994) provides a summary of the biological organisms evaluated for Canada thistle control up to about 1993. The problem with many of the organisms that have been effective on Canada thistle is that they are pests of important crop species, which limits their use on commercial farms. Concern also has been raised regarding the impact of foreign biological control agents on native thistle species (Peschken 1981). Native North American insects such as larvae of the painted lady butterfly (*Cynthia cardui*) will defoliate thistle (Moore 1975). It also was documented years ago that the American goldfinch will eat Canada thistle seed (Detmers 1927).

Some organisms show promise for controlling Canada thistle. Three insects— Canada thistle stem weevil, Canada thistle bud weevil, and thistle stem gall fly—are currently available commercially. Canada thistle stem weevil (*Ceutorhynchus litura*) is a stem-boring beetle introduced into the United States in 1972 (Rees 1990, Rees et al. 1996). Larvae of this insect feed on new leaf, stem, root crown, and root tissue as they emerge in the spring. The greatest damage is to the stems and root crown. The larvae will also chew an exit hole below the soil surface that increases the plant's susceptibility to secondary damage from other arthropods, nematodes, and pathogens. The effect of this insect is usually most apparent the following spring because the underground shoots have rotted. Canada thistle bud weevil (*Larinus planus*) was accidentally introduced to North America. The larvae of this beetle feed in the flower heads and the adults feed on the foliage, however it is the larvae that cause the most problem by reducing seed production (Rees et al. 1996). It will also feed on other *Cirsium* species, as well as *Carduus*, *Onopordum*, and *Arctium* species. The thistle stem gall fly (*Urophora cardui*) is a stem and shoot gall fly introduced into the United States in 1977. As larvae develop within the thistle stem, a gall forms which interrupts nutrient flow in the plant. This reduces the plant's vigor, making it less competitive and more susceptible to attack by other insects or pathogens. Gall size ranges from marble to walnut size, depending on the number of larvae within the gall. Other insects that will attack Canada thistle include *Cassida rubiginosa* and *Rhinocyllus conicus*.

Two pathogenic rust fungi have been tested for control of Canada thistle. *Sclerotinia sclerotiorum* attacks the crown and roots, and has been shown to cause 20% to 80% death of Canada thistle shoots (Brosten and Sands 1986). The drawback to this organism is that it is not host specific and can attack desirable broadleaf crops. More recently, *Puccinia punctiformis*, an obligate rust fungus parasite specific to Canada thistle, has been found to be effective in reducing flowering and vegetative reproduction. It is now found in British Columbia, Ontario, and Quebec in Canada (Harris 1996), but only rarely found in the prairie provinces.

• Long-Term Control Strategies

Integrated Canada thistle control strategies have been recommended for many years to achieve successful control (Strand 1982, Trumble and Kok 1982). According to Trumble and Kok, development of an integrated management program requires information on the crop system or plant community, life cycles of weeds and potential biological control agents, and the various control techniques intended for inclusion in the program.

For successful, sustainable, long-term management of Canada thistle, it is necessary to integrate management strategies. Combinations of mowing, grazing, and herbicides have shown great promise for reducing Canada thistle populations in pasture and rangeland conditions (Beck and Sebastian 1993, Schreiber 1967, and Welton et al. 1929). In undisturbed areas with small infestations, it is recommended that pulling, cutting, and judicious spot herbicide applications be used (Sather 1987). Implementation of these practices with continued research (Hunter 1995) will make sustainable long-term management successful.

171

Literature Cited

Amor, R.L., and R.V. Harris. 1974. Distribution and seed production of *Cirsium arvense* (L.) Scop. in Victoria, Australia. Weed Res. 14:317-23.

Amor, R.L., and R.V. Harris. 1975. Seedling establishment and vegetative spread of *Cirsium arvense* (L.) Scop. in Victoria, Australia. Weed Res. 15:407-11.

Bakker, A. 1960. A comparative life-history study of *Cirsium arvense* (L.) Scop. and *Tussilago farfara* L., the most troublesome weeds in the newly reclaimed polders of the former Zuiderzee. *In:* J.L. Harper (ed.), The Biology of Weeds. Blackwell Scientific Publishers, Ltd., Oxford, England, 205-22.

Bakker, D. H.A. Roberts, and R.J. Chancellor. 1979. Periodicity of seedling emergence and achene survival in some species of *Carduus, Cirsium,* and *Onopordum*. J. Appl. Ecol. 16:641-47.

Beck, K.G., and J.R. Sebastian. 1993. An integrated Canada thistle management system combining mowing with fall-applied herbicides. West. Soc. Weed Sci. Proc. 46:102-104.

Bixler, A., W. Cooley, and V.F. Carrithers. 1991. Canada thistle control at two stages of plant growth with clopyralid. Proc. West. Soc. Weed Sci. 44:44-47.

Brosten, B.S. ,and D.C. Sands. 1986. Field trials of *Sclerotinia sclerotiorum* to control Canada thistle. Weed Sci. 34:377-80.

Burrows, F.M. 1975. Wind-borne seed and fruit movement. New Phytol. 75:405-18.

Derscheid, L.A., and R.E. Schultz. 1960. Achene development of Canada thistle and perennial sow thistle. Weeds. 8:55-62.

Detmers, F. 1927. Canada thistle, *Cirsium arvense* Tourn. Ohio Agric. Exp. Sta. Bull. 414.

Dewey, H.L. 1901. Canada thistle. U.S. Dep. Agric. Bur. Bot. Circ. 17.

Dewey. S.A. 1991. Weedy thistles of the western United States. *In:* L.F. James, J.O. Evans, M.H. Ralphs, and R.D. Child (eds.), Noxious Range Weeds. Westview Press. Boulder, CO, 247-53.

Donald, W.W. 1990. Management and control of Canada thistle (*Cirsium arvense*). Rev. Weed Sci. 5:193-250.

Donald, W.W. 1994. The biology of Canada thistle (*Cirsium arvense*). Rev. Weed Sci. 6:77-101.

Erickson, L. C. 1983. A review of early introductions of field (Canada) thistle (*Cirsium arvense*) (L.) Scop. to North America and its present distribution. West. Soc. Weed Sci. Proc. 36:200-204.

Friesen, H.A. 1968. Trend in Canadian research to control Canada thistle. Proc. Northeast Weed Control Conf. 22:27-36.

Hamdoun, A.M. 1970. The anatomy of subterranean structures of *Cirsium arvense* (L.) Scop. Weed Res. 10:284-87.

Hamdoun, A.M. 1972. Regenerative capacity of root fragments of *Cirsium arvense* (L.) Scop. Weed Res. 12:128-36.

Hansen, A.A. 1918. Canada thistle and methods of eradication. U.S. Dep. Agric. Farmers Bull. 1002.

Harris, P. 1996. Status of introduced and main indigenous organisms on weeds targeted for biocontrol in Canada. Agriculture and Agri-Food Canada, Lethbridge Research Centre, Alberta.

Hayden, A. 1934. Distribution and reproduction of Canada thistle in Iowa. Amer. J. Bot. 21:355-73.

Henderson, L.F. 1898. Twelve of Idaho's worst weeds. Univ. Idaho Bull. 14.

Hitchcock, C.L., and A. Cronquist. 1994. Flora of the Pacific Northwest. University of Washington Press. Seattle, WA.

Hodgson, J.M. 1958. Canada thistle (*Cirsium arvense* (L.) Scop.) control with cultivation, cropping, and chemical sprays. Weeds. 6:1-11.

Hodgson, J.M. 1968. The nature, ecology and control of Canada thistle. U.S. Dep. Agric. Tech. Bull. 1386.

Holm, L.G., D.L. Plucknett, J.V. Pancho, and J.P. Herberger. 1991. The World's Worst Weeds. Distribution and Biology. Krieger Publ. Co., Malabar, FL.

Hunter, J.H. 1995. Effect of bud vs rosette growth stage on translocation of 14C-glyphosate in Canada thistle (*Cirsium arvense*). Weed Sci. 43:347-51.

Hunter, J.H., and L.W. Smith. 1972. Environment and herbicide effects on Canada thistle ecotypes. Weed Sci. 20:163-67.

Kigel, J. and D. Koller. 1985. Asexual reproduction of weeds. *In:* S.O. Duke (ed.), Weed physiology. Vol. 1. Reproduction and ecophysiology. CRC Press, Boca Raton, FL, 65-100.

Korsmo, E. 1930. Unkrauter im ackerbau der neuzeit. Springer, Berlin.

Lewis, W.H., and M.P.F. Elvin-Lewis. 1977. Medical Botany. John Wiley and Sons, Inc., New York.

Madsen, S.B. 1962. Germination of buried and dry stored seeds III, 1934-1960. Proc. Int. Seed Test. Assoc. 27:920-928.

Marriage, P.B. 1981. Response of Canada thistle to herbicides. Proc. North Cent. Weed Control Conf. 36:162-67.

Maw, M.G. 1976. An annotated list of insects associated with Canada thistle (*Cirsium arvense*) in Canada. Can. Entomol. 108:235-44.

Moore, R.J. 1975. The biology of Canadian weeds. 13: *Cirsium arvense* (L.) Scop. Can. J. Plant Sci. 55:1033-48.

Nadeau, L.B., and W.H. Vanden Born. 1989. The root system of Canada thistle. Can. J. Plant Sci. 69:1199-1206.

Peschken, D.P. 1981. Biological control of Canada thistle. Proc. North Cent. Weed Control Conf. 36:169-73.

Reece, P.E., and R.G. Wilson. 1983. Effect of Canada thistle (*Cirsium arvense*) and musk thistle (*Carduus nutans*) control on grass herbage. Weed Sci. 31:488-92.

Reed, C.F., and Hughes, R.D. 1970. Selected weeds of the United State. U.S. Dep. Agric., Agric. Res. Ser., Agric. Handbook 366. USDA-ARS. Washington, D.C.

Rees, N.E. 1990. Establishment, dispersal, and influence of *Ceutorhynchus litura* on Canada thistle (*Cirsium arvense*) in the Gallatin Valley of Montana. Weed Sci. 38:198-200.

Rees, N.E., P.C. Quimby, Jr., G.L. Piper, E.M. Coombs, C.E. Turner, N.R. Spencer, and L.V. Knutson (eds.). 1996. Biological control of weeds in the west. West. Soc. Weed Sci., USDA-ARS, Montana St. Univ., Bozeman, MT.

Rogers, C.F. 1928. Canada thistle and Russian knapweed and their control. Colo. Agric. Exp. Sta. Bull. 434.

Sather, N. 1987. *Cirsium arvense*. Element stewardship abstract prepared for the Midwest Regional Office, The Nature Conservancy, Minneapolis, MN.

Schreiber, M.M. 1967. Effect of density and control of Canada thistle on production and utilization of alfalfa pasture. Weeds 15:138-42.

Seely, C.I. 1952. Controlling perennial weeds with tillage. Idaho Agric. Exp. Sta. Bull. 188.

Sheldon, J.C., and F.M. Burrows. 1973. The dispersal effectiveness of the achene-pappus units of selected Compositae in steady winds with convection. New Phytol. 72:665-75.

Sheley, R.L., B.H. Mullin, and P.K. Fay. 1995. Managing riparian weeds. Rangelands. 17:154-57.

Stachon, W.J., and R.L. Zimdahl. 1980. Allelopathic activity of Canada thistle (*Cirsium arvense*) in Colorado. Weed Sci. 28:83-86.

Stevens, A. 1846. Extirpation of Canada thistles. N.Y. Agric. Soc. Trans. Pp. 406-28.

Strand, O.E. 1982. An integrated approach for Canada thistle control on no-cropland. Proc. North Cent. Weed Control conf. 37:113-14.

Thrasher, F.P., C.S. Cooper, and J. M Hodgson. 1963. Competition of forage species with Canada thistle, as affected by irrigation and nitrogen levels. Weeds 11:136-38.

Trumble, J.T., and L.T. Kok. 1982. Integrated pest management techniques in thistle suppression in pastures of North America. Weed Res. 22:345-59.

Welton, F.A., V.H. Morris, and A.J. Hartzler. 1929. Organic food reserves in relation to the eradication of Canada thistle. Ohio Agric. Exp. Sta. Bull. 441.

Willard, C.J., and R.D. Lewis. 1939. Eradicating Canada thistle. Ohio State Univ. Agric. Coll. Ext. Serv. 146.

Wilson, R.G., Jr. 1979. Germination and seedling development of Canada thistle (*Cirsium arvense*). Weed Sci. 27:146-51.

Wilson, R.G., Jr. 1980. Dissemination of weed seeds by surface irrigation water in western Nebraska. Weed Sci. 28:87-92.

Zimdahl, R.L., and G. Foster. 1993. Canada thistle (*Cirsium arvense*) control with disking and herbicides. Weed Technol. 7:146-49.

Cheatgrass

Jeffrey C. Mosley, Stephen C. Bunting, and Mark E. Manoukian

Cheatgrass (*Bromus tectorum*) is a non-native annual grass that occurs throughout most of the United States and Canada and in northern Mexico. It is an adaptive species with a wide ecological amplitude, occurring from the sagebrush semi-desert biome in the southern Great Basin into the coniferous forest zone of the Rocky Mountains (Hull and Pechanec 1947). Cheatgrass is a major herbaceous species particularly in the sagebrush steppe and Pacific bunchgrass region, where it is a dominant on more than 100 million acres (40 million ha) (Whisenant 1990). Cheatgrass is also dominant in localized areas within the northern mixed-grass prairie of eastern Montana, eastern Wyoming, southern Alberta, and southwestern Saskatchewan. Cheatgrass can significantly alter native rangeland vegetation composition through competitive exclusion of native species reproduction and the facilitation of wildfires.

Identification

Cheatgrass is usually a winter annual. Its seeds typically germinate in the early fall and overwinter as small seedlings. Plants resume growth early in the spring. Plants range from 2 inches (5 cm) to 2 feet (60 cm) tall, depending largely upon water availability and plant density. A cheatgrass plant can be comprised of one or two tillers or as many as 20 tillers per plant. Cheatgrass has a panicle inflorescence with five to eight florets per spikelet. Its fibrous root system is concentrated in the upper 12 inches (30 cm) of soil (Klemmedson and Smith 1964). Most cheatgrass leaves have numerous fine hairs that give them a soft, downy feeling when the plant is green and actively growing, thus the common name of "downy brome." When wheat yields of pioneer farmers were reduced by this grass, the farmers felt they were being cheated, and labeled the weed "cheatgrass." Other common names include cheat, downy chess, downy cheat, broncograss, and six-weeks grass.

Cheatgrass can usually be distinguished from most other grasses in the seedling stage by its very hairy blades and sheaths. In the seedling stage the most similar rangeland species is Japanese brome (*Bromus japonicus*). Japanese brome tends to grow in more moist sites than does cheatgrass, but the two species can occur together. These two species are more readily distinguished once their inflorescences are visible. Cheatgrass has an open, drooping panicle with long straight awns attached to the florets; the awns of Japanese brome are usually twisted. Glumes and lemmas of cheatgrass are typically hairy, whereas those of Japanese brome are not. Lemmas of cheatgrass taper into two narrow teeth. At maturity, cheatgrass plants are reddish-purplish in color.

Origin, History, and Distribution

Cheatgrass apparently began its evolution in southwestern Asia. It was probably introduced into North America independently several times (Mack 1981). One of the original vectors of cheatgrass seed into North America was the soil used as ballast in ships traveling from Eurasia. The first introductions are believed to have been in St. Louis, Missouri, from ballast dumps in about 1850. The first report of cheatgrass in western North America was in 1890 at Spences Bridge, British Columbia. In 1892, cheatgrass was recorded along the Northern Pacific Railroad near Ritzville, Washington. Cheatgrass was collected near Provo, Utah, in 1893, and was first reported in Nevada in 1906.

Early infestations were commonly found near wheat cropland and railroads. Wheat seed was often contaminated with cheatgrass seed, and straw infested with cheatgrass seed was used as packing material for merchandise transported by the railroads. Straw used as bedding in train cars carrying livestock was discarded along the railroad rights-of-way. Once introduced, cheatgrass spread rapidly into adjacent rangeland vegetation.

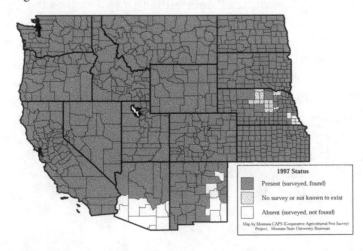

Distribution of cheatgrass by county in the western United States. Data were collected by surveying various weed authorities in each state.

Cheatgrass expansion was especially rapid through the intermountain West. One reason was that cheatgrass was pre-adapted to the environmental conditions of the region. Also, the introduction of cheatgrass followed a period of excessive livestock grazing in the late 1800s that had reduced the competitiveness of the indigenous vegetation. The agricultural depression following World War I forced the abandonment of countless dryland homesteads in the Great Basin and the abandoned cropland fields were eventually colonized by cheatgrass (Piemeisel 1951).

Cheatgrass became widespread and locally abundant by 1914 in the intermountain region and occupied much of its current range by 1930 (Mack 1981). By 1980, nearly every county in Montana, Idaho, and Wyoming had infestations (Forcella and Harvey 1981). The first cheatgrass infestations in Alberta and Saskatchewan occurred in the 1960s, but since 1980 the number of infestations in these provinces has increased exponentially. Cheatgrass infestations in the northern mixed-grass prairie of eastern Montana and eastern Wyoming have also increased dramatically since 1980.

Potential Invasion

Cheatgrass can invade rangelands throughout western North America, and the seed is so ubiquitous that the potential for invasion is virtually unlimited. The most susceptible sites have deep, loamy soils, south-facing slopes, and 12 to 22 inches (30 to 56 cm) of annual precipitation that peaks in late winter or early spring. Populations of cheatgrass are genetically very similar (Pyke and Novak 1994). However, high plasticity allows the species to thrive under a wide variety of site conditions. It can be found in salt desert shrub communities that receive 6 inches (15 cm) of annual precipitation, and in high elevation coniferous forests that exceed 25 inches (64 cm) of annual precipitation.

Cultivation and subsequent land abandonment, excessive livestock grazing, and repeated fires can all interact to proliferate cheatgrass (Hull and Pechanec 1947). But cheatgrass also thrives in areas that have never been cultivated or grazed by domestic livestock (West 1991, Svejcar and Tausch 1991, Kindschy 1994, Tausch et al. 1994). Cheatgrass can outcompete many otherwise competitive plant species. These include halogeton (*Halogeton glomeratus*), tumble mustard (*Sisymbrium altissimum*), and Russian thistle (*Salsola iberica*) (Piemeisel 1951, DeFlon 1986). Cheatgrass may also outcompete seedlings of perennial species and prevent their establishment. Many seedings have failed through excessive cheatgrass competition.

Cheatgrass can spread into uninfested areas through several pathways. Cheatgrass is spread short distances by wind. Animals (wild and domestic) carry cheatgrass in their feces, hooves, hair, feathers, and tails. Cropland intermingled with rangeland is another seed source, particularly fields of winter wheat and dryland hay. Humans may transport cheatgrass seeds in vehicles and clothing. Also, although most cheatgrass seeds generally do not survive longer than one year on rangelands (Hulbert 1955), cheatgrass seeds may remain viable for several years when stored dry within

177

bales of hay or straw. Subsequently transporting and feeding these bales to livestock in uninfested areas can spread cheatgrass.

Cheatgrass is a more significant weed in drier environments. For example, on well-managed mountain foothill sites cheatgrass generally does not compete effectively with perennial grasses. It is dominant only in early successional stages, and is rapidly replaced by perennial species. Cheatgrass often remains as a minor species in the later successional stages on these moist sites, occupying the interspaces between perennial plants. In the drier prairie and steppe environments, however, cheatgrass is very competitive and can rapidly increase whenever perennial plants are stressed by dought, fire, or excessive grazing. Cheatgrass progressively colonizes these drier sites in a pulse-like fashion, whereby the cheatgrass population exhibits a surge whenever growing conditions favor cheatgrass. The stand may remain relatively unchanged for several years until another pulse or surge occurs. The critical point of expansion is when cheatgrass populations become so vast that they begin to influence the occurrence of wildfires.

Cheatgrass can dramatically influence plant community composition by its effect on fire regimes. This effect is most evident in the sagebrush steppe where, prior to European settlement, fire-free intervals probably varied from about 20 to 25 years in higher elevation mountain big sagebrush (*Artemisia tridentata vaseyana*) habitat types, to 50 or 100 years in the drier Wyoming big sagebrush (*Artemisia tridentata wyomingensis)* habitat types that dominated the Snake River Plain (Burkhardt and Tisdale 1976, Wright and Bailey 1982). Much of the Snake River Plain today burns at intervals of five years or less because cheatgrass has increased the continuity of fine-textured fuels which promotes more frequent and larger fires. These cheatgrass-influenced fires also burn the landscape more uniformly, so that islands of unburned areas do not remain. Fine-fuel continuity influences fire frequency more so than the amount of fine fuel (Whisenant 1990), although the amount of fine fuel is also important (Boltz 1994). Fires are especially harmful to desert shrubs such as winterfat (*Ceratoides lanata*) and shadscale (*Atriplex confertifolia*). Fortunately, cheatgrass is usually a minor species in desert shrub communities, but following two or more years with above-average precipitation, sufficient mulch may be present to sustain a wildfire and convert the plant community to cheatgrass indefinitely (Pellant and Reichert 1984).

The cheatgrass-fire cycle is self-promoting (Pellant 1990, Peters and Bunting 1994). Increased fire frequency quickly removes nonsprouting shrubs such as Wyoming big sagebrush. More fire-tolerant, sprouting shrubs may persist for awhile, but they also cannot tolerate the short fire-free intervals common today. Continued increases in fire frequency eventually remove and exclude all perennial shrubs, grasses, and forbs from the landscape, and cheatgrass competition prevents their reestablishment. These altered fire regimes and subsequent changes in botanical composition can occur without any influence from livestock grazing (Cottam and Evans 1945).

Impacts

Cheatgrass is very persistent across its range of habitats (Daubenmire 1940, 1975; Cline et al. 1977) and in some places is so dense that few perennial grasses or shrubs are present. Cheatgrass replaces many other annual species, too, resulting in lower plant species diversity (Rosentrater 1994). Plant production on cheatgrass-infested rangeland is substantially less and much more variable from year to year than the indigenous vegetation (Murray et al. 1978). Also, the length of time each year that cheatgrass is green and actively growing is shorter than the indigenous perennial vegetation. The green period during spring-summer can be as brief as six to eight weeks. This brief period reduces the green feed period for foraging animals.

Although the range livestock industry has adapted to the presence of cheatgrass, range livestock production probably would be enhanced if cheatgrass could be replaced with palatable and nutritious perennial grasses. One problem with cheatgrass is the susceptibility of its florets to ergot, a fungus that is poisonous to livestock. Some livestock producers prefer to use cheatgrass-dominated range for winter grazing (DeFlon 1986) because in winter cheatgrass is more palatable and just as nutritious as crested wheatgrass (*Agropyron cristatum* and *A. desertorum*). Livestock that graze dry cheatgrass in winter need energy and protein supplements to meet nutrient requirements. Livestock should also be observed closely, because eating dry cheatgrass herbage can greatly increase the incidence of lumpy-jaw infections in cattle and cause severe eye injuries from the sharp seeds.

Most range livestock producers utilize cheatgrass-dominated rangeland in spring. At this time of year cheatgrass is palatable and nutritious with crude protein levels of 15% to 18%. Livestock perform well on cheatgrass diets in spring. Yearling ewes can gain 0.3 lbs/day (0.1 kg/day) and yearling steers 1.7 to 2.0 lbs/day (0.8 to 0.9 kg/day). These gains are equivalent to those attainable from most rangeland vegetation during spring. Livestock gains decline to near zero once the cheatgrass plants mature (Cook and Harris 1952, Murray 1971, Murray et al. 1978).

Biology and Ecology

• Seed Production and Seed Viability

Cheatgrass is a prolific seed producer. Even during years with unfavorable growing conditions, cheatgrass produces enough seed to perpetuate itself (Tisdale and Hironaka 1981). In extremely dry conditions, cheatgrass may reach only 1 or 2 inches (2.5 to 5.0 cm) in height, and yet these small plants can produce viable seed. With adequate soil moisture, cheatgrass seed production can exceed 1 billion seeds per acre (2.6 billion seeds/ha). Individual cheatgrass plants grown in high densities may produce about 25 seeds each, but a single cheatgrass plant with abundant tillers and grown with abundant moisture and sunlight can easily produce 5,000 seeds (Young et al. 1987).

Cheatgrass plants are largely self-pollinated (Pyke and Novak 1994), and seeds become germinable soon after they mature. Seeds are viable when the fruits have barely started to turn purple and are still mostly green. However, some viable seed is produced even when the inflorescences are clipped before any purple coloration appears (Hulbert 1955). Ungerminated cheatgrass seeds do not usually remain viable for longer than one year under rangeland conditions (Hulbert 1955).

• Seed Germination and Seedling Establishment

A fluffy seedbed favors establishment of cheatgrass. It can establish on the soil surface in the mesic portions of the northern mixed-grass prairie and Pacific bunchgrass biomes, but in drier environments cheatgrass must be covered by soil or litter. The annual seed mass of cheatgrass often provides sufficient mulch for some of its seeds to germinate (Young et al. 1987).

Cheatgrass seeds germinate quickly and at very high rates (up to 95%) as soon as moisture conditions are favorable (Hulbert 1955, Hull and Hansen 1974). In the Great Plains where summer rains can occur, cheatgrass seeds produced in spring have an afterripening requirement that inhibits their germination until autumn. As a winter annual, cheatgrass seeds typically germinate in the early fall when moisture becomes sufficient, and then cheatgrass plants grow rapidly until cold temperatures arrive. Above-ground growth sometimes continues during winter if the weather is warm and rainy (Klemmedson and Smith 1964). The root system of cheatgrass often continues to develop during winter, when above-ground growth is minimal (Harris 1967, 1970). Cheatgrass roots can grow at much cooler soil temperatures than many other plant species.

Plants from seeds that germinated in the fall or early winter begin rapid growth in late winter or early spring when ambient temperatures warm. Plants grow and develop rapidly in the spring and usually flower, develop seeds, and become fully dried within two to three months (Stewart and Hull 1949, Hulbert 1955, Klemmedson and Smith 1964). Corresponding stages are about six weeks later in bluebunch wheatgrass (*Agropyron spicatum*). Soil water depletion by cheatgrass is one of its principal mechanisms for successfully competing with perennial grasses (Melgoza et al. 1990, Melgoza and Nowak 1991), especially during the establishment of perennial grass seedlings. Cheatgrass seeds can also germinate in the spring, but spring-germinated plants are often less numerous and vigorous, and seedling emergence is slow (Stewart and Hull 1949, Hull and Hansen 1974). Spring-germinated plants produce fewer inflorescences, and when seeds germinate late in the spring no inflorescences are produced (Hulbert 1955). Some cheatgrass plants that do not produce seed during the spring-summer growing season will overwinter and produce seed the following spring. Additions of nitrate fertilizer stimulate dormant cheatgrass seeds to germinate. In a long, wet spring with abundant nitrification, many cheatgrass seeds lose dormancy and germinate.

Cheatgrass adds considerable organic matter to the soil surface (Stewart and Hull 1949). This facilitates the establishment of cheatgrass seedlings, which are favored by large amounts of plant mulch (Evans and Young 1970). Mulch also results in less cover of lichens and mosses on the soil surface, and substantial mulch may inhibit establishment of perennial bunchgrasses (Svejcar and Tausch 1991, Tausch et al. 1994).

Management

Any attempts to control undesirable plants should be approached cautiously. The idea of suppressing cheatgrass to enhance desirable perennials assumes that desirable perennials will fill the temporary void left by cheatgrass. But in many areas, desirable perennials may be outcompeted for site resources by plant species often considered more undesirable than cheatgrass, such as medusahead (*Taeniatherum asperum*) or yellow starthistle (*Centaurea solstitialis*) (Larson and Sheley 1994). A thorough reconnaissance and site evaluation should be completed before initiating any form of plant control (J.A. Young 1983).

As mentioned earlier, cheatgrass is very persistent wherever it becomes established, and thus eradication of cheatgrass is not a reasonable goal in most situations. But the extent to which cheatgrass dominates a plant community greatly determines the appropriate suppression strategy. Large areas that are mostly devoid of perennials and have fire-free intervals of five or fewer years have probably crossed a threshold, and the cheatgrass community probably represents a relatively stable "steady state" (Laycock 1994). This steady state exists on many depleted sites within Wyoming big sagebrush habitat types of the Snake River Plain and Columbia Basin. Cheatgrass control in these areas can be implemented without concern for perennial plants. In other seral plant communities where cheatgrass is present but herbaceous perennials remain abundant, cheatgrass control measures should consider the needs of the perennial plants. Cheatgrass control is effective only when combined with other techniques that establish perennial plants. Control without replacement by desirable perennials will only result in the reestablishment of cheatgrass or some other undesirable species.

• *Mechanical Control*

Disking and other mechanical control methods applied alone are often ineffective for controlling cheatgrass on rangeland (Pellant 1990). The disturbed soil and fluffy seedbed favor cheatgrass. To be effective, the tillage must be 4 to 6 inches (10 to 15 cm) deep to bury cheatgrass seeds in the soil and prevent them from germinating. More than one tillage treatment is usually needed, because the tillage equipment does not consistently cut deep enough to bury the seed (Mattise and Scholten 1994). Mechanical treatments that increase soil bulk density will inhibit cheatgrass germination and establishment. However, such treatments should be applied cautiously to avoid undesirable impacts to other plant species and resource values.

181

• *Chemical Control*

Chemical control can be used effectively when desired perennial plants are still abundant in the plant community. One year of chemical application will only temporarily thin the cheatgrass population and actually increase cheatgrass seed production. Treatment must be repeated from two to five years consecutively.

Paraquat and glyphosate should be applied in the spring when the cheatgrass seedheads begin to emerge. Paraquat at 0.5 lbs active ingredient per acre (0.6 kg ai/ha) applied at either the vegetative growth stage or the early dough stage will control cheatgrass if repeated for two consecutive years (Evans et al. 1967, Whitson and Koch 1998). Surfactants increase the effectiveness of paraquat for cheatgrass control (Evans and Eckert 1965). Glyphosate at 0.5 lbs ai/ac (0.6 kg ai/ha) for three consecutive years will control cheatgrass when applied at either the vegetative growth stage or the early dough stage. Glyphosate at 0.4 lbs ai/ac (0.4 kg ai/ha) will also control cheatgrass if applied at the early dough stage. Glyphosate at 0.6 lbs ai/ac (0.7 kg ai/ac) will kill desirable perennial grasses (Whitson and Koch 1998).

• *Prescribed Fire*

Burning is an ineffective method for controlling cheatgrass. Although fire greatly reduces the density of cheatgrass plants the next growing season, the plants that do establish produce so much more seed per plant that total post-burn seed production for a site may actually increase by a factor of 100 (J.A. Young 1983). Unless an artificially seeded species becomes established and outcompetes cheatgrass, density of cheatgrass plants may exceed pre-burn levels within one to five years (Wright et al. 1979).

• *Biological Control*

No insects or fungi are available as biological control agents for cheatgrass as of 1998. Developmental research is exploring the use of soil bacteria from the genus *Pseudomonas* (Kennedy 1994). These microbes colonize the seeds and roots of cheatgrass and produce a toxin that appears to inhibit cell elongation in the growing plant. The toxin may also inhibit cheatgrass germination and seedling vigor. One particular strain, *Pseudomonas fluorescens* D7, inhibits several biotypes of cheatgrass from throughout the world. This strain of bacteria also attacks other species in the *Bromus* genus (Stubbs 1996).

• *Prescribed Livestock Grazing*

Livestock grazing can be purposely manipulated to control cheatgrass (Megee 1938, Daubenmire 1940, Vallentine and Stevens 1994, Mosley 1996). This tool is probably best suited to localized areas, either for protecting existing stands of perennial plants from fire, or for aiding the artificial seeding of severely depleted sites.

Grazing should begin in spring when cheatgrass has grown tall enough to become accessible and palatable to livestock. The goal is to graze the cheatgrass plants before they turn purple in color, and thus prevent the seeds from reaching the soft dough stage. This will prevent most cheatgrass seeds from becoming viable. At least two defoliations are needed in the spring of each year. Plants should be grazed in late spring during the boot stage, then allowed to regrow for three to four weeks before re-grazing. Grazing is required for a minimum of two consecutive years to control cheatgrass. Grazing intensity should be light enough so that desirable grasses have a residual stubble height of 3 inches (8 cm) or more. Residual stubble heights can be lower if the site is in a cheatgrass-dominated steady state (Mosley 1996).

Grazing of dried cheatgrass during winter dormancy will help reduce mulch accumulations and enhance seedling establishment of perennials. Grazing intensities in winter can be moderately heavy without damaging desirable plants, provided soils are dry and firm.

Prescribed grazing also can help diminish the fire hazard from cheatgrass by disrupting fine-fuel continuity and reducing fine-fuel loads. This will extend fire-free intervals and enhance the competitiveness of perennial plants. Grazed firelines should be at least 250 feet (75 m) wide (Wright et al. 1979, Wright and Bailey 1982).

• *Greenstripping*

Greenstripping is a method of drill-seeding fire breaks to impede the flow of wildfires and thereby increase fire-free intervals (Pellant 1994). These fuelbreaks are 30 to 100 feet (10 to 30 m) wide, and are usually seeded with bunchgrasses because the openings between individual bunchgrass plants help to disrupt the continuity of the fine fuels created by cheatgrass. The bunchgrasses also remain green and moist later into summer, which makes greenstripping vegetation less flammable. Crested wheatgrass and forage kochia (*Kochia prostrata*) have been the most successful species in greenstrips. Wildfire speed often slows when entering a greenstrip, thus allowing fire suppression crews to extinguish the fire. Some wildfires burn into greenstrips and extinguish themselves. Greenstripping is an effective control method to use after the cheatgrass/wildfire cycle has become established.

• *Revegetation*

Although disking, herbicides, or prescribed fire applied alone often do not control cheatgrass on rangeland, each of these treatments can be used effectively to prepare sites for revegetation. One strategy is first to disk the soil to stimulate germination of cheatgrass. Once the cheatgrass seedlings emerge, the site is disked again or glyphosate is applied to kill the new seedlings. The site is then drill- or broadcast-seeded. Another strategy is to use prescribed fire in autumn to prepare sites for seeding the following spring (R.P. Young 1983, Bunting et al. 1987, Rasmussen 1994). Disking, herbicides, or prescribed livestock grazing can then be used in late spring to reduce vigor and seed production of the cheatgrass plants that establish after the fire. Seeding the site can be delayed until after the disking, herbicide, or prescribed grazing treatment, or the site can be broadcast-seeded immediately before the grazing treatment, in order to use the livestock to trample in the seed. Prescribed livestock grazing can similarly be used to suppress cheatgrass on sites that have been initially treated with herbicides prior to seeding, or sites pre-treated with disking.

The atrazine-fallow technique was an effective chemical method for preparing cheatgrass-infested rangeland for revegetation (Eckert and Evans 1967, Eckert et al. 1974). Atrazine application was followed about one year later by seeding with a deep-furrow drill, and sometimes 2,4-D was also applied to control broadleaf weeds (Evans and Young 1977). The atrazine-fallow technique is no longer an option for controlling cheatgrass on U.S. rangeland, because atrazine is no longer registered for this use. Glyphosate at 0.5 lbs ai/ac (0.6 kg ai/ha) or paraquat at 0.5 lbs ai/ac (0.6 kg ai/ha) can substitute for atrazine in the chemical fallow technique (Whitson and Koch 1998). These herbicides can be applied in early summer when cheatgrass is actively growing, and the sites can then be reseeded later that fall or the next spring. Deep-furrow drills are not necessary when reseeding sites that have been summer-fallowed with either glyphosate or paraquat.

Sulfometuron is another herbicide that has shown potential for cheatgrass control when followed with seeding. After applying sulfometuron as a preemergent or early postemergent spray in spring or fall, perennial grasses can be seeded into the area one year later. Sulfometuron should be applied at 0.1 lb ai/ac (0.1 kg ai/ha) in areas with 20 inches (50 cm) or less of annual precipitation. A rate of 0.2 to 0.3 lb ai/ac (0.2 to 0.3 kg ai/ha) should be used in areas that receive more than 20 inches (50 cm) of annual precipitation. Sulfometuron is a more selective herbicide than glyphosate or paraquat. Established plants of Sandberg bluegrass (*Poa sandbergii*) may be harmed by sulfometuron, but wheatgrasses (*Agropyron* spp.) and needlegrasses (*Stipa* spp.) are reasonably resistant (S. Jirik 1998, personal communication).

Hycrest crested wheatgrass, Sodar streambank wheatgrass (*Agropyron riparium*), and Luna pubescent wheatgrass (*Agropyron trichophorum*) are competitive cultivars that can outcompete cheatgrass in new seedings following cheatgrass control (Whitson and Koch 1998). These grasses are suitable for revegetating most non-irrigated sites where annual precipitation is 12 to 16 inches (30 to 40 cm). All three of these wheatgrasses should be seeded at 7 to 10 lbs Pure Live Seed (PLS) per acre (7.8 to 11.2 kg PLS/ha) with row spacings of 12 to 14 inches (30 to 35 cm).

Areas receiving less than 12 inches (30 cm) of annual precipitation should be seeded with drought-resistant species such as Hycrest created wheatgrass or Siberian wheatgrass (*Agropyron sibericum*). Perennial grass seedlings on such dry sites are especially vulnerable to cheatgrass competition, and often fail to establish.

Regar meadow bromegrass (*Bromus biebersteinii)* and orchardgrass (*Dactylis glomerata*) are suitable for reseeding most cheatgrass-infested rangeland where annual precipitation exceeds 16 inches (40 cm). Regar meadow bromegrass should be seeded at 7 to 10 lbs PLS/ac (7.8 to 11.2 kg PLS/ha), and orchardgrass should be seeded at 3 to 4 lbs PLS/ac (3.4 to 4.5 kg PLS/ha). Row spacings of 6 to 12 inches (15 to 30 cm) are appropriate for these two species.

Acknowledgments

The authors thank Tom Whitson and four anonymous reviewers for their helpful comments.

Literature Cited

Boltz, M. 1994. Factors influencing postfire sagebrush regeneration in south-central Idaho. *In*: S.B. Monsen and S.G. Kitchen (eds.), Proc. Ecology and management of annual rangelands. USDA For. Ser. Gen. Tech. Rep. INT-GTR-313, 281-90.

Bunting, S.C., B.M. Kilgore, and C.L. Bushey. 1987. Guidelines for prescribed burning sagebrush-grass rangelands in the Northern Great Basin. USDA For. Ser. Gen. Tech. Rep. INT-231.

Burkhardt, J.W., and E.W. Tisdale. 1976. Causes of juniper invasion in southwestern Idaho. Ecology 76:472-84.

Cline, J.F., D.W. Uresk, and W.H. Rickard. 1977. Plants and soils of a sagebrush community on the Hanford Reservation. Northwest Sci. 51:60-70.

Cook, C.W., and L.E. Harris. 1952. Nutritive value of cheatgrass and crested wheatgrass on spring ranges of Utah. J. Range Manage. 5:331-37.

Cottam, W.P., and F.R. Evans. 1945. A comparative study of the vegetation of grazed and ungrazed canyons of the Wasatch Range, Utah. Ecology 26:171-81.

Daubenmire, R.F. 1940. Plant succession due to overgrazing in the *Agropyron* bunchgrass prairie of southeastern Washington. Ecology 21:55-64.

Daubenmire, R. 1975. Plant succession on abandoned fields, and fire influences in a steppe area in southeastern Washington. Northwest Sci. 49:36-48.

DeFlon, J.G. 1986. The case for cheat grass. Rangelands 8:14-17.

Eckert, R.E., Jr., J.E. Asher, M.D. Christensen, and R.A. Evans. 1974. Evaluation of the atrazine fallow-technique for weed control and seedling establishment. J. Range Manage. 27:288-92.

Eckert, R.E., Jr., and R.A. Evans. 1967. A chemical-fallow technique for control of downy brome and establishment of perennial grasses on rangeland. J. Range Manage. 20:35-41.

Evans, R.A., and R.E. Eckert, Jr. 1965. Paraquat-surfactant combinations for control of downy brome. Weed Sci. 13:150-51.

Evans, R.A., R.E. Eckert, Jr., and B.L. Kay. 1967. Wheatgrass establishment with paraquat and tillage on downy brome ranges. Weed Sci. 15:50-55.

Evans, R.A., and J.A. Young. 1970. Plant litter and establishment of alien annual weed species in rangeland communities. Weed Sci. 18:697-703.

Evans, R.A., and J.A. Young. 1977. Weed control-revegetation systems for big sagebrush-downy brome rangelands. J. Range Manage. 30:331-36.

Forcella, F., and S.J. Harvey. 1981. New and exotic weeds of Montana. II. Migration and distribution of 100 alien weeds in northwestern USA, 1881-1980. Mont. Weed Survey, Mont. State Univ., Bozeman.

Harris, G.A. 1967. Some competitive relationships between *Agropyron spicatum* and *Bromus tectorum*. Ecol. Monogr. 37:89-111.

Harris, G.A. 1970. Competition for moisture among seedlings of annual and perennial grasses as influenced by root elongation at low temperature. Ecology 51:529-34.

185

Hironaka, M. 1994. Medusahead: Natural successor to the cheatgrass type in the northern Great Basin. *In*: S.B. Monsen and S.G. Kitchen (eds.), Proc. Ecology and management of annual rangelands. USDA For. Ser. Gen. Tech. Rep. INT-GTR-313, 89-91.

Hulbert, L.C. 1955. Ecological studies of *Bromus tectorum* and other annual bromegrasses. Ecol. Monogr. 25:181-213.

Hull, A.C., and W.T. Hansen. 1974. Delayed germination of cheatgrass seed. J. Range Manage. 27:366-68.

Hull, A.C., and J.F. Pechanec. 1947. Cheatgrass—a challenge to range research. J. For. 45:555-64.

Kennedy, A.C. 1994. Biological control of annual grass weeds. *In*: S.B. Monsen and S.G. Kitchen (eds.), Proc. Ecology and management of annual rangelands. USDA For. Ser. Gen. Tech. Rep. INT-GTR-313, 186-89.

Kindschy, R.R. 1994. Pristine vegetation of the Jordan Crater kipukas: 1978-91. *In*: S.B. Monsen and S.G. Kitchen (eds.), Proc. Ecology and management of annual rangelands. USDA For. Ser. Gen. Tech. Rep. INT-GTR-313, 85-88.

Klemmedson, J.O., and J.G. Smith. 1964. Cheatgrass (*Bromus tectorum* L.). Bot. Rev. 30:226-62.

Larson, L.L., and R.L. Sheley. 1994. Ecological relationships between yellow starthistle and cheatgrass. *In*: S.B. Monsen and S.G. Kitchen (eds.), Proc. Ecology and management of annual rangelands. USDA For. Ser. Gen. Tech. Rep. INT-GTR-313, 92-94.

Laycock, W.A. 1994. Implications of grazing vs. no grazing on today's rangelands. *In*: M. Vavra, W.A. Laycock, and R.D. Pieper (eds.), Ecological implications of livestock herbivory in the West. Soc. for Range Manage., Denver, CO, 250-80.

Mack, R.N. 1981. Invasion of *Bromus tectorum* L. into western North America: An ecological chronicle. Agro-Ecosystems 7:145-65.

Mattise, S.N., and G. Scholten. 1994. Mechanical control of undesirable annuals on the Boise Front, Idaho. *In*: S.B. Monsen and S.G. Kitchen (eds.), Proc. Ecology and management of annual rangelands. USDA For. Ser. Gen. Tech. Rep. INT-GTR-313, 190-93.

Megee, C.R. 1938. Wild oats or downy brome. Mich. Agr. Exp. Sta. Occas. Bull. 20(3):153-56.

Melgoza, G., and R.S. Nowak. 1991. Competition between cheatgrass and two native species after fire: implications from observations and measurements of root distribution. J. Range Manage. 44:27-33.

Melgoza, G., R.S. Nowak, and R.J. Tausch. 1990. Soil water exploitation after fire: competition between *Bromus tectorum* (cheatgrass) and two native species. Oecologia 83:7-13.

Mosley, J.C. 1996. Prescribed sheep grazing to suppress cheatgrass: A review. Sheep and Goat Res. J. 12:74-81.

Murray, R.B. 1971. Grazing capacity, sheep gains: cheatgrass, bunchgrass ranges in southern Idaho. J. Range Manage. 24:407-10.

Murray, R.B., H.F. Mayland, and P.J. Van Soest. 1978. Growth and nutritional value to cattle of grasses on cheatgrass range in southern Idaho. USDA For. Ser. Res. Pap. INT-199.

Pellant, M. 1990. The cheatgrass-wildfire cycle—Are there any solutions? *In*: E.D. McArthur, E.M. Romney, S.D. Smith, and P.T. Tueller (eds.), Proc. Symp. on cheatgrass invasion, shrub die-off, and other aspects of shrub biology and management. USDA For. Ser. Gen. Tech. Rep. INT-276, 11-18.

Pellant, M. 1994. History and applications of the Intermountain greenstripping program. *In*: S.B. Monsen and S.G. Kitchen (eds.), Proc. Ecology and management of annual rangelands. USDA For. Ser. Gen. Tech. Rep. INT-GTR-313, 63-68.

Pellant, M., and L. Reichert. 1984. Management and rehabilitation of a burned winterfat community in southwestern Idaho. *In*: A.R. Tiedeman, E.D. McArthur, H.C. Stutz, et al., compilers. Proc. Symp. on the biology of *Atriplex* and related chenopods. USDA For. Ser. Gen. Tech. Rep. INT-172, 281-85.

Peters, E.F., and S.C. Bunting. 1994. Fire conditions pre- and postoccurrence of annual grasses on the Snake River Plain. *In*: S.B. Monsen and S.G. Kitchen (eds.), Proc. Ecology and management of annual rangelands. USDA For. Ser. Gen. Tech. Rep. INT-GTR-313, 31-36.

Piemeisel, R.L. 1951. Causes affecting change and rate of change in a vegetation of annuals in Idaho. Ecology 32:53-72.

Pyke, D.A., and S.J. Novak. 1994. Cheatgrass demography—establishment attributes, recruitment, ecotypes, and genetic variability. *In*: S.B. Monsen and S.G. Kitchen (eds.), Proc. Ecology and management of annual rangelands. USDA For. Ser. Gen. Tech. Rep. INT-GTR-313, 12-21.

Rasmussen, G.A. 1994. Prescribed burning considerations in sagebrush annual grassland communities. *In*: S.B. Monsen and S.G. Kitchen (eds.), Proc. Ecology and management of annual rangelands. USDA For. Ser. Gen. Tech. Rep. INT-GTR-313, 69-70.

Rosentrater, R. 1994. Displacement of rare plants by exotic grasses. *In*: S.B. Monsen and S.G. Kitchen (eds.), Proc. Ecology and management of annual rangelands. USDA For. Ser. Gen. Tech. Rep. INT-GTR-313, 170-75.

Stewart, G., and A.C. Hull. 1949. Cheatgrass (*Bromus tectorum* L.)—an ecologic intruder in southern Idaho. Ecology 30:58-74.

Stubbs, T. 1996. Biological control of annual grass weeds, control options effectiveness. *In*: Proc. of the downy brome symp., NOVA Gas Transmission Ltd., Medicine Hat, Alberta, 46-51.

Svejcar, T., and R. Tausch. 1991. Anaho Island, Nevada: A relict area dominated by annual invader species. Rangelands 13:233-36.

Tausch, R.J., T. Svejcar, and J.W. Burkhardt. 1994. Patterns of annual grass dominance on Anaho Island: Implications for Great Basin vegetation management. *In*: S.B. Monsen and S.G. Kitchen (eds.), Proc. Ecology and management of annual rangelands. USDA For. Ser. Gen. Tech. Rep. INT-GTR-313, 120-25.

Tisdale, E.W., and M. Hironaka. 1981. The sagebrush-grass region: A review of the ecological literature. Idaho For., Wildl. and Range Exp. Sta. Bull. 33.

Vallentine, J.F., and A.R. Stevens. 1994. Use of livestock to control cheatgrass—a review. *In*: S.B. Monsen and S.G. Kitchen (eds.), Proc. Ecology and management of annual rangelands. USDA For. Ser. Gen. Tech. Rep. INT-GTR-313, 202-6.

West, N.E. 1991. Grazing may be unfairly blamed for changes in pristine areas. Utah Sci. 53:50.

Whisenant, S.G. 1990. Changing fire frequencies on Idaho's Snake River Plains: Ecological and management implications. *In*: E.D. McArthur, E.M. Romney, S.D. Smith, and P.T. Tueller (eds.), Proc. Symp. on cheatgrass invasion, shrub die-off, and other aspects of shrub biology and management. USDA For. Ser. Gen. Tech. Rep. INT-276, 4-10.

Whitson, T.D., and D.W. Koch. 1998. Control of downy brome (*Bromus tectorum*) with herbicides and perennial grass competition. Weed Technol. 12 12:391-96.

Wright, H.A., and A.W. Bailey. 1982. Fire ecology: United States and southern Canada. John Wiley and Sons, New York.

Wright, H.A., L.F. Neuenschwander, and C.M. Britton. 1979. The role and use of fire in sagebrush-grass and pinyon-juniper plant communities: A state-of-the-art review. USDA For. Ser. Gen. Tech. Rep. INT-58.

Young, J.A. 1983. Principles of weed control and plant manipulation. *In*: S.B. Monsen and N. Shaw (eds.), Managing intermountain rangelands—Improvement of range and wildlife habitats. USDA For. Ser. Gen. Tech. Rep. INT-157, 6-10.

Young, J.A., R.A. Evans, R.E. Eckert, Jr., and B.L. Kay. 1987. Cheatgrass. Rangelands 9:266-70.

Young, R.P. 1983. Fire as a vegetation management tool in rangelands of the Intermountain Region. *In*: S.B. Monsen and N. Shaw (eds.), Managing intermountain rangelands—Improvement of range and wildlife habitats. USDA For. Ser. Gen. Tech. Rep. INT-157, 18-31.

Common Crupina

Donald C. Thill, Cindy T. Roché, and David L. Zamora

Common crupina (*Crupina vulgaris* var. *typica* and var. *brachypappa*) is a non-indigenous species in North America that has invaded more than 63,000 acres (25,506 ha) of disturbed land in California, Idaho, Oregon, and Washington. It forms dense populations that dominate forages on disrupted or depleted rangelands, and it reduces the biodiversity of many native ecosystems where it occurs in the Pacific Northwest and California (Prather et al. 1991). Common crupina is designated "noxious" by federal and state laws and must be controlled.

The biology, distribution, and control of common crupina have been studied extensively as part of a 10-year pilot eradication project (1979 to 1989) conducted by researchers from the University of Idaho and the USDA Animal Plant Health Inspection Service (APHIS). French researchers also have studied the species extensively. Biology studies focused mainly on seed production, dissemination, germination, and viability, because common crupina is an annual and only reproduces by seed. Survey methods were developed to locate new infestations, identify the boundaries of known infestations, and to appraise the effectiveness of various eradication tactics. Methods were developed to eradicate common crupina from all types of land use where it was found (Zamora et al. 1989b). Sites included steep canyons, riparian areas, and infestations adjacent to homes. Eradication tactics included aerial and ground application of herbicides, hand-weeding, quarantine, and revegetation. After nearly 10 years of research, Thill and Zamora concluded in 1989 that eradication of common crupina was biologically feasible.

In 1990 through 1992, efforts were undertaken by Idaho, Oregon, and Washington Departments of Agriculture, the University of Idaho, and APHIS to eradicate common crupina infestations in the Pacific Northwest. Eradication projects were initiated in Oregon and Washington, but not Idaho. The Idaho Department of Agriculture conducted additional detection surveys during this time. At least two new and discrete infestations were identified in Idaho during the 1991-92 surveys. Efforts in Oregon and Washington were unsuccessful because funding for the projects was limited. In Washington, the methods for eradication were limited because much of the infestation was on lands administered by the U.S. Forest Service in rugged,

189

remote, wilderness areas along Lake Chelan. Earlier eradication efforts in California also were not successful.

APHIS prepared an extensive environmental assessment (EA) for eradication of common crupina in Idaho (Anonymous 1991a). Several public meetings were held to discuss the eradication project and the EA. Ultimately, the project was abandoned by APHIS because of concerns expressed by some stakeholders regarding the use of herbicides (especially aerial application of picloram), identification of a large, previously unknown infestation along the Salmon River, and lack of funding to complete the project successfully. Research on biological control of common crupina began in 1991, and selecting and screening agents for release continues in 1998.

Identification

Common crupina, a herbaceous annual, is a member of the sunflower (Asteraceae) family and the thistle (Cynareae) tribe. This tribe contains several other troublesome rangeland weeds including yellow starthistle (*Centaurea solstitialis*), diffuse knapweed (*C. diffusa*), and spotted knapweed (*C. maculosa*). Couderc-LeVaillant (1984) provides a detailed description of the taxonomy of the genus *Crupina*.

Common crupina var. *typica* seedlings appear aboveground as two oblong, sessile cotyledons $1/2$ to 1 inch (1.3 to 2.5 cm) in length (Prather et al. 1991). The prominent midvein of the cotyledons is usually purple or red, especially when plants are grown at cool temperatures. The midvein and the large, fleshy cotyledons distinguish common crupina from associated species. Rosette leaves develop above the cotyledons, and progress from entire (smooth margins) to toothed to lobed to finely dissected as the plant grows. Rosette leaves can be up to 3 inches (7.6 cm) long. The finely divided, lace-like leaflets are produced alternately along the elongating stem. Older leaves develop short, stiff trichomes (epidermal "hairs") that are prickly to the touch.

Each plant usually has one main flowering stem, 6 inches to 3 feet (0.2 to 0.9 m) tall, that can branch near the top into five to 15 branches under good growing conditions. One or more flower heads appear at the end of each branch. Under poor growing conditions, or where common crupina plants are very crowded, each plant produces only one to three branches with flower heads. Flowers are rose to purple in flower heads that are $1/2$ inch (1.3 cm) long. Seeds (achenes) are reported to be $1/8$ to $1/4$ inch (0.3 to 06 cm) long, cone-shaped, and tapered to a blunt point (Prather et al. 1991). A dense circle of $1/4$ inch (0.6 cm)-long barbed hairs (pappus) surrounds the wide end of the seed. Seeds are black or silvery beige and are covered with fine hairs.

The habit of common crupina var. *brachypappa* is identical to var. *typica* (Couderc-LeVaillant 1984). However, the number of cauline leaves is large, about 40, and the internodes shorter in var. *brachypappa* than in var. *typica*. Seeds from four North American populations (Sonoma County, California; Harpster Grade-Idaho County, Idaho; Umatilla County, Oregon; and Chelan County, Washington) of common crupina were collected and examined to determine their taxonomic traits (Couderc-LeVaillant and Roché 1993). The seeds of the two varieties were easily distinguished. In common crupina var. *typica*, the length of the seeds varies from $^5/_{32}$ to $^7/_{32}$ inch (0.4 to 0.5 cm) and the length of the longest pappus bristle ranged from $^9/_{32}$ to $^{11}/_{32}$ inch (0.7 to 0.9 cm). This varies somewhat from traits reported by Prather et al. (1991). In common crupina var. *brachypappa*, seeds were $^1/_8$ to $^7/_{32}$ inch (0.3 to 0.5 cm) in length and the longest pappus bristle was $^5/_{32}$ to $^1/_4$ inch (0.4 to 0.6 cm). According to Couderc-LeVaillant and Roché (1993), pappus length provided a clear differentiation, while mean seed length also distinguished the two varieties, even though the range of values overlapped. The size differences were not influenced by biotype or environmental conditions. Seeds of common crupina collected in Sonoma County, California, Idaho County, Idaho, and Umatilla County, Oregon, were identified as var. *typica* using these traits, while seeds from Washington were identified as var. *brachypappa*.

Origin, History, and Distribution

Common crupina is native to the Mediterranean region of Europe where it is widely distributed (Marcet 1953, U.S. Dept. of Agric. 1973, Flora Europaea 1976, Couderc-LeVaillant 1984). It is found everywhere in central and southern Europe, including France, Spain, Italy, Romania, Switzerland, the former Yugoslavia, Greece, Bulgaria, Luxembourg, Russia, Turkey, Iran, Morocco, and Algeria (Couderc-LeVaillant 1984), where herbarium samples dated back to 1837 (Roché 1990, personal communication). Common crupina has been reported to be an economic pest in rangelands of southern Russia (U.S. Dept. of Agric. 1973).

How common crupina was first introduced into the United States is unknown. It was discovered in 1968 in Idaho (Stickney 1972), California in 1975 (Davis and Sherman 1991), Washington in 1984, Oregon in 1985, and additional, discrete locations in California in 1989 and 1991 (Couderc-LeVaillant and Roché 1993). Among the five varieties of common crupina recognized in its native range (Couderc-LeVaillant 1984), two varieties, *typica* and *brachypappa*, occur in the U.S. (Couderc-LeVaillant and Roché 1993). *Crupina vulgaris* var. *typica* is found in California, Idaho, and Oregon. The Washington infestation and one infestation in California were identified as *C. vulgaris* var. b*rachypappa* (Couderc-LeVaillant and Roché 1993). Random amplified polymorphic DNA (RAPD) procedures were used to confirm that common crupina in Idaho and Oregon were the same varieties (Personal communication 1994, Alfonso Susanna, Senior Researcher, Laboratory of Molecular Systematics, Botanical Institute of Barcelona, Spain). The occurrence of two varieties of common crupina in the Pacific Northwest demonstrate that multiple introductions occurred, with at least two distinct origins (Couderc-LeVaillant and Roché 1993). Hybridization between common crupina varieties is possible. However, in Europe, populations of common crupina usually are isolated and homogeneous (Couderc-LeVaillant and Roché 1993).

191

Common crupina occurs in a wide range of habitats in the Pacific Northwest. The primary habitat is southern slopes on steep canyon grasslands (Prather et al. 1991). Open forested areas also can support this weed. Common crupina has been reported rarely in annually tilled cropland, but occurs along field edges and in improved pastures, hayfields, grass seed fields, and Conservation Reserve Program plantings. The weed frequently infests gravel pits, roadsides, railroad embankments, and other rights-of-way (Prather et al. 1991).

Stickney (1972) reported that common crupina appeared to dominate a 45-acre (18-ha) area of rangeland along State Highway 13, 6 miles (9.6 km) east by northeast of Grangeville, Idaho. A University of Idaho survey in 1979 indicated 8,000 acres (3,238 ha) were infested within a 40,000-acre (16,194 ha) area that included parts of Clearwater, Idaho, and Lewis counties (Baysinger and Lee 1980). 1980 and 1981 surveys of the three counties indicated that 22,900 acres (9,271 ha) were infested with common crupina (Miller and Thill 1983). Results of a 1990 survey showed 49,500 acres (20,040) in Idaho County and 5,500 acres in Clearwater and Lewis counties were infested with common crupina (Idaho Dept. Agric. 1990). About 80% of the infestation was contained within the boundaries of the Nez Perce Indian Reservation (Anonymous 1991a). There were approximately 234 discrete, infested sites ranging in size from 10 to 4,800 acres (4 to 1,943 ha). Infested sites were equally divided between public and private lands. The 1991 EA (Anonymous 1991a) contained maps showing infested areas and a table listing legal description, county, land status (private or public), watershed, acres, and soil type. In 1992, about 100 acres (40 ha) of common crupina-infested rangeland were reported in The Nature Conservancy Garden Creek Preserve in Nez Perce County, Idaho (Personal communication 1992, Janice Hill, Botanist, Garden Creek Preserve, TNC, Deary, ID).

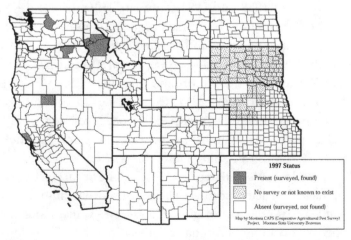

Distribution of common crupina by county in the western United States. Data were collected by surveying various weed authorities in each state.

In Washington State, about 480 acres (194 ha) along the north shore of Lake Chelan between Prince Creek and Hunts Bluff were infested with common crupina (Anonymous 1991b). About 210 acres (85 ha) were found on private land, 255 acres (103 ha) were within the Lake Chelan-Sawtooth Wilderness Area, and 15 acres (6 ha) were on national forest land outside of the wilderness. Originally, common crupina was reported to infest about 500 acres (202 ha) of rangeland along Dry Creek in Umatilla County, Oregon (Thill 1985, personal communication). In 1990, 4,000 to 8,000 acres (1,619 to 3,239 ha) of rangeland in Umatilla County were reported to be infested with common crupina (Anonymous 1990). Detailed maps of the Oregon and Washington infestations are contained in the Washington and Oregon EAs (Anonymous 1990, 1991b). The Santa Rosa, California, infestation (Davis and Sherman 1991) of common crupina was reported to be about 300 acres (121 ha) scattered over 1,200 acres (486 ha) of rangeland in and around Annadel State Park (Zamora 1990, personal communication). Other infestations in Sonoma and Modoc Counties were reported in 1991 (Couderc-LeVaillant and Roché 1993). Apparently, an infestation of common crupina near Pincher Creek in British Columbia, Canada, was eradicated sometime between 1980 and 1986 (Zamora, personal communication from R. Cranston, British Columbia Ministry of Agric. and Food, Surrey, B.C., 1987).

Potential for Invasion

Common crupina has invaded more than 63,000 acres (25,506 ha) of disturbed land in California, Idaho, Oregon, and Washington. Primary habitats include steep canyon grasslands, sparsely forested areas, pastures, hay and grass seed fields, waste areas, and rights-of-way. Although not considered weedy in its native habitat in Europe (Roché 1990, personal communication), common crupina is considered weedy in the western United States (Thill et al. 1985). Common crupina appears to be adapted to a wide range of edaphic and climatic conditions (Patterson and Mortenson 1985) and is capable of establishing solid stands (Lee et al. 1980, Belles et al. 1981). For example, the South Atlantic states, parts of the Midwest, the intermountain region, and the Pacific Coast States are agroclimatic analogues of native populations of common crupina in the eastern hemisphere and are susceptible to invasion (Patterson and Mortensen 1985).

The weed infests sites where downy brome (*Bromus tectorum*), wheatgrass (*Thinopyrum* sp.), fescue (*Festuca* sp.), lupine (*Lupinus* sp.), and arrowleaf balsamroot (*Balsamorhiza sagittata*) occur. Ponderosa pine (*Pinus ponderosa*) and Douglas-fir (*Pseudotsuga menziesii*) are associated with common crupina, as are oceanspray (*Holodiscus* sp.), smooth sumac (*Rhus glabra*), and poison ivy (*Toxicodendron radicans*). Common crupina likely will continue to invade previously noninfested, disturbed areas throughout the western United States, because the plant is adapted to a wide range of habitats, and seed can be moved naturally and through human activities.

193

Impacts

Considerable benefits have accrued from the presence of many non-indigenous species in the United States. Other introduced species have caused considerable harm (U.S. Congress 1993). The Office of Technology Assessment estimated that economic loss from the introduction of 79 harmful non-indigenous species between 1906 and 1991 was $97 billion (1991 dollars), which likely is a fraction of the total cost for the period (U.S. Congress 1993). Only 14% of the non-indigenous species were included in the estimate, because economic loss information was not available for the other species. Fifteen species of plants included in the estimate caused $603 million in economic loss.

APHIS prepared an economic analysis of potential impacts for known infestations of common crupina in 1980 (Wallace 1980). Most of the winter rangelands in north central Idaho, northeastern Oregon, and southeastern Washington were considered at risk for invasion by common crupina. If all these high risk rangelands became infested with common crupina within 25 years (by 2005), the net present value to society of eradicating the weed was estimated to be $2.6 million. Only the economic effect of common crupina infestations on livestock production was considered in the estimate. Effects on native plant communities, rare or endangered species, wildlife habitat, recreation, crops, and on soil and water quality were not included in the 1980 APHIS economic analysis.

Potential impacts of common crupina also were included in the EAs for the eradication of common crupina in Idaho, Oregon, and Washington (Anonymous 1990, 1991a, 1991b). Common crupina eradication was not attempted in Idaho and was not fully implemented in Oregon and Washington. No documentation exists for eradication of any infestation of common crupina. Thus, the "no action" portions of the EAs were used here to assess potential impacts of common crupina.

Common crupina poses a threat to rangelands, native plant populations, export-marketed agricultural commodities, and wildlife (Anonymous 1991a). Environmental and economic impacts will increase as common crupina continues to spread in California, Idaho, Oregon, and Washington. Without control, common crupina is expected to invade 230,400 acres (93,279 ha) of suitable habitat in the Clearwater Basin by 2001. Common crupina is expected to replace native plant communities and form pure stands of up to 4,000,000 unpalatable plants per acre (9,900,000 per ha) on these 230,400 acres (93,279 ha). This will reduce biodiversity within the infested area. This is contrary to previous observations in which common crupina was grazed by sheep and goats in Europe (Roché 1990, personal communication). At the estimated spread rates, valuable agricultural crops will be contaminated. Without effective commercial monitoring or plant quarantine, seeds will be shipped in hay and other forage to all western states. Irretrievable resource losses include grazing for livestock and winter range for deer and elk, which reduce farm income and income associated with hunting, respectively. The loss in forage is estimated to exceed $19 million in 1990 discounted dollars.

In Washington, uncontrolled infestations of common crupina may adversely affect water quality in Lake Chelan through increased soil erosion, because the weed is less effective in stabilizing soils compared with the perennial bunchgrasses that it displaces (Anonymous 1991b). About one-half of the common crupina infestation is within wilderness boundaries near Lake Chelan. Land disturbance within this wilderness area has been caused mostly by fire, grazing by horses, and soil erosion associated with spring snowmelt and summer storms. As common crupina spreads, it will change the plant composition of this wilderness ecosystem and reduce native plant and wildlife habitat. Forage productivity on private and public lands will continue to be reduced by common crupina. Forage losses were estimated to be $2.25 per acre ($5.56 per ha) (Anonymous 1991b). Uncontrolled common crupina infestations in Oregon are expected to spread, resulting in reduced forage production and degraded habitat for both livestock and some wildlife species (Anonymous 1990). This will cause adverse affects on livestock and wildlife. Effects from common crupina infestation in California should be similar to those reported for Idaho, Oregon, and Washington.

Biology and Ecology

In France, common crupina is found along coastal regions and in valleys up to 3,600 feet elevation (Couderc-LeVaillant 1984). A reconnaissance investigation of common crupina in Greece, Italy, and France was conducted in 1990 (Roché 1990, personal communication). Populations of common crupina were found in all three countries where it was considered nonweedy. Common crupina occupied dry, bunchgrass grasslands where it grew nonaggressively in openings between the grass clumps. It was obviously palatable to sheep and goats, and probably palatable to cattle. Common crupina populations appeared to decrease over time with grazing, being replaced by species more tolerant of grazing. No common crupina was found in croplands, recently abandoned fields, vacant lots, trampled areas around livestock watering sites, or in heavily grazed pastures.

Common crupina is in the same botanical tribe (Cynareae) as the knapweeds, and is closely related to *Centaurea* species; its former name was *Centaurea crupina* (Couderc-LeVaillant 1984). Common crupina is a facultative winter annual with many of the ecological habits of other winter annual *Centaurea* species (Prather et al. 1991). For example, seeds usually germinate in late summer or fall after rains begin (Thill et al. 1987). However, spring germination has been observed often. Seeds germinate best at day/night temperatures ranging from 77° to 84° F (25° to 29° C) and 39° to 59° F (4° to 15° C), respectively (Patterson and Mortensen 1985), when soil moisture is ample (Kambitsch 1983). Germination has been observed at constant temperatures as low as 37° F (3° C) (Roché et al. 1997b).

In Idaho studies, seedling densities of common crupina ranged from 15 to 30 per square foot (161 to 323 per square meter) with low mortality from seedling to adult stage (Zamora 1988). Plants overwinter as rosettes, usually bolt in April and begin flowering during May and June (Zamora 1988, Roché et al. 1997a). Common crupina planted in February developed 35- and 22-inch (89- and 56-cm)-long roots by May when grown in full sun and 10% to 20% full sun, respectively (Roché 1996). It is facultatively self-fertile with no apparent reduction in fecundity under conditions of self-pollination (Roché 1996).

Flower initiation has been reported to respond to lengthening days in spring, although flowering occurred in photoperiods as short as eight hours (Patterson and Mortensen 1985). Common crupina var. *typica* and *brachypappa* planted in August through October flowered synchronously in May (Roché 1996, Roché et al. 1997a). When not vernalized, var. *brachypappa* required more time to flower than var. *typica*. Most plants of var. *brachypappa* planted in June and all plants planted in July remained vegetative or died during the summer. Flowering is indeterminate and continues until frost, as long as soil moisture is available. Rangeland plants produced one or two seeds per flower head and two to 23 seeds per plant in summer (Zamora 1988). Average 100-seed weight was 0.13 and 0.07 ounces (3.6 and 1.9 g) for common crupina var. *typica* and *brachypappa*, respectively (Roché et al 1997a). Seeds produced from rangeland-grown plants were about 96% viable. Approximately 85% of these seeds germinated that autumn (Zamora 1988). Most seeds fall from the plant at maturity, but a few remain on the plant through the winter. Common crupina seeds exhibit seasonal dormancy (Thill et al. 1985), which includes an after-ripening requirement that prevents summer germination (Roché et al. 1997a). Also, common crupina seeds respond to priming; following wetting and drying they germinate faster than unprimed seeds. Seeds that do not germinate during the first year survive in the soil for as long as 32 months (Thill et al. 1985).

Wind may spread common crupina seeds up to 5 feet (1.5 m) from the source plant (Prather et al. 1991). Rodents have carried seeds up to 50 feet (15 m) and cattle and deer have carried seeds on hooves and hair at least 300 feet (91 m) from the source plant. Streams and rivers can disseminate floating common crupina seeds (Thill and Zamora 1986, personal communication). Common crupina seeds have survived passage through the digestive system of cows, horses, upland game birds, and deer, but not sheep (Thill et al. 1986). Livestock may transport seeds for great distances before excreting them. Recreational vehicles, such as motorcycles and all-terrain vehicles, also can spread seeds over long distances (Prather et al. 1991). Viable seeds of common crupina can last 25 to 32 months in soil in the field (Thill et al. 1985, Zamora and Thill 1989). In fact, it has been estimated that it would take more than 12 years to deplete viable common crupina seeds from the soil at an annual control level of 97% (Zamora et al. 1989a). Thus, transporting soil from infested to noninfested areas may spread this species.

Common crupina grows in sandy or loamy soils that are well drained and slightly acidic (Lee et al. 1980). The precipitation where this weed occurs is 15 to 30 inches (38 to 76 cm) per year, and the mean annual temperature is 46° to 54° F (8° to 12° C). Common crupina has been found at elevations between 1,000 and 3,200 feet (305 to 975 m) in the Pacific Northwest (Prather et al. 1991).

Management

Various management strategies for common crupina are discussed in the Idaho, Oregon, and Washington EAs (Anonymous 1990, 1991a, and 1991b), and by Burnworth (1991), Zamora et al. (1989b), Zamora and Thill (1988), Miller (1982), and Belles et al. (1981). Control methods include manual, mechanical, biological, thermal, cultural, and chemical methods. Successful management of common crupina requires site-specific integration of several control methods. The best combination of tactics will depend on such factors as the level of infestation, degree of control desired, land use and ownership, topography, associated vegetation, use of adjacent property, economics, and local regulations governing the use of certain tactics.

• *Preventative Techniques*

Preventing immigration of common crupina seed into uninfested sites is the most important weed management strategy. Immigration is prevented by knowing how common crupina seed can move from site to site. For example, it is suspected that a rancher in Idaho moved livestock from a site infested with common crupina to Umatilla County, Oregon. Seed may have been attached to vehicles or livestock, contained in contaminated hay or bedding material, or excreted by the livestock. Dispersal within an infested area to adjacent noninfested areas can be caused by human activities (farming and ranching operations and recreational uses) and by livestock and wildlife. New infestations often are found along trails or primitive roads used by humans and domestic and wild animals. Long-distance spread is most always associated with human activities. Although unpopular, quarantine of plants and animals, plant and animal products, and any other article or material associated with the infestation can effectively prevent spread.

• *Manual and Mechanical*

Hand-pulling, hoeing, or other tillage is frequently the best treatment in and adjacent to home sites, gardens, urban areas, and some herbicide-sensitive crops, or where infestations consist of only a few plants and can be inspected frequently (Prather et al. 1991). Infested sites should be inspected every two to four weeks each spring and summer to find and remove all common crupina plants before they flower. All plants must be controlled to prevent seed production. Sites must be reinspected for at least three years following the last year that seed is produced, to control plants arising from seed already in the soil.

• *Biological*

No insects, herbivores, or special pathogens were available for release in 1998. However, biological control of common crupina is being investigated at the USDA European Biological Control Laboratory in Montpellier, France. Four potential agents had been selected for study: a moth, a mite, a weevil (Sobhian et al. 1996), and a fungus, all of which attack common crupina seedlings. Any effective biological control agent identified should be tested on both varieties of common crupina that infest the Pacific Northwest and California.

• *Cultural*

Competitive grasses can suppress common crupina and are readily available for sites where planting is possible (Prather et al. 1991, Lass et al. 1998). Competitive grasses effectively resist invasion of other weeds in addition to common crupina, and they are themselves productive. The importance of healthy communities cannot be overemphasized. The best grasses to use are those best adapted to the target site. Obviously, they must be appropriate for the land's expected use. To determine which grasses are most appropriate to a site, contact the university Extension office or the Natural Resources Conservation Service office in your county.

Unfortunately, little information is available on using competitive grasses to manage common crupina infestations. However, information is available for yellow starthistle, which infests most sites where common crupina is found (Lass et al. 1998). In general, where yellow starthistle is well adapted, intermediate wheatgrass (*T. intermedium* subsp. *intermedium*), tall wheatgrass (*T. ponticum*), pubescent wheatgrass (*T. intermedium* subsp. *Barbulatum*), tall oatgrass (*Arrhenatherum elatius*), and bunch-type fescue (*Festuca* spp.) are adapted.

For sites that have shallow soils or are inaccessible for renovation operations, such as steep canyon lands, pasture yield should not be the primary expectation. Instead, ecological stability should be the goal in order to prevent weed disruption. In such circumstances, a bunch-type fine fescue, such as sheep fescue (*F. ovina* var. *ovina*), hard fescue (*F. ovina* var. *duriuscula*), or Idaho fescue (*F. idahoensis*) should be considered. These grasses develop extensive root systems that prevent roots of seedling yellow starthistle or common crupina from establishing. Although the grasses are slow to establish, they can also persist under semiarid conditions and provide significant ecological stabilization.

For sites with moderate soil depth and some potential productivity as rangeland, and on land that is accessible with tractors and tillage equipment, grasses of higher potential yield should be considered. These may include sod-forming species such as intermediate or pubescent wheatgrass. These grasses are able to spread within the stand using rhizomes, thus successfully competing against all annual grasses and weedy forbs such as yellow starthistle or common crupina.

Where soil productivity is high and precipitation is generally above 16 inches (41 cm) per year, or moisture is available from subsoil flow such as at the base of steep slopes, tall oatgrass, tall wheatgrass, and streambank wheatgrass (*Agropyron riparium*) may be suitable. These sites are typically potential cropland fields that would produce marginal to fair crop yields, but would be excellent pastures. Competition among these grasses and yellow starthistle appears to be based on limiting light to seedling yellow starthistle plants. Excessive grazing does not allows sufficient time for grasses to regrow and shade seedling yellow starthistle plants in the fall and spring, and thus reduces the competitive effects of these grasses. The same is likely to be true for common crupina seedlings.

Grasses adapted to pastures as high-yielding forage, such as smooth brome (*B. inermis*), meadow brome (*B. biebersteinii*), timothy (*Phleum pratense*), orchardgrass (*Dactylis glomerata*), redtop (*Agrostis alba*), creeping red (*F. rubra*) and tall fescues (*F. arundineae*), and bluegrass, will establish on a yellow starthistle and common crupina-infested site, but are not very competitive in a typical infested site.

Burnworth (1991) prepared a comprehensive revegetation management plan for the proposed common crupina project in Idaho. She concluded that reseeding may not be required following herbicide treatment to rehabilitate some plant communities. She did state that reseeding practices in certain infested sites would be beneficial to restore the health of the rangeland plant community. This would improve the quality and productivity of livestock and wildlife forage. Also, grazing practices would need to change to prevent reinvasion by common crupina or other noxious weeds.

• *Thermal*
Prescribed burning has been used to destroy common crupina plants invading specific sites along Lake Chelan (Anonymous 1991a). However, the success of this treatment has not been evaluated long-term. It reduced seedling numbers in the season following the burn.

• *Herbicides*
Control of common crupina in most infested sites has depended mainly on use of herbicides, because many infestations are in rugged, semi-remote areas (Zamora et al. 1989b, Prather et al. 1991). Effective herbicides for control of common crupina include clopyralid, 2,4-D, dicamba, and picloram. Always read product labels to verify use is legal, and to comply with safety requirements.

A combination of biological control, reduced grazing pressure, selective use of some herbicides, and revegetation is likely to provide the best long-term control of common crupina. Any single management practice by itself will be less effective or fail. Prevention is always the first line of defense. If you do not have the problem, avoid it by using land management practices that prevent the introduction of common crupina or other noxious weeds. Eradication is an alternative that should be considered, especially with small, isolated infestations.

Literature Cited

Anonymous. 1990. Environmental assessment for eradication of common crupina in Umatilla County, Oregon. USDA-APHIS, Portland, OR.

Anonymous. 1991a. Eradication of common crupina in Idaho. Environmental Assessment. USDA-APHIS, Hyattsville, MD.

Anonymous. 1991b. Eradication of common crupina in Washington. Environmental Assessment. USDA Forest Service, Wenatchee National Forest, Chelan Ranger District.

Baysinger, O.K., and G.A. Lee. 1980. 1979 survey of exotic noxious weeds in Idaho. Univ. of Idaho Misc. Ser. No. 57.

Belles, W.S., D. W. Wattenbarger, and G.A. Lee. 1981. Chemical control of Crupina vulgaris, a new range weed in Idaho and the United States. J. Range Mange. 34:468-70.

Burnworth, R.S. 1991. Revegetation management plan for common crupina eradication project Idaho. USDA-SCS, Moscow, ID.

Couderc-LeVaillant, M. 1984. L'amphiploidie dans le genre Crupina DC. Essai de systematique synthetique. Thése Docteur en Sciences, Univ. Paris-Sud, Orsay.

Couderc-LeVaillant, M., and C.T. Roché. 1993. Evidence of multiple introduction of *Crupina vulgaris* in infestations in the western United States. Madrono 40:63-65.

Davis, L.H., and R.J. Sherman. 1991. *Crupina vulgaris* Cass. (Asteraceae: Cynareae), established in Sonoma County, California, at Annadel State Park. Madrono 38:296.

Flora Europaea. 1976. Cambridge Univ. Press, London, England. 4:301.

Idaho Dept. of Agric. 1990. Common crupina survey. Boise, ID.

Kambitsch, D.L. 1983. The longevity, germination and dissemination of common crupina (*Crupina vulgaris* Cass.) achenes. M.S. Thesis, Univ. of Idaho, Moscow, ID.

Lass, L.W., J.P. McCaffrey, D.C. Thill, and R.H. Callihan. 1998. Yellow starthistle biology and management in pastures and rangeland. Bull. 805. Univ. of Idaho, Moscow, ID.

Lee, G.A., D.W. Wattenbarger, T.L. Miller, and W.J. Schumacher. 1980. Common crupina. Univ. of Idaho Curr. Info. Ser. No. 542.

Marcet, A.F. 1953. Flora Montserratina. Bol. Soc. Espanols Hist. Nat. 47:307-33.

Miller, T.L. 1982. Biology and control of common crupina (*Crupina vulgaris* Cass.). M.S. Thesis, Univ. of Idaho, Moscow, ID.

Miller, T.L., and D.C. Thill. 1983. Today's Weed: Common crupina. Weeds Today. 14(3):10-11.

Patterson, D.T., and D.A. Mortensen. 1985. Effects of temperature and photoperiod on common crupina (*Crupina vulgaris*). Weed Sci. 33:333-39.

Prather, T.S., R.H. Callihan, and D.C. Thill. 1991. Common crupina biology, management, and eradication. Univ. of Idaho Curr. Info. Ser. No. 880.

Roché, C.T. 1996. Development biology in common crupina (*Crupina vulgaris* Pers.) and yellow starthistle (*Centaurea solstitialis* L.). Ph.D. Dissertation, Univ. of Idaho, Moscow, ID.

Roché, C.T., D.C. Thill, and B. Shafii. 1997a. Prediction of flowering in common crupina (*Crupina vulgaris*). Weed Sci. 45:519-28.

Roché, C.T., D.C. Thill, and B. Shafii. 1997b. Estimation of base and optimum temperature for seed germination in common crupina (*Crupina vulgaris*). Weed Sci. 45:529-33.

Sobhian, R, L. Knutson, and J.S. Rodier. 1996. Biology and host specificity notes on *Styphlus penicilus* Gyllenhal (Coleoptera: Curculionidea), examined as a biological control agent for *Crupina vulgaris* in the United States with remarks on its host plant. Proc. Entomol. Soc. Wash. 98(2):317-23.

Stickney, P.F. 1972. *Crupina vulgaris* (Compositae: Cynareae), new to Idaho and North America. Madrono 21:402.

Thill, D.C., D.L. Zamora, and D.L. Kambitsch. 1985. Germination and viability of common crupina (*Crupina vulgaris*) achenes buried in the field. Weed Sci. 33:344-48.

Thill, D.C., D.L. Zamora, and D.L. Kambitsch. 1986. The germination and viability of excreted common crupina (*Crupina vulgaris*) achenes. Weed Sci. 34:237-41.

Thill, D.C., D.L. Zamora, and D. Kidder. 1987. Common crupina identification and biology. Univ. of Idaho Curr. Info. Ser. No. 542 (rev.).

U.S. Congress, Office of Technology Assessment. 1993. Harmful non-indigenous species in the United States. OTA-F-565. Washington, D.C.: U.S. Government Printing Office, 51-76.

U.S. Dept. Agric. 1973. Economically important foreign weeds. Potential problems in the United States. U.S. Dept. Agric. Handbook No. 498.

Wallace, W.S. 1980. Economic analysis of *Crupina vulgaris* eradication. USDA-APHIS report. Hyattsville, MD.

Zamora, D.L. 1988. The reproductive biology and seed longevity of *Crupina vulgaris* as it pertains to an eradication strategy for plant invasions. Ph.D. Dissertation. Univ. of Idaho, Moscow, ID.

Zamora, D.L., and D.C. Thill. 1988. The compatibility of rhodamine B dye with herbicides for common crupina (*Crupina vulgaris*) control. Weed Technol. 2:16-19.

Zamora, D.L., and D.C. Thill. 1989. Seed bank longevity of common crupina (*Crupina vulgaris*) in natural populations. Weed Technol. 3:166-69.

Zamora, D.L., D.C. Thill, and R.E. Eplee. 1989a. An eradication plan for plant invasions. Weed Technol. 3:2-12.

Zamora, D.L., D.C. Thill, and R.E. Eplee. 1989b. Eradication manual for common crupina (*Crupina vulgaris* Cass.). Univ. of Idaho Bull. No. 701.

Dalmatian and Yellow Toadflax

Sherry Lajeunesse

Dalmatian toadflax (*Linaria dalmatica*) and yellow toadflax (*Linaria vulgaris*) are aggressive, highly competitive noxious weeds that were introduced into North America as ornamentals, fabric dyes, and folk remedies. They soon escaped from gardens to infest farmland, pastures, and rangeland across the United States and Canada, displacing native plant species, livestock forage, and wildlife habitat. Unfortunately, toadflax continues to be sold commercially as an ornamental plant, and is recommended for xeriscape (low-water-use) landscape plantings. Toadflax thrives in a wide range of habitat types and climatic zones and is expensive and extremely difficult to manage, partly because of localized biotypes that respond differently to herbicides and other management tactics.

Much of the information in this chapter applies to both Dalmatian and yellow toadflax; however, the differences between the species can be important in management efforts, and should be considered when designing a management program.

Identification

Three species of toadflax are considered noxious weeds in the United States and Canada: broad-leaved Dalmatian toadflax, narrow-leaved Dalmatian toadflax (*Linaria genistifolia*), and yellow toadflax, all in the plant family Scrophulariaceae (Hartl 1974, Davies 1978). These toadflax species are herbaceous perennials with stems that are robust and somewhat woody at the base and smooth toward the top.

Stems of broad-leaved Dalmatian toadflax grow 2 to 3 feet (0.6 to 0.9 m) or taller. Both leaves and stems are waxy with a whitish or bluish cast. The leaves are usually heart-shaped, but can vary from broad to lanceolate shape, especially on lower portions of the plant, and the bases tend to wrap around the stem. Flowers grow at the bases of upper leaves and are bright yellow with an orange center and a spur on the end that is approximately as long as the rest of the flower combined (Cronquist et al. 1984). The mature narrow-leaved Dalmatian toadflax plant is quite similar in appearance to broad-leaved Dalmatian toadflax, except for the somewhat narrower leaves and smaller flowers (Cronquist et al. 1984). Management information for broad-leaved Dalmatian toadflax will apply to both Dalmatian species in this text.

Yellow toadflax stems are usually 1 to 3 feet (0.3 to 0.9 m) high, are somewhat woody at the base and smooth toward the top, sparingly branched. Leaves are narrow, linear, somewhat pointed at both ends, and 1 to 2 inches (2.5 to 5 cm) long. Stems and leaves are pale green. Flowers, which grow at the bases of upper leaves, are bright yellow with an orange center, with a spur that is approximately as long as the rest of the flower combined (Cronquist et al. 1984). Flowers occur at first in clusters near the ends of the stems, becoming more widely spaced along the stem as the season progresses.

Hybrids between yellow toadflax and Dalmatian toadflax can be produced in the laboratory, so natural occurrence of this hybrid in the field should be considered (Saner 1994).

Origin, History, and Distribution

The Dalmatian toadflaxes are native to the Mediterranean region. In Europe the broad-leaved species has been cultivated as an ornamental for nearly four centuries, and was brought to the west coast of North America as an ornamental about 1874 (Alex 1962). Currently, heaviest North American infestations of this escaped ornamental are found in the northwestern states, British Columbia, and Alberta, and in widely scattered locations in other regions (Lajeunesse et al. 1993). In North America, broad-leaved Dalmatian toadflax is the most widely distributed of the two Dalmatian species. However, in the native Eurasian habitat narrow-leaved Dalmatian toadflax is the more widely distributed of the two, possibly indicating its potential to become widely distributed in North America as well (Smith 1959, Harris 1988). Narrow-leaved Dalmatian toadflax is currently found in several locations in western Oregon, northwestern Washington, and occasionally in British Columbia.

Yellow toadflax originated in south-central Eurasia (Meusel et al. 1978) where it has been used for centuries as a folk remedy and fabric dye (Mitich 1993). It was introduced into New England in the late 1600s as an ornamental and folk remedy (Fernald 1905, Rousseau 1968) and continues to be sold today in nurseries and seed catalogs as "Butter and Eggs," "Jacob's Ladder," or "Wild Snapdragon"; infestations still originate from these escaped plantings. By the 1950s yellow toadflax had spread throughout North America (Saner 1991). Distribution is most common throughout the northeastern United States and southeastern Canada, and localized in other parts of the continent, particularly the western Canadian provinces.

In their regions of origin, Dalmatian and yellow toadflax evolved in plant communities that are grazed moderately to intensely, primarily by sheep and goats, and by cattle to a lesser extent. Because much of the land is arable in the region of origin, many populations have evolved with, and are adapted to, the periodic disturbances of agriculture.

203

Potential for Invasion

The toadflax species are unpredictable, variable weeds. High genetic variability enables these weed to adapt to a wide variety of conditions. Characteristic plant communities associated with toadflax species are those associated with disturbed, open habitats (Arnold 1982).

Dalmatian and yellow toadflax are often found in well-drained, relatively coarse-textured soils varying from coarse gravels to sandy loams, but are also sometimes found in heavier soils. They are highly competitive in areas where summers tend to be dry. Areas of low competition between species, sparsely vegetated soils, and drier, open areas on rangeland seem susceptible to invasion in some cases, particularly south- and southeast-facing slopes. Sites where Dalmatian and yellow toadflax can establish include roadsides, areas near dwellings, vacant lots, cemeteries, gravel pits, fields, waste areas, and other disturbed sites, spreading to valleys, sagebrush flats, and overstocked spring-grazed pastures (Lange 1958, Alex 1962, Reed and Hughes 1970, Robocker 1974, Parker and Peabody 1983).

Minimum and no-till farming methods could enable yellow toadflax to invade or re-invade areas where regular tillage has kept populations at acceptable levels (McClay 1992). Wet or dark conditions appear to limit yellow toadflax (Zilke 1954, Saner 1994), although it is often found on well-drained gravelly or rocky river banks. Roadsides, dry fields, grainfields, waste areas, gravel pits, pastures and rangeland, clearings, clearcuts, vacant lots, and railroad yards are sites typically colonized by yellow toadflax (Reed and Hughes 1970, Frankton and Mulligan 1970, Lorenzi and Jeffrey 1987, Saner 1994).

Even in pristine areas and on rangeland in excellent condition, new infestations of Dalmatian or yellow toadflax can establish in naturally-occurring disturbances or small openings. Once the highly competitive vegetative growth begins, the condition of the rangeland will probably do little to slow expansion of the infestation.

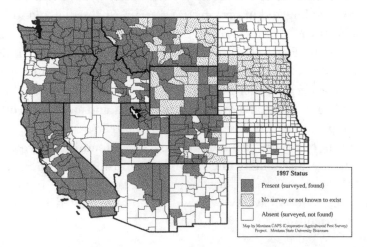

1997 Status

■ Present (surveyed, found)

░ No survey or not known to exist

□ Absent (surveyed, not found)

Map by Montana CAPS (Cooperative Agricultural Pest Survey) Project. Montana State University-Bozeman

Distribution of Dalmatian toadflax by county in the western United States. Data were collected by surveying various weed authorities in each state.

Impacts

• *Ecological and Environmental*

Dalmatian and yellow toadflax can displace existing plant communities and associated animal life. Loss of forage can impact big game species, especially on winter ranges. Although deer have been observed to browse Dalmatian toadflax and seed is used by some species of birds and rodents, it is not known to be heavily used by any native species (Robocker 1970). All toadflax species can provide cover for smaller animals. Where sod-forming or bunch grass communities are replaced by toadflax, soil erosion, surface runoff, and sediment yield can be increased. However, on harsh, sparsely vegetated sites toadflax can actually help stabilize soil.

• *Economic*

The toadflaxes can displace desirable plants, cause loss of forage for domestic livestock and some big game, and habitat loss for associated animal life. The economic impact of Dalmatian toadflax is primarily on rangeland and wildlife habitat, while yellow toadflax can be a serious problem on cultivated land as well as on rangeland.

Economic data specific to Dalmatian toadflax are scarce, but direct management costs averaged $40 per acre ($99 per ha) in 1992 on a Montana ranch of which 30% of the 1,064 acres (431 ha) was severely infested with Dalmatian toadflax (25% to 100% vegetative cover) (James 1994). Reduction in cattle-carrying capacity and reduction in the appraised value of infested ranch land increases the economic impact (Lacey and Olsen 1991).

Cattle will sometimes casually browse flowering shoots (Harris and Carder 1971). Occasional cases of mild poisoning have been reported for cattle (Kingsbury 1964, Mitich 1993), but the toadflaxes are usually avoided by cattle, and such cases are rare. Sheep will use Dalmatian toadflax as a major food source, showing no apparent

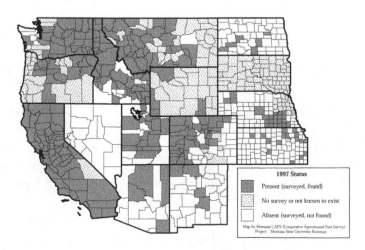

Distribution of yellow toadflax by county in the western United States. Data were collected by surveying various weed authorities in each state.

205

ill effects (Barnett 1992, personal communication; James 1994, personal communication).

Yellow toadflax can be a serious weed in cultivated crops, as well as on rangeland, and is believed to become more prominent in reduced-tillage farming operations. This problem is compounded by the ineffectiveness of herbicides in many cases (Ontario Ministry of Agriculture and Food 1993). In Alberta, a 1987 survey showed an estimated 69,188 acres (28,000 ha) infested with the weed. Twenty percent of the infestation was on rangeland and non-agricultural land, and 30% was in annual crops and forages. These infestations cost more than $360,000 per year, or about $5.25 per acre ($13 per ha) per year (McClay 1987).

In New York state, the root system has been found to provide an overwintering site for cucumber mosaic virus and broad bean wilt virus, disease pests of cultivated crops (Rist and Lorbeer 1989).

Yellow toadflax is reported to be mildly poisonous to cattle (Parker and Peabody 1983, Saner 1994), but reported cases of poisonings are rare, probably due to avoidance of the plant.

Biology and Ecology

Dalmatian and yellow toadflax are deep-rooted, short-lived herbaceous perennials that reproduce by seeds and by vegetative buds on the roots.

• Seeds

A single Dalmatian toadflax plant can produce up to 500,000 seeds (Robocker 1970) beginning in late June or early July in northern climates, and continuing until September or early October. In one study, 97% of total seed production occurred in the first five weeks (Lange 1958). Seed production can begin on lower portions of the stems while upper portions are still in various stages of bloom (Parker and Peabody 1983).

Yellow toadflax seed capsules usually contain 10 to 40 seeds (Arnold 1982), with numbers being highly variable. Number of seeds produced per plant is considerably less than Dalmatian toadflax and have been estimated at 15,000 to 30,000 (McClay 1992).

Dalmatian toadflax seed dispersal begins as early as July in northern climates and continues into winter. Dried floral stalks can remain standing for two years, retaining some seeds but dispersing most during the first year (Robocker 1974). Yellow toadflax seed dispersal begins in August or September in northern climates and continues into winter (Lewis 1954, Clements and Cavers 1990, Nadeau and King 1991, Saner 1994).

206   Although wind has been considered a major means of seed dispersal for toadflax species (Lange 1958, Alex 1962, Robocker 1970, Robocker 1974), it may be less a factor than previously thought. Studies of yellow toadflax have shown that 80% to 90% of seeds fall within 18 inches (0.5 m) of the parent plant, and "very few" seeds

fall more than 5 feet (1.5 m) from the parent plant (Nadeau and King 1991). Average seed size is similar for yellow and Dalmatian toadflax. Dalmatian toadflax seeds are slightly heavier (Robocker 1970), and the papery wing surrounding both kinds of seeds is less developed on Dalmatian toadflax. This suggests Dalmatian toadflax seeds may also fall within short distances of the parent plant. However, when seeds of either species fall from upright dried floral stems onto crusted snow they can be blown by wind across the surface (Lowe 1992, personal communication; Saner 1994).

Some Dalmatian toadflax seed germination occurs in fall, but most occurs the following spring, with peaks in April and May. Germination rates are as high as 75% for Dalmatian toadflax, and seeds can remain dormant at least 10 years (Robocker 1970). These dormant seeds can rapidly reinfest a site following control applications, even when pre-emergent herbicides are used, because only a portion of the seeds will germinate in any given year.

Yellow toadflax seed germination rates vary greatly but are usually low, often below 10% (Clements and Cavers 1990). The remaining seeds are either dormant or non-viable. Most seeds produced by yellow toadflax are dormant (Lewis 1954, Zilke 1954, Nadeau and King 1991) and can remain dormant up to 10 years (Carder 1963) or more. Seed viability is often low also, 40% to 50% in one study (Nadeau and King 1991). These characteristics result in low seedling establishment, an important consideration in integrated weed management.

Although seeds and root sections appear to be equally capable of initiating new yellow toadflax infestations, seeds appear more important in initiating new infestations, especially in non-arable lands. (In arable land, root pieces can cling to farm implements, easily initiating new infestations.) Established infestations of yellow toadflax appear to expand mainly by vegetative reproduction by roots rather than by seed (Nadeau et al. 1992). New infestations of Dalmatian toadflax usually originate from seed, also; however, established patches of Dalmatian toadflax expand by both seed and vegetative reproduction, an important difference between the species.

• *Seedlings*

Seedlings are vulnerable to dehydration, competition from other species, and certain control strategies. This period of vulnerability is an important consideration in development of toadflax management strategies. Survival of toadflax seedlings after emergence often depends on precipitation or lack of competition from other plants. Seedlings are easily outcompeted by plants in areas with good groundcover, and in particular by well-adapted perennials (Robocker 1970). Dalmatian toadflax seedlings are also outcompeted by cheatgrass (*Bromus tectorum*) when soil moisture is limited.

Seedlings of Dalmatian toadflax emerge in spring and autumn, usually first on south- or southeast-facing 40% slopes, because soils on these sites warm first in the spring and remain warm later in the fall. Such slopes can be useful for monitoring seasonal development of infestations. In eastern Washington, spring emergence on slopes usually begins the first or second week in March, and lasts until the first or second week in April. Emergence on level ground occurs two to three weeks later.

207

Rather weak floral stems and some seed can be produced by seedlings the first year (Robocker 1970).

Yellow toadflax seedling emergence begins in early to mid-May in Alberta (Nadeau and King 1991), and earlier in warmer regions of the United States. Most seedlings emerge in spring, then a second, smaller flush occurs in the fall.

• Vegetative Regeneration

The extreme competitiveness of established Dalmatian and yellow toadflax is partly due to early spring regeneration from vegetative buds on root stock. These vegetative shoots are not particularly dependent on soil moisture and are not vulnerable to competition from other plants. They are highly efficient in competing for available soil moisture and consequently moisture is seldom a limiting factor in vegetative regrowth. Because they usually emerge before desirable species in spring and use existing moisture so efficiently, vegetative shoots are extremely competitive even on rangeland in excellent condition.

Yellow toadflax vegetative shoots usually emerge in early to mid-April in Canada, when soil temperatures reach 42° to 50° F (5° to 10° C) (Saner 1994), and possibly in mid- to late March in warmer regions of the United States.

• Rosette Stage

In early autumn, first-year Dalmatian toadflax plants often produce prostrate stems, forming a mat-like rosette that survives into the following spring. The rosette stage stores carbohydrates. The following spring, floral stems grow from the rosette. The fall regrowth of rosettes is thought to be a vulnerable stage for Dalmatian toadflax, a consideration in management strategy. The sequence also occurs to a lesser degree in mature plants (Robocker 1974). With the exception of these prostrate stems, topgrowth dies back in the fall and is regenerated from the root system each spring.

• Adult Stage

Floral stems of the mature Dalmatian toadflax plant are apparently produced only after a winter's dormancy and exposure to temperatures between 50° and 68° F (10° and 20° C). The lack of abundant seed production by plants not exposed to the required low temperatures may be a factor in the geographical distribution of Dalmatian toadflax. Because of the relatively short life of a plant, the ultimate survival of a stand of Dalmatian toadflax probably depends most on floral stem and seed production, which is usually associated with the rosette-like prostrate stems from the previous autumn (Robocker 1974). Strategies that prevent rosettes from producing floral stems and seeds may be an important component of long-term Dalmatian toadflax management.

208

Flowering of Dalmatian toadflax normally begins in June and continues until September or October (Lange 1958), but can occur earlier in warm seasons or warmer habitats.

When flowering stems of yellow toadflax are 16 to 24 inches (40 to 60 cm) tall, branching begins, which signals the beginning of flower bud formation. Flowering begins in May and continues until October, and is variable. Seed dispersal can begin on lower portions of the stem while flower buds are still forming on upper portions (Parker and Peabody 1983). Aboveground portions of the plant die at freezing, but the woodier stems may remain standing through the winter. Due to lower seed production and viability, the ultimate survival of a stand of yellow toadflax probably depends most on vegetative reproduction of roots, indicating the importance of limiting vegetative spread in a management program for yellow toadflax.

• Roots

For the first several weeks after seed germination, roots of seedlings are poor competitors for soil moisture, and are easily outcompeted by many annual and perennial plants. After this initial period of vulnerability, however, the toadflax is an extremely effective competitor, and competition from other species usually has little effect.

Nine weeks after Dalmatian toadflax seed germination, vertical roots can reach depths of 20 inches (50 cm) or more. Lateral roots are usually 1 to 4 inches (2 to 10 cm) beneath the soil surface. Within nine weeks, seedling roots have vegetative buds that can give rise to new, independent plants (Bakshi and Coupland 1960). Yellow toadflax seedlings can begin producing vegetative shoots from root buds two to three weeks after germination, giving rise to their own independent daughter shoots (Zilke 1954, Nadeau et al. 1992).

Mature toadflax plants have extensive, well-developed root systems with vegetative buds that can produce shoots which can later become independent plants. Roots of mature Dalmatian toadflax plants reach depths of 4 to 10 feet (1.2 to 3 m). Vegetative buds have been found as deep as 6 feet (1.8 m) (Robocker 1974), but most plants produced from vegetative buds occur on lateral roots, which are normally found in the upper 2 to 8 inches (5 to 20 cm) of the soil, and can extend 10 feet (3 m) or more from the parent plant (Lange 1958, Reed and Hughes 1970, Robocker 1974, Cronquist et al. 1984).

Roots of mature yellow toadflax can penetrate into the soil 3 feet (0.9 m) or more, while lateral roots can be several yards (meters) long (Saner 1994). Most vegetative reproduction is associated with the lateral roots, which usually grow in the top 2 to 8 inches (5 to 20 cm) of soil.

• Longevity

Dalmatian toadflax is a short-lived perennial, with individual plants living an average of three to five years. Individual patches can persist for 13 years or more under favorable conditions. Dalmatian toadflax stands frequently disappear for several years, then re-establish, either from buried seeds, or perhaps from vegetative root buds (Robocker 1974).

Individual yellow toadflax plants live up to four years, but it is difficult to determine which parts are individual plants, because of vegetative reproduction characteristics. Theoretically, a stand may persist indefinitely.

Management

Successful management of the toadflax species requires integrating as many management strategies as possible into the program design. This is especially important for the toadflaxes, because of their high genetic variability. This variability results in localized populations that respond differently to the same herbicides, management methods, biological control agents, and environmental conditions. Management programs for Dalmatian toadflax should emphasize equally the prevention of seed formation and vegetative spread (spreading by roots), while programs for yellow toadflax can emphasize control of vegetative spread with secondary emphasis on prevention of seed formation.

• Prevention

Because the toadflax species are expensive, labor-intensive, and extremely difficult to manage once infestations become established, preventing infestations from occurring is time- and cost-effective. New infestations of toadflax originate from seeds or vegetative buds on root pieces; therefore, keeping seeds and contaminated materials or equipment out of uninfested areas is good strategy.

Farm operations, outdoor recreation, and other human activities can transport seeds into uninfested areas. Seeds can be transported in mud on motorbikes, construction equipment, vehicles, agricultural machinery, tires, and in mud on feet of livestock. Check and clean equipment before moving it into uninfested areas or before bringing it in from infested areas. Fill dirt and landscaping soil can also bring seeds into an uninfested area.

When moving livestock from infested to uninfested areas, hold them in corrals or small pastures until viable seeds have had time to pass through the digestive tract. Hold cattle six days and sheep 11 days. Monitor these holding areas regularly for seedling establishment. Avoid purchasing feed or seed that could be contaminated with weed seeds.

• Mechanical and Physical Controls

Pulling toadflax by hand can be effective for small infestations, especially in sandy soils or when soils are moist. Pulling each year for five to six years is necessary to deplete the remaining root system of reserves. Try to follow lateral roots to their ends. The site must be revisited for 10 to 15 years to remove seedlings produced from dormant seeds.

Because established infestations of yellow toadflax spread mainly by roots, physical removal (especially around perimeters) can limit spread.

Mowing is not recommended for any of the toadflax species, since it does not affect root reserves or buried seeds, nor is it feasible on most sites. Hand-removal of the flowering tops from the plants is a marginal strategy even for very small infestations.

Cultivation, where feasible, will control toadflax. Using sweep-type cultivators, begin cultivation early in June in the northwestern states, perhaps earlier in warmer climates, and repeat every 7 to 10 days. Eradication requires at least two years, with four to five cultivations the second year (Parker and Peabody 1983). Once begun, tillage to control yellow toadflax should be consistent; irregular tilling can spread infestations, because root pieces as short as $^1/_2$ inch (1.5 cm) can produce new shoots that can expand to a patch more than 3 feet (0.9 m) in diameter in a single season (Nadeau et al. 1992). Care must be taken not to transport toadflax root pieces to clean fields on machinery.

• Cultural Controls

Active growth of the toadflax species during all seasons (if temperature and moisture are adequate) enhances the competitiveness of the species. However, the initial vulnerability of toadflax seedlings often enables well-adapted plant species to outcompete toadflax seedlings. Through seeding and/or land management, encourage combinations of desirable species that can provide competition throughout the season and at all levels of the soil profile (with roots that grow to shallow, intermediate, and deeper levels), providing competition with toadflax through time and space. After initial weed control activities, areas should be seeded with a mixture of competitive, well-adapted species. Rangeland in excellent condition can outcompete many (but not all) germinating toadflax seedlings.

Low viability of yellow toadflax seeds can make competition by desirable species an even more effective strategy for this species. In cropland, barley appears to inhibit vegetative reproduction of yellow toadflax to some extent (Nadeau and King 1991).

• Grazing

Overgrazing by cattle in spring can increase establishment of toadflax because seedlings can more effectively compete with grazed plants for soil moisture and other resources. Timing of grazing is important in developing and maintaining competitive, desirable plant communities that reduce establishment of seedlings for long-term toadflax management. Grazing strategies will be less effective in restricting expansion of established stands, because of toadflax's deep, competitive roots.

However, preliminary results of field trials in Montana show that sheep can be used to help suppress stands of Dalmatian toadflax and limit seed production. In these studies, 1,000 ewes and lambs were placed in a hilly rangeland area of moderate to heavy infestations with Dalmatian toadflax densities of 25% to 100% of existing vegetative coverage. Approximately 35% to 45% of the toadflax foliage was stripped,

211

including the terminal 6 to 10 inches (15 to 25 cm) of plant stems. Although initially the sheep only nibbled at the plants, in two to three weeks they were consuming Dalmatian toadflax regularly, even though other forages were present. In these studies, the sheep did well and showed good weight gain (Barnett 1992, personal communication; James 1994, personal communication).

Burning is usually not effective for the toadflaxes, because root buds and buried seeds are unaffected. Burning can increase the competitiveness of the toadflax by removing desirable plants. Removal of top growth can also stimulate production of vegetative shoots. However, scorching of floral stalks using propane burners can help prevent seed production.

• *Biological Controls*

As of 1998, several species of insects have been released on Dalmatian toadflax, including a defoliating moth (*Calophasia lunula*), an ovary-feeding beetle (*Brachypterolus pulicarius*), and two seed capsule-feeding weevils (*Gymnaetron antirrhini* and *Gymnaetron netum*). A stem-boring weevil (*Mecinus janthinus*) and a root-boring moth (*Eteobalea intermediella*) have been released in Canada and the United States.

All insect species that have been released for control of Dalmatian toadflax also attack yellow toadflax (McClay 1987, Harris and Carder 1971, Saner 1991), and are intended as biological control agents for that species also. The ovary-feeding beetle and the two seed-capsule-feeding weevils have been found most frequently on yellow toadflax. *B. pulicarius* and *G. netum* are believed to be partially responsible for declines in infestations of yellow toadflax in western Canada (McClay 1992, Darwent et al. 1975).

Although each species has some degree of impact on individual plants or on seed production (Jeanneret and Schroeder 1992, Nowierski 1992, Moeller-Schroeder 1994, Saner 1994), it is too soon to know what cumulative effect these biological control agents will have in a toadflax management program. When these insects are used as part of an integrated weed management program, place them on carefully selected sites to which the insects are adapted, then protect them from pesticide applications (including herbicides), disturbance by livestock, and other disruptions.

• *Chemical Controls*

Effectiveness of herbicides used to manage the toadflax species is highly variable, partly because of the plants' high genetic variability. The waxy leaf surface of Dalmatian toadflax provides a protective barrier that hinders herbicide uptake in some cases. Soil type is also important, since it is more likely that the herbicide will leach below the plant root zone on sites with sandy soils or soils low in organic matter, especially in stands in which lateral roots are very close to the soil surface.

Even when herbicide treatment of toadflax appears to be effective, long-term control may not be achieved and reinvasion may occur. It is necessary to treat an infestation every three to four years for as long as 12 years to eradicate the weed on sites where herbicides are effective.

After plants such as spotted knapweed (*Centaurea maculosa*) and St. Johnswort (*Hypericum perforatum*) are removed from plant communities that include the toadflaxes, releases (sudden germination of dormant seeds) of toadflax may occur, and higher herbicide rates are often needed (Duncan 1992, personal communication; Lange 1958). For fall herbicide applications, 1 to 3 inches (3 to 8 cm) of fall regrowth indicates roots are storing energy for winter. This may be used as an indicator for fall application of herbicides or other control methods.

For Dalmatian toadflax on some sites fall applications of picloram at a rate of 1 pound active ingredient per acre (ai/ac) (1.12 kg ai/ha) has given up to 98% control for up to three years. Spring applications on these sites gave less control. On other sites, picloram at 0.5 lb ai/ac (0.56 kg ai/ha) applied in fall has given comparable control of Dalmatian toadflax for two years (Sebastian and Beck 1989, Sebastian et al. 1990). However, on other sites, picloram has not been effective (Hanson et al. 1989). Picloram at the higher rate will kill many broadleaf species, possibly injuring desirable plant species. Under dry conditions, picloram is not moved into the soil, and significant losses of the herbicide can occur from exposure to sunlight in three or four weeks.

In one Dalmatian toadflax study, dicamba applied prebloom at a rate of 4 lb ai/ac (4.83 kg ai/ha) gave excellent control for one year. In another study, a tank mix of picloram plus 2,4-D (0.5 lb + 1 lb ai/ac [0.56 kg + 1.12 kg ai/ha]) applied prebloom or in the fall provided 90% to 100% control (Sebastian and Beck 1989, Sebastian et al. 1990), but in other studies 2,4-D was ineffective when applied in a tank mix of triclopyr and 2,4-D amine (1.5 lb + 3 lb ai/ac [1.68 kg + 3.36 kg ai/ha]). Triclopyr and fluroxypyr alone or in combination were ineffective in those studies (Ferrell and Whitson 1989).

Effectiveness of herbicides is also highly variable on yellow toadflax. In one study, fair to good control of yellow toadflax was obtained for only one year using high rates of picloram plus fluroxypyr (1 lb + 1 lb. ai/acre [1.12 kg + 0.89 kg ai/ha]) applied pre-bloom (Sebastian and Beck 1989). On other sites, picloram has had variable results. Yellow toadflax has not been effectively controlled by 2,4-D, MCPA, 2,4-DB, MCPB, or mecoprop. If yellow toadflax infestations develop in low-till cropland, spot treatments with labeled rates of glyphosate, amitrole, diquat, or picloram can be used. Glyphosate has been used in conjunction with cultivation to reduce yellow toadflax effectively for up to two years (Saner 1994).

• *Reseeding and Revegetation*

After initial weed control, areas should be seeded with a mixture of competitive, well-adapted species that include deep-rooted species. The mixture should include plants with roots that extend as deep or deeper into the soil than the roots of the toadflax species to maximize competition for water, nutrients, and space. Also include species with roots that extend to intermediate depths, and others with shallow root systems to compete with toadflax seedlings.

Strive for combinations of species that will provide growth activity and desirable competition for as much of the year as possible. Include species that grow actively in early season, some in midseason, and others in late season. To help outcompete germinating toadflax seedlings, include winter and summer annuals and shallow-rooted perennials. Although mature toadflax plants are particularly competitive against winter annuals and shallow-rooted perennials, it may be useful to include plants in those groups in reseeding and revegetation mixtures to crowd out germinating seedlings. Species recommendations for revegetation efforts can be obtained from regional Plant Materials Centers of the Natural Resources Conservation Service.

• Long-Term Management

Determining effective and economical methods for a particular toadflax population will require annual re-mapping and evaluation. It is only with this information that the site-specific management efforts required for this highly variable weed can be adjusted to determine the best combination of strategies for a site. Persistent implementation of those strategies will be required.

Literature Cited

Alex, J.F. 1962. The taxonomy, history, and distribution of *Linaria dalmatica*. Can. J. Bot. 40: 295-307.

Arnold, R.M. 1982. Pollination, predation, and seed set in *Linaria vulgaris* (Scrophulariaceae). Am. Midl. Nat. 107:360-69.

Bakshi, T.S., and R.T. Coupland. 1960. Vegetative propagation in *Linaria vulgaris*. Can. J. Bot. 38:243-49.

Barnett, K. 1992. Personal communication.

Carder, A.C. 1963. Control of yellow toadflax (*Linaria vulgaris*) by grass competition plus 2,4-D. Weeds 11: 13-14.

Clements, D.R., and P.B. Cavers. 1990. Seasonal seed viability patterns and the role of incomplete seed development in the life history strategy of *Linaria vulgaris*. Le Naturaliste Canadien. 117:189-98.

Cronquist, A., A.H. Holmgren, N.H. Holmgren, J.L. Reveal, and P.K. Holmgren. 1984. Intermountain flora-vascular plants of the intermountain west. NY Bot. Garden, New York. Vol.4, 459-60.

Darwent, A.L., W. Lobay, W. Yarish, and P. Harris. 1975. Distribution and importance in Northwestern Alberta of toadflax and its insect enemies. Can. J. Plant Sci. 55:157-62.

Davies, P.H. 1978. Flora of Turkey, Vol. 6. Edinburgh University Press, Edinburgh.

Fernald, M.L. 1905. Some recently introduced weeds. Trans. Mass. Hort. Soc. Part 1:11-22.

Ferrell, M.A., and T.D. Whitson. 1989. Dalmatian toadflax control in rangeland. Res. Prog. Rep., West. Soc. of Weed Sci., pp. 46 and 95.

Frankton, C., and Mulligan, G.A. 1970. Weeds of Canada. Canada Dept. of Agric., Pub. 948, Ottawa, Ontario, 150.

Hanson, D.E., K.G. Beck, and J.R. Sebastian. 1989. Dalmatian toadflax control in rangeland, Livermore, California. Res. Prog. Rep., West. Soc. of Weed Sci., 31-32.

Harris, P. 1988. The biocontrol of Dalmatian toadflax—a cooperative project report for 1988, Agric. Canada, British Columbia (BCMA, BCFS, and Cattlemen's Assoc.), and Montana.

Harris, P., and A.C. Carder. 1971. *Linaria vulgaris* Mill., yellow toadflax, and *L. dalmatica* (L.) Mill., broad-leaved toadflax (Scrophulariaceae). *In:* Biological control programmes against insects and weeds in Canada 1959-1968. Commonwealth Agriculture Bureaux, Slough, England, 94-97.

Hartl, D. 1974. Family Scrophulariaceae, in Issustrierte Flora von Mittel-Europa VI, 1. Teil (Hegi, G., ed.). Carl Hauser Verlag, Munchen, 73-94.

James, S. 1994. Personal communication.

Jeanneret, P., and D. Schroeder. 1992. Biology and host specificity of *Mecinus janthinus* Germar (Col.: Curculionidae), a candidate for the biological control of yellow and Dalmatian toadflax, *Linaria vulgaris* (L.) Mill. and *Linaria dalmatica* (L.) Mill. (Scrophulariaceae) in North America. Sci. and Tech. 2:25-34.

Kingsbury, J.M. 1964. Poisonous plants of the United States and Canada. Prentice-Hall, Inc. Englewood Cliffs, NJ.

Lacey, J., and B. Olsen. 1991. Environmental and Economic Impacts of Noxious Range Weeds. *In:* Noxious Range Weeds, L.F. James et al. (eds.), Westview Press, Boulder, CO, 5-16.

Lajeunesse, S., P.K. Fay, D. Cooksey, J.R. Lacey, R.M. Nowierski, and D. Zamora. 1993. Dalmatian and yellow toadflax: weeds of pasture and rangeland. Montana State Univ. Ext. Serv., EB 115. Bozeman, MT.

Lange, A.W. 1958. Dalmatian toadflax—A possible rival of goatweed as a serious range weed. Weed Sci. 6: 68-70.

Lewis, N.G. 1954. Germination of toad flax, *Linaria vulgaris.* Hill. Ass. Off. Seed. Anal. News. 28: 9-10.

Lorenzi, H.J., and L.S. Jeffery. 1987. Weeds of the United States and their control. Van Nostrand Reinhold, New York.

McClay, A. 1987. Proposal to screen further agents for biological control of Common Toadflax, *Linaria vulgaris* (L.) Mill. Project proposal, Alberta Environmental Centre, Vegreville, Alberta, unpublished report.

McClay, A.S. 1992. Effects of *Brachypterolus pulicarius* (L.) (Coleoptera: Nitidulidae) on flowering and seed production of common toadflax. Can. Entomol. 124:631-36.

Meusel, H., E. Jager, S. Rauschert, and E. Weissert. 1978. Vergleichende Chorologie der zentraleuropaischen Flora. VEB Gustav Fischer Verlag, Jena, Germany. Band 2.

Mitich, L.W. 1993. Intriguing world of weeds—Yellow Toadflax. Weed Technol. 7:791-93.

Nadeau, L.B., and J.R. King. 1991. Seed dispersal and seedling establishment of *Linaria vulgaris* Mill. Can. J. Plant Sci. 71:771-82.

Nadeau, L.B., J.R. King, and K.N. Harker. 1992. Comparison of growth of seedlings and plants grown from root pieces of yellow toadflax (*Linaria vulgaris*). Weed Sci. 40:42-47.

Nowierski, R.M. 1992. Dalmatian toadflax, *Linaria genistifolia* spp. *dalmatica* (L.) Maire and Petitmengin (Scrophulariaceae). *In:* L.A. Andres, J.W. Beardsley, R.D. Goeden, and G. Jackson (eds.), Biological Control in the U.S. Western Region: Accomplishments and Benefits of Regional Research Project W84 (1964-1989). UC Press, Berkeley, CA.

Ontario Ministry of Agriculture and Food. 1993. Guide to weed control. Publication 75. Queens Printer, Toronto, Ontario.

215

Parker, R., and D. Peabody. 1983. Yellow toadflax and Dalmatian toadflax. Pacific Northwest Coop. Ext. Bull. 135. Wash. State Univ., Pullman, WA.

Reed, C.F., and R.O. Hughes. 1970. Selected Weeds of the United States, USDA Agric. Res. Serv. Dover Publications, Inc., New York.

Rist, D.L., and J.W. Lorbeer. 1989. Occurrence and overwintering of cucumber mosaic virus and broad bean wilt virus in weeds growing near commercial lettuce fields in New York. Phytopath. 79:65-69.

Robocker, W.C. 1970. Seed characteristics and seedling emergence of Dalmatian toadflax. Weed Sci 18:720-25.

Robocker, W.C. 1974. The history, ecology, and control of Dalmatian toadflax. Wash. Ag. Exp. Sta. Bull. 1330. 20 pp.

Rousseau, C. 1968. Histoire, habitat et distribution de 220 plantes introduites au Quebec. Le Naturaliste Canadien. 95:49-169.

Saner, M.A. 1991. Interactions of *Linaria vulgaris* Mill. and *L. dalmatica* (L.) Mill. (Scrophulariaceae) with insect herbivores. Dissertation (Ph.D thesis), Universite, Basel, Switzerland.

Saner, M.A. 1994. The Biology of Canadian Weeds: 105. *Linaria vulgaris* Mill. Can. J. Plant Sci. 75:525-37.

Saner, M.A., and H. Moeller-Schroeder. 1994. Impact of *Eteobalea* spp. on clonal growth and sexual reproduction of common toadflax, *Linaria vulgaris* Mill. Weed Res. 34: 199-204.

Sebastian, J. R., and K. G. Beck. 1989. Yellow toadflax control with fluroxypyr and picloram on Colorado rangeland. Res. Prog. Rep., West. Soc. of Weed Sci., 33-34.

Sebastian, J.R., K.G. Beck, and D.E. Hanson. 1990. Dalmatian toadflax control with fluroxypyr and picloram on Colorado rangeland. Res. Prog. Rep., West. Soc. of Weed Sci., 99-100.

Smith, J.M. 1959. Notes on insects, especially *Gymnaetron* spp. (Coleoptera: Curculionidae) associated with toadflax, *Linaria vulgaris* Mill. (Scrophulariaceae) in North America. Can. Entomol. 91:116-21.

Zilke, S. 1954. Some aspects of the ecological life history of *Linaria vulgaris* Mill. M.A. thesis, University of Saskatchewan, Saskatoon, Saskatchewan.

Diffuse Knapweed

Ben F. Roché, Jr., and Cindy Talbott Roché

The genus to which our knapweeds belong, *Centaurea*, originated in the eastern Mediterranean region and evolved into numerous species following the retreat of the last major glaciation, some 10,000 years ago (Small 1919). As the glaciers retreated, watering the plains, the knapweeds moved into the glacially disturbed area. Over time, different species of knapweeds were provided the opportunity to fit themselves to the many types of disturbed sites created (Prodan 1930). This all preceded the decision by humans, about 7,000 years ago, to settle in the same general area, develop the first planned cropping systems, and domesticate grazing animals (Lowdermilk 1953). The two scenarios, one by a genus of potential weeds (rapidly evolving, early successional species) and the other by humans creating disturbances, provided the *Centaurea* species ample opportunity to become preadapted to similar disturbances in similar environments in our region. Differences in the ecologic amplitude among the invasive knapweeds reflects evolutionary habitats and selection pressures in their native distributions. This chapter will explore the specific adaptations of diffuse knapweed (*Centaurea diffusa*) and the implications these have for management of the sites it invades.

Identification

Diffuse knapweed grows 1 to 3 feet (0.3 to 0.9 m) tall from a deep taproot. Upright stems have numerous spreading branches, which give the plant a ball-shaped appearance and tumble-weed mobility when broken off. The basal leaves, which form rosettes on a central crown, are borne on short stalks and are deeply divided into lobes on both sides of the midrib. Stem leaves are stalkless, becoming progressively smaller and less divided higher up the stem, with the uppermost small leaves bractlike. Urn-shaped flower heads are $3/16$ to $1/4$ inch (3 to 6 mm) in diameter, and $5/16$ to $7/16$ inch (8 to 11 mm) long, excluding spines and flowers. Heads are solitary or borne in clusters of two or three at the ends of the branches. Bracts surrounding the flower heads are yellowish green with a buff or brown margin. Each bract is edged with a fringe of spines and ends with a longer spreading spine, about $1/8$ inch (3 mm) long, at the tip. Most plants have white flowers, but rose-purple and lavender flowered plants are not uncommon. Colored flowers also have spots of color on the bracts. This causes confusion with spotted knapweed (*C. maculosa*), but the significant difference is the longer, spreading spine at the tip of the diffuse knapweed bract. Flowering occurs from June to September, continuing later into the season as permitted by adequate moisture and mild temperatures. Seeds are buff to dark brown, about $^1/_8$ inch (3 mm) long, having a plume of bristle-like hairs that varies from scalelike to $^1/_8$ the length of the seed (Roché and Roché 1993).

Origin, History, and Distribution

Diffuse knapweed, native to grasslands and shrub steppes of the eastern Mediterranean and western Asia, occurs from the southern former USSR to western Germany (Müller and Schroeder 1989), and is especially common in the Ukraine and Crimea (Popova 1960). The climate is characterized by an arid period in the summer, and soils are fertile Chernozems (Harris and Cranston 1979). In the eastern part of its geographical distribution, diffuse knapweed is associated with continental steppe and silvo-steppe vegetation: grassland and fringe forest (Dostal 1976).

The earliest record of diffuse knapweed in western North America is from an alfalfa field at Bingen, Washington, in 1907 (Howell 1959, Roché and Talbott 1986). It may have been introduced with Turkestan alfalfa seed from the Caspian Sea region (Harris and Myers 1979). Maddox (1979) implicates alfalfa seed from Asia Minor-Turkmenistan or hybrid alfalfa seed from Germany as sources. Howell (1934) reported that diffuse knapweed at The Dalles, Wasco County, Oregon, 1931, was the first naturalized colony in the United States. Renney (1959) reported that diffuse knapweed grew in British Columbia before 1930. The 1930s appear to be the decade of rapid movement of diffuse knapweed to widely scattered locations along roadsides and railroads in British Columbia, Washington, Oregon, and Idaho (Roché and Talbott 1986). By 1967 it was reported in 12 eastern Washington counties, and considered a serious range weed problem in Chelan, Ferry, Kittitas, Klickitat, Okanogan, Spokane, and Stevens counties (Roché 1967). The problem escalated in the 1970s, as diffuse knapweed moved from initial introduction sites along travel corridors onto adjacent pasture and rangeland. This was the response predicted by Cade (1968) when he wrote that the "very first plant or seed of a bad weed is...the slow motion equivalent of the tiny flame that could eventually burn the house down." The spread of diffuse knapweed has been like a wildfire: sending out fingers along roads, establishing spot infestations in disturbed sites, and then eventually coalescing the spots.

Lacey (1989) reported approximately 3.1 million acres (1.2 million ha) in the western United States infested with diffuse knapweed: 30,000 acres (12,000 ha) in Colorado, 1.4 million acres (567,000 ha) in Idaho, 10,000 acres (4,000 ha) in Montana, 1.2 million acres (486,000 ha) in Oregon, 1000 acres (400 ha) in South Dakota, 25 acres (10 ha) in Utah, 427,000 acres (173,000 ha) in Washington, and 5,000 acres (2,000 ha) in Wyoming. It also grows in Nevada and California.

Since the 1989 summary, diffuse knapweed has been expanding rapidly. For example, by 1997 it occupied an estimated 100,000 acres (40,000 ha) in Colorado (G. Beck 1997, personal communication) and the infested area in Utah increased to between 100 and 200 acres (40 to 80 ha), mainly in the foothills east of Moab and along the Wasatch Front near Ogden (S. Dewey 1997, personal communication). In California, diffuse knapweed primarily occurs as single plants or small patches, and is under eradication in most areas except Trinity County (Joley and Woods 1996).

Estimates of infestation size are extremely subjective because survey groundrules vary. Area reported varies from actual area occupied to total area exposed to the invader. Note the discrepancies in this paper: Lacey (1989) reported 1.4 million acres in Idaho and 1.2 million in Oregon. Callihan and Sanders (1994) suggested that the 100,000 acres (567,000 ha) in Blaine County plus the 487 acres (200 ha) reported by other counties adequately estimated Idaho's diffuse knapweed. In Oregon, diffuse knapweed's distribution range increased from 53 townships in 1982 to 252 townships in 1992 (Isaacson 1993). That represents 5.8 million gross acres (2.3 million ha), not actual infestations. In Oregon, the weed board has classified diffuse knapweed as a "B" weed. Since this listing attaches no special priority to control this weed over the other 44 weeds similarly listed, more detailed surveys are not available.

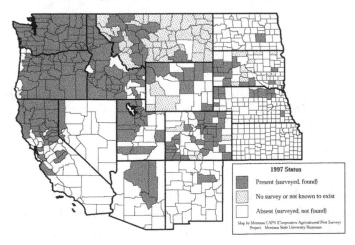

Distribution of diffuse knapweed by county in the western United States. Data were collected by surveying various weed authorities in each state.

Potential Invasion

In eastern Washington, diffuse knapweed has demonstrated a wide ecologic amplitude: elevational range was sea level to over 5,000 feet (1,500 m), all aspects and slope positions, flat to over 60%, in a wide spectrum of soil properties. Average annual precipitation ranged from 6 to 35 inches (15 to 90 cm), and 26 habitat types were recorded (Talbott 1987). However, its zone of maximum competitiveness is in the shrub steppe, with superior invasiveness in the bitterbrush/bunchgrass communities (*Purshia tridentata/Agropyron spicatum* with or without *Stipa comata*) commonly found on light, well-drained soils such as sandy or gravelly loams or loamy fine sands. Diffuse knapweed is less competitive on shallow soils (less than 15-inch [38 cm] depth) and very coarse-textured soils such as sand or loamy coarse sand, although it thrives on these sites when disturbance removes other vegetation. Diffuse knapweed does not grow in dense shade or on poorly drained soils.

In eastern Washington, major areas of diffuse knapweed dominance include the bitterbrush habitat types, with or without an overstory of ponderosa pine (*Pinus ponderosa*), and ponderosa pine/ or Douglas-fir/shrub habitat types (*Pseudotsuga menziesii/ Symphoricarpos albus* or *Physocarpus malvaceus*), which have been cleared and are either successional tree regeneration or converted to pasture. Most of the dry pasturelands or meadows were dominated by Kentucky bluegrass (*Poa pratensis*) prior to diffuse knapweed invasion. At middle and lower elevations of the east slope of the Cascades, invaded habitat types range from ponderosa pine and bunchgrass (with or without a shrub union) into the big sagebrush (*Artemisia tridentata*)/ bunchgrass types. In British Columbia, all of the interior natural grasslands (1.1 million ha) and fringe forest areas are considered vulnerable to diffuse knapweed invasion (Harris and Cranston 1979). The bitterbrush/bunchgrass shrub steppe in Utah is also highly susceptible to invasion (S. Dewey 1997, personal communication). In Colorado, diffuse knapweed has invaded the short grass steppe along the Front Range, including the glaciated foothills. Areas susceptible to invasion include the remaining short grass steppe, the adjacent montane zones, and the lower elevation pinyon-juniper-oak brush zones in southwestern Colorado (G. Beck 1997, personal communication). In Montana, the largest infestations are on natural grasslands and fringe forest areas near Helena, Big Timber and Ennis (Lacey et al. 1992).

Impacts

Diffuse knapweed has been credited with reducing biodiversity, increasing soil erosion, (Sheley et al. 1997), reducing land value (B. Roché 1988a), increasing cost of roadside maintenance (Korpi 1988), threatening Natural Area Preserves (Schuller 1992), and replacing wildlife and livestock forage on rangeland and pasture. Thus, losses due to diffuse knapweed invasion are both environmental and economic. Economic losses are often calculated in terms of forage loss. For example, a 1988 estimate of loss of forage production on knapweed-infested range and pasture in eastern Washington was 130 million lbs (59 million kg) of feed annually, which was translated into $951,000 as pasture or rangeland AUMs, or $2.9 million in hay as replacement feed (C. Roché 1988).

Forage is defined as "all browse and herbaceous foods that are available to grazing animals" (Kothman 1974). The knapweeds and starthistles have traditionally been considered poor forage, that is, less desirable, less palatable, and without food value. The first two descriptives are rated or ranked according to what else is available. Neither desirability nor palatability is as important to a foraging animal as availability. Forage value varies with developmental stages of the plant as well as the season. Although diffuse knapweed has commonly been rated as very poor forage, crude protein levels were found to be 18% in the rosette stage, 11% in bolting plants, 8% during bud and flowering stages, and 7% at seed ripe (B. Roché 1989, unpublished data).

Miller (1990) reported that diffuse and spotted knapweeds were important forage for mule deer, white-tailed deer, and California bighorn sheep in the West Kootenays of British Columbia. Prior to snowfall, bighorn sheep diet was 80% grass, 18% forbs, and 2% shrubs. As the snow receded in January and February, knapweed rosettes comprised 80% of the diet, while grass contributed 18%. Nutritional analyses from the Robson/Syringa Park area in mid-December showed that crude protein levels in knapweed rosettes (18%) and seedheads (8%) were higher than in the associated grasses, which ranged from 2% to 7.5% (Miller 1990). In addition to use by deer and elk, sheep and cattle will graze diffuse knapweed, at least through the bolting stage.

Diffuse knapweed is a source of pollen and nectar for honeybees during mid- to late summer when alternative sources are in short supply; however, diffuse knapweed honey is not as good as yellow starthistle honey, because it tends to have a bitter flavor. Because diffuse knapweed remains green later in the season than most associated vegetation, it sometimes serves as emergency rations for pest grasshopper species during outbreaks (Fielding et al. 1996).

Biology and Ecology

Diffuse knapweed is normally a biennial, but may behave as an annual or short-lived perennial (Watson and Renney 1974). Seeds germinate in the fall or spring and develop into low-lying, tap-rooted rosettes. Seasonal precipitation is critical in determining knapweed seedling survival (Myers and Berube 1983). Continuous seed rain and sequential establishment ensure occupancy of available safe sites for seed germination, and minimizes competition between cohorts (Sheley and Larson 1996). Germination polymorphism also distributes seed germination over time (Nolan and Upadhyaya 1988). Crowded plants often die before flowering, or live for several years as a rosette and then die after producing seed (Powell 1990). A few of the early germinating plants flower the first year, and others continue to grow after producing seed to flower again in the following year, behaving as a short-lived perennial (Schirman 1981, Roché and Roché 1990). Seed production by diffuse knapweed in

221

northeastern Washington averaged 11,200 to 48,100 seeds per square meter (Schirman 1981).

Diffuse knapweed is ideally suited to spread by vehicles and by tumbling in the wind. It evolved to spread by the wind blowing the ball-shaped plants in the same manner as tumble mustard (*Sisymbrium altissimum*). The seeds, held in urn-shaped heads which do not open widely, are lost gradually, giving the plant the advantage of distant distribution. This technique adapts extremely well to hitchhiking on the frames of vehicles and colonizing the bare shoulders of roads. Plants bearing seeds are also carried in rivers and irrigation systems.

The answer to the question of whether or not invasion by diffuse knapweed depends on overgrazing is not simple. It is clear that disturbance greatly increases the rate and final density of invasion for diffuse knapweed, and allows it to invade a wider range of habitats. Disturbances need not be recent and they may be as small as rodent activity or a single hailstorm (Lacey et al. 1990, Schuller 1992). Fletcher and Renney (1963) reported that diffuse knapweed was allelopathic, producing and releasing into the environment toxic substances that were poisonous to other plants, leading to rapid spread and dominance by the weed. Tucker (1990) claimed, in a cleverly written piece entitled "The Myths of Knapweed," that managers were hiding behind allelopathy to excuse their lack of management action. Kelsey and Bedunah (1989) provided evidence that, although a chemical (cnicin) could be isolated from the aerial tissues of knapweed species that would, at a range of concentrations, reduce seedling development of selected species, the source material (knapweed foliage) when applied at three times normal litter production provided no appreciable reduction in grass growth. Nevertheless, it is likely that these knapweeds are our best symptom of range degradation. They fill the niches created by disturbances and in some instances, due to differences among and between species and their varying abilities to compete, are able to dominate the site indefinitely. The results are decreased forage production, increased surface runoff, and reduced rain-use efficiency. That is desertification: ". . .the sustained decline and/or destruction of biological productivity of arid and semi-arid lands caused by manmade stresses, sometimes in conjunction with natural extreme events" (Sabadell et al. 1982).

Management

Weeds are best defined as those species having a negative value in a given management system. This recognizes the necessity for a definable value system and accepts the premise that all resources are or should be subject to predetermined managerial objectives.

At the 1984 Knapweed Symposium, Roy Cranston proposed that the long-term solution to knapweed included a combination of proper grazing management, judicious use of herbicides, development of effective biological control agents, and a high level of public awareness (Cranston 1984). This approach has proven to be appropriate where diffuse knapweed is well entrenched. However, places still exist where prevention and local eradication remain options. Alberta has maintained a successful exclusion program for diffuse knapweed for several decades, and California is pursuing eradication in most counties. A similar approach of early detection and destruction, of constant vigilance against new introductions, can be applied at a smaller scale for counties or individual ranches.

• *Manual and Mechanical*

Pulling or digging is often feasible for control of scattered diffuse knapweed plants, either as new invaders or those persisting following herbicide treatments. It may also be appropriate in areas where other treatments are not feasible or are prohibited. While labor-intensive, it is effective if enough of the taproot is removed to discourage sprouting. Rosettes cut just below the crown regrew 38% of the time, while only 4% of those cut 2 to 4 inches (5 to 10 cm) below the crown resprouted (Roché and Roché 1990). A single "pulling" event, even on an annual basis, is not enough to achieve control. Youtie and Soll (1994) recommended that handpulling be done three times per year, for as many years as additional knapweed appears. The first pulling is in spring when soil moisture allows extraction of a lethal portion of the taproot, followed by a second pulling in late June of bolting plants, and a third pulling just before seed dispersal to catch any plants overlooked previously. Plants from the last pulling should be burned, or otherwise carefully disposed of, to destroy viable seeds.

Mowing is not a control method, but can be used to reduce seed production or alter phenological development. Twenty-two percent of plants mowed to a 2-inch (5-cm) height each month of the growing season of April through October were still growing four years later (Roché and Roché 1990). Diffuse knapweed can produce viable seeds even if the parent plant is cut the same day that florets emerge from the bud (Roché and Roché 1990). Although diffuse knapweed requires pollination to produce seeds (Harrod and Taylor 1995), energy remaining in the cut plants is adequate for seeds to develop. Diffuse knapweed mowed early in the flowering period produces few viable seeds from the cut stalks. If an abundant seed bank already exists, a few additional seeds are insignificant. In contrast, a few seeds produced by newly-established plants in isolated locations maintain and expand new populations. However, diffuse knapweed mowed in the early flowering stage will usually regrow and produce abundant late-season seeds. Late-season flowerheads are likely to escape attack by biological control insects, as they will be out of synchrony with the normal development timing. When mowing stimulates new growth, the rosette stage of maximum herbicide effectiveness is extended. Mowing can also reduce weed competition during establishment of newly-seeded grass.

In rehabilitating diffuse knapweed-infested pastures, cultivation can control weeds and prepare an optimal seedbed. Best results may be obtained by planting an annual grain or cover crop before reseeding to perennial grass. Perennial grasses were established after a single disking of a diffuse knapweed stand in northeastern Oregon

(Larson and McInnis 1989), but a single cultivation may stimulate weed populations and follow-up treatments should be planned.

• *Fire*

Mixed results have been reported following burning of diffuse knapweed stands (Lacey et al. 1992, Renney and Hughes 1969). Diffuse knapweed resprouts following fire, even if burned by intense wildfire at bolting to flowering stage. Fire may damage residual desirable grasses while stimulating knapweed populations. However, fire may be used to remove plant debris and improve herbicide efficacy (Lacey et al. 1992). Wildfire may present the opportunity for rehabilitation of infested areas, using herbicides and seeding of perennial grasses.

• *Chemical*

Several herbicides are registered for control of diffuse knapweed on rangeland, with varying degrees of residual activity for control of later germinants. In order of decreasing residual effects, the following herbicides control diffuse knapweed: picloram at 0.25 to 0.5 lb active ingredient (ai) per acre (0.28 to 0.56 kg/ha); clopyralid at 0.25 to 0.5 lb ai/ac (0.28 to 0.56 kg/ha); Curtail® at 2 to 5 qts/ac; and 2,4-D at 1 to 2 lb ae/ac (1.12 to 2.24 kg ae/ha) (William et al. 1997). Herbicides may be applied in spring or fall, and best results are normally obtained when the knapweed is in the rosette stage of growth. Picloram or clopyralid may be used to reduce weed competition during establishment of grass seedlings (Fagerlie 1989; D. Gaiser 1997, personal communication). Herbicides are among the most effective means to control diffuse knapweed in reseeding programs (Hubbard 1975).

• *Biological Control*

Twelve of the insects established in the western United States for classical biological control of knapweeds attack diffuse knapweed (Piper et al. 1996, Rees et al. 1996). Seedhead feeders include the two gall-forming flies, *Urophora affinis* and *U. quadrifasciata*, introduced in the early 1970s, the peacock fly, *Chaetorellia acrolophi*, and two weevils, *Bangasternus fausti* and *Larinus minutus*. *Terellia virens* (fly), *Larinus obtusus* (weevil) and *Metzneria paucipunctella* (moth), three seedhead feeders released for spotted knapweed, use diffuse knapweed to a lesser extent. Larvae of the buprestid beetle, *Sphenoptera jugoslavica*, the weevil *Cyphocleonus achates*, and the moths *Pterolonche inspersa* and *Agapeta zoegana*, damage plants by feeding in the roots. While the *Urophora* flies have demonstrated high occupancy rates in diffuse knapweed heads, the reduction in seed production has had no apparent effect on diffuse knapweed populations. On hot, dry sites, *Sphenoptera jugoslavica* has been reported to weaken diffuse knapweed rosettes in competition with perennial vegetation. High populations of this beetle alter diffuse knapweed population dynamics by reducing rosette survivorship, delaying reproduction, and lowering seed production (Powell and Myers 1988). It infested nearly 80% of the roots in a diffuse knapweed population in Trinity County, California, in 1996 (Joley and Woods

1996). *Sphenoptera* is expected to be less damaging on cooler, moister sites, because female adult beetles require five days of 86° F (30° C) or higher temperatures to lay eggs, and dry conditions favor larval survival (Rees et al. 1996). Of the new agents, *Larinus minutus* is expected to be especially damaging to diffuse knapweed, because a single larva can destroy the entire contents of a seedhead. *Bangasternus* and *Larinus* larvae are aggressive, killing other insects in the same seedhead. More time will be needed to see the full effects of all the agents in combination (Story 1984). Two fungal pathogens, not yet cleared as agents, can be quite destructive to diffuse knapweed under certain conditions. *Puccinia jaceae* attacks the leaves, and *Sclerotinia sclerotiorum* attacks the crowns (Piper et al. 1996).

• *Grazing*

Methods of utilizing diffuse knapweed can be patterned after programs designed for spotted knapweed and yellow starthistle (Wallander et al. 1992, Thomsen et al. 1989, 1993). Timing relative to the development stages of both the weed and associated vegetation is critical to achieve the desired selectivity. Wallander et al. (1992) timed sheep grazing on spotted knapweed to minimize use of Idaho fescue by grazing the pastures first in mid-June when the spotted knapweed was bolting and the Idaho fescue was going dormant, followed by a September grazing before fall growth of the cool-season grasses. Thomsen et al. (1989) found that proper timing of grazing is also critical to suppressing yellow starthistle, and that the first grazing should be timed to the bolting, pre-spiny stage. Subsequent grazings are generally required; local conditions (the moisture regime) determine the number. Timing was more important than class of animal, although sheep and goats accepted yellow starthistle more readily than cattle. A major consideration in planning a grazing regime is whether conditions exist to utilize diffuse knapweed without damaging forage species or soil resources. If not, utilization levels should key on the desirable forage species to maintain vigor in competition with ungrazed weedy species. Diffuse knapweed is more likely to be grazed under the following conditions: 1) by sheep; 2) when it is green and succulent (rosette through bud stage), especially when the associated vegetation is dry (mature or dormant); or 3) when it is the only plant available.

• *Integrated Management*

Integrated management of diffuse knapweed is of necessity site specific, and must be designed for particular conditions and objectives. As an example, results from research plots on a typical diffuse knapweed site in northeastern Washington (Roché 1988) will be used to discuss a few potential alternatives. The effects of five treatments, alone and in combination (burning, cultivation, picloram, seeding of smooth brome [*Bromus inermis*], and nitrogen fertilization [40 lb/ac, 45 kg/ha]) were evaluated for eight years. The site was formerly cropland, used for hay and pasture, before invasion by diffuse knapweed. Native vegetation was a Douglas-fir forest supported by a deep silt loam soil, with average annual precipitation of 18 inches (46 cm).

225

Establishment of a competitive grass cover is imperative for long-term control. Because precipitation levels were high enough, fertilizer increased vigor in residual grasses and extended herbicidal effectiveness. Fertilization alone stimulated both grass and knapweed, producing more diffuse knapweed than grass. Picloram alone dramatically decreased the diffuse knapweed population for two years, but by the third year, the knapweed had rebounded. However, grass production remained higher on sprayed plots than on the control plots through the fifth year. When rainfall was at or above normal, the combination of a herbicide and fertilizer produced maximum weed control and forage production. As single treatments, burning or cultivation provided fleeting control of diffuse knapweed, resulting in weed production equal to or greater than untreated controls after the first year. When all five treatments were combined, grass production reached a peak the third year after treatment, then declined during three subsequent drought years. Best results in seeding were obtained with seedbed preparation and control of weed competition during establishment.

The discouraging aspect of this study was the total reestablishment of diffuse knapweed on all of the plots only eight years after treatment, in the absence of grazing or clipping. In this case we attribute the reinvasion to three years of drought, and the small plot size. Smooth brome was less competitive when annual precipitation dropped from 21 to 15 inches (53 to 38 cm). In the absence of adequate moisture, the fertilizer did not benefit the grass. At this site, Durar hard fescue (*Festuca ovina* var. *duriuscula*) limited diffuse knapweed reinvasion more effectively than smooth brome or orchardgrass (*Dactylis glomerata*). Plots were 6 by 10 feet (1.8 m by 3 m), allowing abundant reseeding of diffuse knapweed from adjacent weedy plots. Regardless of the vigor of native or introduced grass stands, diffuse knapweed seed-rain results in establishment of some weed seedlings (Sheley et al. 1997). Residual herbicides can be used to establish a perennial grass, but do not last as long as weed seed reserves (Lacey et al. 1992). Selection of the most competitive grasses, forbs and/or shrubs for a given site is crucial to long-term management. The best case scenario that one can hope to achieve is selection and establishment of competitive forage species that can, with the help of effective biological control agents and carefully prescribed grazing practices, maintain the diffuse knapweed at low levels.

"Being asked to contribute to this text is an honor. Being given the opportunity to own it as a reference is a privilege. Beyond those two niceties, individual responses will be as variable as are the readers. Weed control, vegetation management, and species manipulation are concepts that reflect the maturity of the program rather than its day-to-day management. Our goal is to offer sufficient insight to place the user in the company of those who understand the species and its ecologic potentials well enough to manipulate (or if you prefer, design and implement) plant communities that will benefit society." —Ben Roché (1924-1997)

Ben wrote the preceding paragraph in 1994, and although this book was not completed in time for him to own it as a reference, the influence of his life's work is evident in more chapters than this one which bears his name. May those who acquire this reference benefit from his legacy.—Cindy Talbott Roché

Literature Cited

Cade, L. 1968. Weeds: biggest danger to range production. Oregon Farmer. February 1, 1968, 20, 22-23.

Callihan, R., and K. Sanders. 1994. Status of Knapweed in Idaho. Knapweed Newsletter, Vol. 8, No. 1. Wash. State Univ. Coop. Ext., Pullman.

Cranston, R.S. 1984. Knapweed in British Columbia. *In*: J.R. Lacey and P.K. Fay (eds.), Proc. Knapweed Symposium, Coop. Ext. Serv. Bull. 1315, Montana State Univ., Bozeman. p. 4-7.

Dostal, J. 1976. *Centaurea* L. *In*: T.G. Tutin, V.H. Heywood, N A. Burges, D.H. Valentine, S.M. Walters and D.A. Webb (eds.), Flora Europaea. Vol 4, 254-301. Cambridge Univ. Press.

Fagerlie, D. 1989. Grass reseeding on sites where picloram has been or will be used. *In*: P.K. Fay and J.R. Lacey (eds.), Proc. Knapweed Symp., April 4-5, 1989. Plant and Soil Sci. Dept .and Coop. Ext. Serv. Montana State Univ., Bozeman. EB45, 220-21.

Fielding, D.J., M.A. Brusven, and L.P. Kish. 1996. Consumption of diffuse knapweed by two species of polyphagous grasshoppers (Orthoptera:Acrididae) in southern Idaho. Great Basin Nat. 56:22-27.

Fletcher, R.A., and A.J. Renney. 1963. A growth inhibitor found in *Centaurea* species. Can. J. Plant Sci. 43:475-81.

Harris, P., and R. Cranston. 1979. An economic evaluation of control methods for diffuse and spotted knapweed in western Canada. Can. J. Plant Sci. 59:375-82.

Harris, P., and J. H. Myers. 1979. *Centaurea diffusa* Lam. and *C. maculosa* Lam. S. lat. diffuse and spotted knapweed (Compositae). *In*: Biological control programmes against insects and weeds in Canada 1969-1980, Commonwealth Agric Bur., 127-37.

Harrod, R.J. and R.J. Taylor. 1995. Reproduction and pollination biology of *Centaurea* and *Acroptilon* species, with emphasis on *C. diffusa*. Northwest Sci. 69:97-105.

Howell, J.T. 1934. *Centaurea diffusa* Lam. Leafl. West. Bot. 1:144.

Howell, J.T. 1959. Distributional data on weedy thistles in western North America. Leafl. West. Bot. 9:1232.

Hubbard, W.A. 1975. Increased range forage production by reseeding and the chemical control of knapweed. J. Range Manage. 28:406-407.

Isaacson, D. 1993. Status and trend of weeds of the *Centaurea* complex in Oregon. Knapweed Newsletter, Vol. 7, No. 3. Wash. State Univ. Coop. Ext., Pullman.

Joley, D.B., and D.M. Woods. 1996. Biological control of diffuse knapweed, *Centaurea diffusa. In:* D.M. Woods (ed.), Biological Control Program Annual Summary, 1996. Calif. Dept. Food and Agric., Div. of Plant Industry, Sacramento, CA.

Kelsey, R.G., and D.J. Bedunah. 1989. Ecological Significance of Allelopathy for *Centaurea* species in the Northwestern United States. *In:* P.K. Fay and J.R. Lacey (eds.), Proc. Knapweed Symp., April 4-5, 1989. Plant and Soil Sci. Dept. and Coop. Ext. Serv. Montana State Univ., Bozeman. EB45, 10-32.

Korpi, J. 1988. Cost of Herbicide Ban. Knapweed Vol 2, No. 1:1. Wash. State Univ. Coop. Ext., Pullman.

Kothman, M.M. (ed.). 1974. A glossary in terms used in range management. 2nd ed. Society for Range Manage., Denver, CO.

Lacey, C.A. 1989. Knapweed management: a decade of change. *In:* P.K. Fay and J.R. Lacey (eds.), Proc. Knapweed Symp., April 4-5, 1989. Plant and Soil Sci. Dept. and Coop. Ext. Service. Montana State Univ., Bozeman. EB45, 1-6.

Lacey, C.A., J R. Lacey, P K. Fay, J.M. Story, and D.L. Zamora. 1992. Controlling Knapweed on Montana Rangeland. Montana State Univ. Coop. Ext. Serv. Circ. 311., Bozeman.

Lacey, J., P. Husby, and G. Handl. 1990. Observations on spotted and diffuse knapweed invasion into ungrazed bunchgrass communities in western Montana. Rangelands 12(1):30-32.

Larson, L.L., and M.L. McInnis. 1989. Impact of grass seedings on establishment and density of diffuse knapweed and yellow starthistle. Northwest Sci. 63:162-66.

Lowdermilk, W.D. 1953. Conquest of the land through seven thousand years. U.S. Dept. Agric., Agri. Info. Bull. No. 99.

Maddox, D. M. 1979. The Knapweeds: their economics and biological control in the western states, U.S.A. Rangelands 1(4):139-41.

Miller, V.A. 1990. Knapweed—A Forage for Big Game in the Kootenays. *In:* B.F. Roché, Jr., and C.R. Roché (eds.), Proc. PNW Range Shortcourse: Range Weeds Revisited. January 25, 1989, Spokane, WA. Wash. State Univ. Coop. Ext., Pullman, Misc. 0143, 35-37.

Müller, H., and D. Schroeder. 1989. The biological control of diffuse and spotted knapweed in North America: what did we learn? *In:* Proc. Knapweed Symp., Montana State Univ., Bozeman, MT, 151-69.

Myers, J.H., and D.E. Berube. 1983. Diffuse knapweed invasion into rangeland in the dry interior of British Columbia. Can. J. Plant Sci. 63:981-87.

Nolan, D.G., and M.K. Upadhyaya. 1988. Primary seed dormancy in diffuse and spotted knapweed. Can. J. Plant Sci. 68:775-83.

Piper, G.L., S.S. Rosenthal, J.M. Story, and N.E. Rees. 1996. Diffuse knapweed (*Centaurea diffusa*). *In:* Biological control of weeds in the West. West. Soc. Weed Sci., USDA Agric. Res. Serv., Montana Dept. Agric., and Montana State Univ., Bozeman.

Popova A.YA. 1960. *Centaurea diffusa* Lam., a steppe-pasture weed in the Crimea. Bot. Zh (Moscow) 45:1207-13.

Powell, R.D. 1990. The role of spatial pattern in the population biology of *Centaurea diffusa*. J. Ecol. 78:374-88.

Powell, R.D., and J.H. Myers. 1988. The effect of *Sphenoptera jugoslavica* Obenb. (Col., Buprestidae) on its host plant *Centaurea diffusa* Lam. (Compositae). J. Appl. Ent. 106:25-45.

Prodan, I. 1930. Centaureele Romaniei. Buletinul Academiei de Cluj Institulul de arte Grafice. Ardealul Strada, Memorandului No. 22 1930.

Rees, N.E., P.C. Quimby, Jr., G.L. Piper, E.M. Coombs, C.E. Turner, N.R. Spencer, and L.V. Knutson (eds.). 1996. Biological control of weeds in the West. West. Soc. Weed Sci., USDA Agric. Res. Serv., Montana Dept. Agric., and Montana State Univ., Bozeman.

Renney, A.J. 1959. *Centaurea* spp. infestation in British Columbia. Proc. Joint meeting, North Central Weed Conference 16 and West Canada Weed Control Conf. 10:18-19.

Renney, A.J., and E.C. Hughes. 1969. Control of knapweed, *Centaurea* species, in British Columbia with Tordon herbicides. Down to Earth 24:6-8.

Roché, B.F., Jr. 1967. Range and pasture weed survey for Washington by county. Wash. State Univ. Coop. Ext., Pullman.

Roché, B.F., Jr. 1988a. Knapweed award upheld by appeals court. Knapweed Newsletter Vol. 2. No. 1:4. Wash. State Univ. Coop. Ext., Pullman.

Roché, B.F., Jr. 1988b. Management Technologies for Diffuse Knapweed Control. Knapweed Newsletter, Vol. 2, No. 4. Wash. State Univ. Coop. Ext., Pullman.

Roché, B.F., Jr. 1994. Status of Knapweeds in Washington. Knapweed Newsletter, Vol. 8, No. 1. Wash. State Univ. Coop. Ext., Pullman.

Roché, B.F., Jr., and C.J. Talbott. 1986. The collection history of *Centaureas* found in Washington State. Agri. Res. Center. Res. Bull. XB0978. Wash. State Univ. Coop. Ext., Pullman.

Roché, C.T. 1988. Cost of knapweeds on rangelands in eastern Washington. Knapweed Newsletter Vol. 2, No. 4:4. Wash. State Univ. Coop. Ext., Pullman.

Roché, C.T., and B.F. Roché, Jr. 1990. Mowing diffuse knapweed: effects on longevity and seed production. *In*: Proceedings from Washington State Weed Conference, Washington State Weed Assoc., Yakima, 19-23.

Roché, C.T., and B.F. Roché, Jr. 1993. Identification of knapweeds and starthistles in the Pacific Northwest. PNW 432.

Sabadell, J.E., E.M. Risley, H.T. Jorgenson, and B.S. Thornton. 1982. Desertification in the United States. U.S. Dept. Interior, Bureau of Land Management.

Schirman, R. 1981. Seed production and spring seedling establishment of diffuse and spotted knapweed. J. Range Manage. 34:45-47.

Schuller, R. 1992. Knapweeds invade natural areas. Knapweed Newsletter Vol. 6, No. 4: 4. Wash. State Univ. Coop. Ext., Pullman.

Sheley, R L., and L.L. Larson. 1996. Emergence date effects on resource partitioning between diffuse knapweed seedlings. J. Range Manage. 49:241-44.

Sheley, R.L., B.E. Olson, and L.L. Larson. 1997. Effect of weed seed rate and grass defoliation level on diffuse knapweed. J. Range Manage. 50:39-43.

Small, J. 1919. The origin and development of the Compositae. New Phytologist 18:1-35.

Story, J.M. 1984. Collection and Redistribution of *Uropha affinis* and *U. quadrifasciata* for Biological Control of Spotted Knapweed. Montana Sate Univ. Coop. Ext. Circ. 308.

Talbott, C.J. 1987. Distribution and ecologic amplitude of selected *Centaurea* species in Eastern Washington. M.S. Thesis. Washington State University, Pullman.

Thomsen, C.D., W.A. Williams, M.R. George, W.B. McHenry, F.L. Bell, and R.S. Knight. 1989. Managing yellow starthistle on rangeland. Calif. Agric. 43(5):4-7.

Thomsen, C.D., W.A. Williams, M. Vayssiéres, F.L. Bell, and M.R. George. 1993. Controlled grazing on annual grassland decreases yellow starthistle. Calif. Agric. 47(6):36-40.

229

Tucker, R. 1990. The myths of knapweed. Knapweed Newsletter, Vol. 4, No. 1. Wash. State Univ. Coop. Ext., Pullman.

Wallander, R., B. Olson, and J. Lacey. 1992. Sheep graze spotted knapweed. Knapweed Newsletter, Vol. 6, No. 2. Wash. State Univ. Coop. Ext., Pullman.

Watson, A.K., and A.J. Renney. 1974. The biology of Canadian weeds. *Centaurea diffusa* and *C. maculosa*. Can. J. Plant Sci. 54:687-701.

William, R.D., D. Ball, T.L. Miller, R. Parker, J.P. Yenish, R.H. Callihan, C. Eberlein, G.A. Lee, and D.W. Morishita. 1997. Pacific Northwest Weed Control Handbook. Oregon State Univ., Corvallis.

Youtie, B., and J. Soll. 1994. Non-chemical control of diffuse knapweed (*Centaurea diffusa*). Knapweed Newsletter Vol. 8, No. 3:2-3. Wash. State Univ. Coop. Ext., Pullman.

Dyer's Woad

Erin G. McConnell, John O. Evans, and Steven A. Dewey

Loathed by land managers and loved by photographers for its showy floral display, dyer's woad (*Isatis tinctoria*) is no exception to the rule that many noxious weeds were at one time introduced as beneficial plants. Dyer's woad is a member of the mustard family (Brassicaceae) and has blue-green leaves and numerous bright yellow flowers in an umbrella-shaped inflorescence, making it easy to identify. Even before the Christian era, dyer's woad was believed to have medicinal attributes, and later it was cultivated as a dye crop (Varga and Evans 1978). Ancient warriors painted themselves with blue extract from dyer's woad to look more ferocious prior to going into battle. Dyer's woad was introduced from Europe and cultivated in the eastern United States as a textile dye crop during the colonial period, and escaped to become a nuisance on range and cropland west of the Missouri River. Though it is still grown occasionally in the eastern states as a dye crop, it is not a threatening weed. Researchers speculate that the rapid invasion of forest and rangeland in the western United States by dyer's woad is because the plant was preadapted to the alkaline soils and arid climate of the West, as well as to the large expanses of public lands, uncommon in the East. Dyer's woad and its control are of extreme concern to public land managers and policymakers throughout the intermountain West.

Identification

Dyer's woad has small, bright yellow flowers, each with four petals and four sepals. Petals are about $1/8$ inch (0.3 cm) wide, and only slightly longer. Flowers are clustered in racemes on upper parts of multi-branched stems. At peak flowering, dense stands of dyer's woad appear bright yellow with a hint of chartreuse green. Each flower produces a teardrop-shaped winged silicle (fruit) that hangs from a small stalk. Fruits are $1/2$ to $3/4$ inch (1 to 2 cm) long and $1/4$ inch (0.6 cm) wide, black or purplish-brown at maturity. The fruit is strongly flattened, with a wing around a thickened center where one seed is held. Occasionally, two seeds exist, though the second seed generally is not viable. Seeds are brownish-yellow and cylinder-shaped. The fruits drop intact from the plant, unlike most mustards whose fruits split to release the seeds. Dyer's woad normally grows 1 to 3 feet (0.3 to 1 m) tall, but may reach over 5 feet (2 m). Typically, it has a 3 to 5 foot (1 to 2 m)-long taproot and some lateral roots in the upper 12 inches (30 cm) of soil. Rosette leaves, attached by a stalk, are widest near the tip and are usually covered with soft fine hairs. Stem leaves are alternate, lance-shaped, and clasp the stem with short basal lobes. Stem leaves usually lack hairs and their margins are mostly entire. All leaves have a cream-colored midrib on the upper surface from the base to the tip, a key identifying feature. The flower stems are branched near the top of the stem, and stiffen into an umbrella-like structure at maturity.

Origin, History, and Distribution

A native of southeastern Russia, dyer's woad has spread on its own or has been transported to many countries. It exists on six continents and grows wild in China, Tibet, and Afghanistan (Varga and Evans 1978). It was imported to North America from Europe by early colonists as a textile dye crop then accidently spread as a crop seed contaminant. Dyer's woad arrived in the West as a contaminant in a shipment of alfalfa seed to California from Ireland (Roché 1992). It was also introduced near Brigham City, Utah, about 1910, again as a contaminant in alfalfa seed (Callihan 1990). Today, dyer's woad persists as a weed in eight western states and threatens to invade others, particularly those with large amounts of rangeland and pasture. Dyer's woad also readily invades alfalfa and small grain fields, orchards, waste areas, and along waterways.

Potential Invasion

Dyer's woad poses a real threat to rangelands, forests, and pastures of the intermountain West because of its ability to dominate plant communities (Farah et al. 1988). Dyer's woad thrives on rocky soil with limited water-holding capacity. Competition begins early in the growing season, probably because of its accelerated growth rate from rosettes to flowering plants. In one experiment, stem growth rate averaged 4 inches (10 cm) per week in April and May. Another reason for dyer's woad's success rests with its root structure. A deep taproot extends into the soil to access the deeper nutrient and moisture reserves, while a shallower set of lateral roots takes advantage of spring moisture and surface nutrients.

Prolific seed production enables dyer's woad to spread at a rapid rate. One infestation south of Dillon, MT, increased from two acres to more than 100 acres in just two years (Aspevig et al. 1985). It is estimated that dyer's woad is spreading at an annual rate of 14% on Bureau of Land Management rangeland in the Pacific Northwest, and reduces grazing capacity by an average of 38% (USDI 1985). The number of infested hectares on national forest lands in the intermountain region increased by a factor of more than 35 between 1969 and 1985 (USDA 1986). On the Cache National Forest of northern Utah, a 1988 study of nearly 150,000 hectares (373,000 acres) concluded that there was potential for a 124-fold increase in the number of dyer's woad populations (Dewey et al. 1991).

Impacts

In 1981 it was estimated that dyer's woad reduced crop and rangeland production in Utah by $2 million (Evans and Chase 1981). The dyer's woad infestation has doubled there in the last decade, and certainly causes several million dollars' loss per year, causing widespread concern among land managers. Dyer's woad can be controlled more easily in cropland than in rangeland and forests. In forests and on rangelands, control is limited by inaccessible terrain, possible undesirable impacts of machinery and chemicals on desirable forage and native plants, and questionable economic returns on weed control investments.

Unlike many weeds in the mustard family, dyer's woad does well in the absence of disturbance. It is capable of encroaching upon and increasing its density on well-vegetated range sites that have not been grazed or disturbed for decades (Roché 1992). A healthy, dense stand of grass and other perennials deters the spread of dyer's woad, but will not stop an invasion.

Biology and Ecology

Dyer's woad behaves as a winter annual, biennial, or short-lived perennial. In the intermountain area it may germinate in the spring or fall. Typically it remains as a rosette of basal leaves during the following summer and winter, flowers in April and May of the second year, and seeds ripen in June and July. Studies have shown that about 1% of fall-germinated plants flowered the first spring, half of the 35% of the plants that survived the second winter flowered the second spring, and 12% did not flower until the third spring (Farah et al. 1988). Winter chilling is necessary for rosettes to bolt and flower. Often up to 20 stalks begin to develop from each rosette, but fewer than eight mature. Plants may produce 350 to 500 seeds each, but selected plants have been known to produce more than 10,000 seeds in one year.

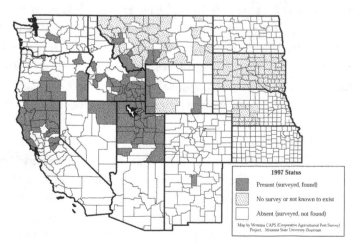

1997 Status

Present (surveyed, found)

No survey or not known to exist

Absent (surveyed, not found)

Map by Montana CAPS (Cooperative Agricultural Pest Survey) Project, Montana State University-Bozeman

233

Distribution of dyer's woad by county in the western United States. Data were collected by surveying various weed authorities in each state.

Dyer's woad spreads to new sites by seed. Seed dispersal studies revealed that 95% of the winged fruits fell within 22 inches (54 cm) of the parent plant (Farah et al. 1988). The greatest distance that fruits were wind-blown from their parents was 8 feet (2.4 m). Wind and rain are important factors in fruit detachment as well as in the direction fruits move away from the parent. Some fruits remain on the plants until winter, when they may blow much greater distances over the surface of crusted snow. Vehicles, flowing water, birds, and contaminated feed, bedding, and crop seed, as well as movement of seed-contaminated soil, are important in long distance dispersal. A fruit pedicel can serve as a hook-type apparatus that can attach to vectors such as animals or people. Long-range seed dispersal is often facilitated by moving water such as canals, streams, and rivers. Anecdotal reports of dyer's woad reappearing after tillage of grasslands suggest that seed may remain viable in the soil for several years, though this has not been verified under controlled conditions.

The fruits of dyer's woad probably contain allelopathic substances, but the chemicals have not been fully characterized. Experiments were conducted in Nevada to determine the allelopathic effects of dyer's woad on itself and on other species (Young and Evans 1971). Fruits of dyer's woad were soaked in water for varying intervals, and the leachate was used to irrigate germinating seeds of several crops as well as other weeds. The water extract inhibited germination of dyer's woad and numerous other species. Dyer's woad seeds separated from the fruits do not exhibit seed dormancy and readily germinate under a variety of conditions, though they do not readily germinate when they remain intact within the fruit (Young and Evans 1971). The inhibitors in the fruit may allow dyer's woad seed to persist over time, correlating with precipitation patterns that leach inhibitors from fruits, insuring that germination occurs under favorable conditions.

Management

Dyer's woad seeds can be transported by vehicles, rail cars, hikers, or passing animals, where they are dropped onto suitable sites. As the seeds germinate, new plants readily grow and produce seeds which spread to neighboring areas or are picked up once again by vectors, and thus the cycle continues. Because roadsides, railways, and trails are such effective avenues of seed dispersal, it is extremely critical that any woad growing in these areas be destroyed or removed. It is especially important not to allow the plants to produce seeds.

Prevention and early detection are paramount in managing dyer's woad invasions. One of the most important methods of containment or control is hand-pulling. The easily-identified, distinct yellow blossoms are readily recognized by individuals who do not have special training in plant identification. Civic volunteer groups such as Boy Scouts, high school students, garden clubs, and other groups such as corrections crews and seasonal employees, can clear large tracts of land in a relatively short time with basic tools. Hand-pulling is very effective in hard-to-reach places, such as along fence lines, canal banks, and wooded areas. It may be the only practical control method in difficult or sensitive terrain.

• Agricultural Settings

In fields where dyer's woad infestations are severe, cultivation and herbicides can be used advantageously. Annual crops and rowcrops are cultivated often enough that tillage itself should eliminate dyer's woad, though occasionally an additional tillage may be required, particularly if cropland is fallowed to conserve moisture. Spring cultivation destroys the vernalized rosettes and effectively stops seed production, provided escapes are dealt with appropriately. Dyer's woad seedlings sometimes appear after spring cultivation, but cannot bolt and blossom until the next year following cold exposure. To remove the competitive effects of immature dyer's woad seedlings in small grain and forage grass fields, selective herbicides such as 2,4-D, metsulfuron, or dicamba can be used.

Dyer's woad is often a problem in perennial crops such as alfalfa, particularly in fields grown without irrigation. Dyer's woad is easily spread to new areas in hay bales transported over great distances in the western states. Animals that consume alfalfa hay contaminated with dyer's woad seeds can further aid the dissemination process to even more remote areas. Domestic animals and wildlife will deposit dyer's woad seed to rangelands where its effects are more severe and more difficult to manage than in alfalfa fields. Hexazinone, metribuzin and 2,4-DB are effective herbicides in controlling dyer's woad in alfalfa fields. Hay stands need to be properly managed for herbicides to control dyer's woad satisfactorily. Fields that display stressed crop plants should be taken out of hay production and rotated to other crops, rather than trying to force herbicidal weed control.

• Rangeland and Forests

There are three major strategies used to manage dyer's woad in rangeland and forests: hand-pulling, herbicide application, and biological control. Grazing experiments using sheep to control dyer's woad in Utah indicated that the stocking rates required to damage dyer's woad populations would likely cause further deterioration of an already depleted range (West and Farah 1989). Sheep did not voluntarily graze significant amounts of dyer's woad, even though the range was in poor condition.

Hand-pulling of individual weeds is probably one of the simplest and most effective methods of dyer's woad control. Hand-pulling is most effective in areas surrounding major infestations, and in areas where the weed has been introduced recently, remote from any other infestations. To be effective, it is generally necessary to wait until the plant bolts and begins flowering before attempting a hand-pulling operation. The distinctive yellow flowers make it easy to locate and identify. Once the plants have been identified, they can be removed by pulling if the ground is wet, or by digging with a hoe or shovel. The fleshy taproot must be removed below the root crown of the plant or regrowth will occur. It is important to remember that there are only four to six weeks from time of flowering until the seeds are mature. Do not let dyer's woad plants go to seed! Breaking or cutting off the flowering stalks does not kill dyer's woad but will encourage it to develop new stems and produce seeds later in the season. Plan to hand-pull dyer's woad two to three times each year for several years. In a demonstration involving youth volunteers, dyer's woad density was reduced by approximately 49% one year after a single hand-pulling event. Subsequent annual hand-pulling operations reduced the density by approximately 75% of the original level after four years, and more than 90% after seven years. Eradication was not achieved on any land unit (because of reintroduction of seed from surrounding infestations), but control reached 97% or greater on most units retained in the program for eight or more years (Dorst et al. 1994). It is essential that the plants be removed as soon as possible after flowering to prevent the possibility of some slipping by and going to seed. Even though the plants have been uprooted, seeds can continue to develop and reach maturity in dyer's woad pulled after green fruits form (Dewey and Squire 1994).

The importance of hand-pulling cannot be overemphasized, especially in those areas that are only lightly infested. Land managers and others should constantly be on the alert for dyer's woad. As they make surveys in May and June, land managers should remove any small isolated patches of the weed.

Metsulfuron in combination with 2,4-D are the herbicides found to be most effective against dyer's woad in pastures and rangeland (Evans and Gunnel, unpublished data). Chlorsulfuron and metsulfuron, with or without 2,4-D, are highly effective in non-crop situations where it is desired to maintain established perennial grasses. Picloram and dicamba have not proven significantly more effective than 2,4-D in controlling dyer's woad. Excellent control of dyer's woad can be obtained by spraying with 2,4-D in the rosette stage. As the plant enters early bud and blossom stages, 2,4-D often does not kill it quickly enough to prevent seed production. Production of fruit and viable seed was reduced little or not at all when dyer's woad plants were treated with 2,4-D in the late bloom stage of development. However, chlorsulfuron or metsulfuron completely prevented viable seed production when applied at that stage. Combining 2,4-D with other herbicides shows more promise to immediately stop dyer's woad growth and seed production. The use of 2,4-D should be confined to those areas where adjacent properties will not be affected by spray drift. Dyer's woad typically invades along the edges of highways, railroads, or canals.

One of the most exciting discoveries with regard to stopping the advance of dyer's woad is a native rust pathogen, tentatively identified as *Puccinia thlaspeos*. First discovered on dyer's woad in 1978 in southern Idaho, the rust has since spread into woad populations throughout Idaho and Utah. Plants usually become infected during spring and early summer, though fall infections are also possible. Symptoms do not appear until three to nine months following infection, and generally only appear when plants are ready to bolt. Initial symptoms appear as stunting, leaf distortion, and chlorosis. The rust itself is visible on the plants as tiny brown or reddish spots on the undersides of the leaves. These structures house the reproductive spores which are rubbed off by vectors or carried on the wind to spread the infection. Fruit and seed production are completely prevented on almost all infected plants (Kropp et al. 1995). Studies are under way to determine optimum conditions for the pathogen and whether rust spores can be hand-disseminated to remote locations where dyer's woad exists. Recent surveys reveal that the rust is naturally spreading to new dyer's woad infestations and significantly slowing the growth and reproduction of many dyer's woad populations.

Literature Cited

Aspevig, K., P. Fay, and J. Lacey. 1985. Dyer's woad: a threat to rangeland in Montana. Montguide. Montana State Univ. Cooper. Ext.. Bozeman, MT.

Callihan, R.H. 1990. Dyer's woad: biology, distribution and control. Current information series No. 857. Univ. Idaho Agric. Exp. Sta. Moscow, ID.

Dewey, S.A., K.P. Price, and D. Ramsey. 1991. Satellite remote sensing to predict potential distribution of dyer's woad (*Isatis tinctoria*). Weed Technol. 5:479-84.

Dewey, S.A., and J.O. Squire. 1994. Germinable seed production of dyer's woad plants hand-pulled at eight stages of flower of fruit development. Proceedings, West. Soc. Weed Sci. 47:52.

Dorst, H.E., S.A. Dewey, and J.O. Evans. 1994. Manual control of dyer's woad on heavily infested rangeland and non-crop sites in northern Utah. Proc., West. Soc. Weed Sci. 47:13-15.

Evans, J.O., and R.L. Chase. 1981. Dyer's woad control. Bull. EL-188. Utah State Univ. Ext. Serv. Logan, UT.

Evans, J.O., and R.W. Gunnel. 1982. Unpublished data. Utah State Univ. Logan, UT.

Farah, K.O., A.F. Tanaka, and N.E. West. 1988. Autecology and population biology of dyer's woad (*Isatis tinctoria*). Weed Sci. 36:186-93.

Kropp, B.R., S. Albee, K.M. Flint, P. Zambino, L. Szabo, and S.V. Thomson. 1995. Early detection of systemic rust infections of dyer's woad (*Isatis tinctoria*) using the polymerase chain reaction. Weed Sci. 43:467-72.

Roché, C. 1992. Dyer's woad. PNW 384. Pacific Northwest Extension Service. WA, ID, OR.

United States Department of Agriculture. 1986. Intermountain region noxious weed and poisonous plant control program environmental impact statement. USDA Forest Service, Intermountain Reg., Ogden, UT.

United States Department of Interior. 1985. Northwest area noxious weed control program environmental impact statement. USDI Bureau of Land Management, Oregon State Office, Portland, OR.

Varga, W.A., and J.O. Evans. 1978. Dyer's woad: from cultivated to cursed. Utah Sci. 39:87-90.

West, N.E., and Farah, K.O. 1989. Effects of clipping and sheep grazing on dyer's woad. J. Range Manage. 42(1):5-10.

Young, J.A., and R.A. Evans. 1971. Germination of dyer's woad. Weed Sci. 19:76-78.

Meadow and Orange Hawkweed

Linda M. Wilson and Robert H. Callihan

Meadow hawkweed (*Hieracium pratense*) (= *Hieracium caespitosum*) and orange hawkweed (*H. aurantiacum*) are among the 11 species of highly invasive hawkweeds introduced into North America from Europe. They are members of the chicory tribe of the sunflower family (Asteraceae), being closely related to dandelion, prickly lettuce, sowthistle, and chicory. Hawkweeds are able to colonize and rapidly dominate new sites, because they possess a wide range of highly successful reproductive strategies: hawkweeds reproduce not only by seeds, but also by rhizomes, stolons, and adventitious root buds. Furthermore, seeds can be produced either sexually or asexually. These reproductive strategies, and a tolerance for low-productivity soils, are probably why hawkweeds have so quickly spread throughout much of the Northwest since their arrival only 30 years ago.

Identification

Meadow and orange hawkweeds are creeping perennials that contain a milky sap and have shallow, fibrous roots. Plants in the vegetative stage are a basal rosette of hairy leaves. The narrow, spatula-shaped leaves are 4 to 6 inches (10.2 to 15.25 cm.) long, dark green above and light green beneath. Each rosette produces 10 to 30 flower stems that are 10 to 36 inches (25.4 to 91.5 cm.) tall (L. Wilson, unpubl. data). Stems have short, stiff hairs, and may have one to three small, clasping leaves below the midpoint of the stem. Each inflorescence consists of 5 to 30 bright yellow (meadow hawkweed) or orange (orange hawkweed) dandelion-like flower heads, from $^1/_2$ to $^3/_4$ inch (1.3 to 1.9 cm.) in diameter. Heads contain all ligulate flowers arranged in a flat-topped cluster. The 12 to 30 tiny, columnar seeds have a tawny tuft of bristles on the flattened end. Flowering plants produce from 4 to 12 leafy stolons that can reach a length of 4 to 12 inches (10.2 to 30.5 cm.).

Hawkweeds are notorious for their complex and confusing classification. Species are difficult to distinguish because they interbreed freely, and many of our hawkweed populations look like hybrids. It is important to distinguish the non-indigenous hawkweeds from the many native *Hieracium* species in the western United States. Native hawkweeds lack stolons, have leafy, branched stems, and bear flowers in open panicles. Taxonomically, the non-indigenous species differ from the native species; the non-indigenous species belong to the subgenus *Pilosella*, whereas the native species belong to either the subgenus *Stenotheca* or the subgenus *Hieracium*.

Origin, History, and Distribution

Meadow hawkweed occurs in the northern, central, and eastern portions of Europe, and is a regular component of the native flora of the foothills of the Alps (Sell 1974). From Europe, it spread to North America (Hulten and Fries 1986) and New Zealand (Grundy 1989). Meadow hawkweed was probably introduced into the United States in 1828 (Britton and Brown 1970). It is now found from Quebec to Ontario, and southward to Georgia and Tennessee (Rickett 1973). The first report of meadow hawkweed in the Pacific Northwest is from June 1969 in Pend Orielle County, WA. (Marion Ownbey Herbarium, Washington State University, Pullman). It is now found in many Inland Northwest counties (Callihan et al. 1997, Wilson et al. 1997). As of 1998, the largest infestations of meadow hawkweed are centered in northern Idaho, northeastern Washington (extending into the Okanogan Highlands), and northwestern Montana (extending to the eastern slopes of the Rocky Mountains). In every location it is considered to be spreading rapidly. It is not reported to be a problem weed in Oregon or California.

In Europe, orange hawkweed comprises a large, diverse group of plants originating from a restricted area in northern and central regions (Sell 1974). It does not occur in southern or eastern Europe. In its native range, orange hawkweed occurs primarily in mountainous meadows and hillsides, although is widely cultivated elsewhere (Skalinska 1976). From Europe, it spread to North America (Hulten and Fries 1986), New Zealand (Grundy 1989), and Japan (Suzuki and Narayama 1977). It was introduced in Vermont in 1875 as an ornamental (Voss and Bohlke 1978). Since

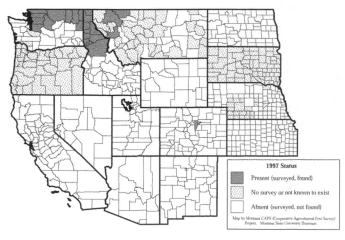

239

Distribution of meadow hawkweed by county in the western United States. Data were collected by surveying various weed authorities in each state.

that time, orange hawkweed has been planted across the northern United States as an ornamental, and is often found escaped from landscape plantings, cemeteries, and gardens. Within 25 years it had spread throughout much of New England, west to Michigan (Voss and Bohlke 1978) and into Canada from New Brunswick to Ontario (Britton and Brown 1970). It now occurs throughout the eastern seaboard and into the Midwest, extending west to Minnesota and Iowa, and south to Virginia and North Carolina (Johnson 1977), and along the eastern slope of the Rocky Mountains in Colorado (Weber 1990). In the western United States, populations of orange hawkweed were first reported from coastal Washington and Oregon (Peck 1941, Abrams and Ferris 1960). It has also been reported from the Lower Mainland of British Columbia (Guppy 1976) where it infests pastures, old fields, and roadsides.

Potential Invasion

In their native range, meadow and orange hawkweeds are ruderal species of pastures, roadside cutbanks, abandoned fields, and meadows (Skalinska 1967). In most cases they are found in small, isolated pockets. Their highest densities are found on recently disturbed areas; they do not persist as dominant members of the early successional community (Skalinska 1967).

Hawkweeds are primarily weeds of moist pastures, forest meadows, abandoned fields, clearcuts, and roadsides. However, they have shown a tendency to invade mid- to high-elevation meadows and abandoned farmland. The potential for spread in the northern Great Plains and the Columbia River region is difficult to predict. However, hawkweeds' occurrence as weeds in the northeastern states during the last century, and their original distribution in northern and central Europe, suggest that they pose the greatest threat to cooler, sub-humid to humid sites in the northern regions of the United States. Based on current infestations, habitats most susceptible

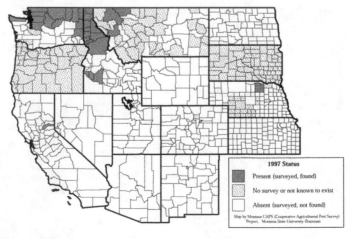

Distribution of orange hawkweed by county in the western United States. Data were collected by surveying various weed authorities in each state.

to invasion range from the lowlands of the northern Pacific Coast to elevations of 5,000 feet (1,524 m) or more in the mountain states. Sites most vulnerable to invasion include roadsides, mountain meadows and clearings in forest zones, permanent pastures, hayfields, cleared timber units, and abandoned farmland where the soil is well drained, coarse textured, and moderately low in organic matter. Consequently, elk habitat, recreation areas, and pristine mountain meadows in areas that have a climate similar to that in their native range are particularly susceptible. Neither species appears to survive in annually-tilled cropland. Meadow hawkweed appears to have a greater potential for spread than does orange hawkweed. Reasons for this are unclear, but may relate to meadow hawkweed's ability to spread vegetatively by root fragments.

Meadow and orange hawkweeds are closely associated with habitats that support oxeye daisy (*Chrysanthemum leucanthemum*), sulfur cinquefoil (*Potentilla recta*), spotted knapweed (*Centaurea maculosa*), gray goldenrod (*Solidago nemoralis*), wild carrot (*Daucus carota*) and dandelion (*Taraxacum* spp.) (Thomas and Dale 1974, Maycock and Guzikowa 1984). None of the non-indigenous species are found in the natural grasslands or shrub steppe of the northern intermountain West, and are not expected to become problem weeds in any dry habitat usually associated with western rangelands.

Impacts

Meadow and orange hawkweeds are tenacious invaders. There is serious concern with the loss of native plant biodiversity in infested areas. Once established, hawkweeds quickly develop into a patch that continues to expand until it covers the site with a solid mat of rosettes. Forage species in pastures and abandoned farmland are choked out by the advancing front of hawkweed. Hawkweeds threaten lawns and gardens too. Hawkweeds have been reported to have allelopathic effects on neighboring vegetation, by exuding toxic chemicals into the soil (Dawes and Maravolo 1973, Makepeace 1976).

Livestock, deer, and elk consume hawkweed foliage and buds. In a study of the nutritive value of hawkweed, leaf protein content ranged from 7% (similar to that in low-quality cured timothy hay) to 11% (equal to high-quality grass hay) (Callihan et al. 1995). The digestibility of hawkweed leaves (74%) was well above that expected of grass pasture (60% to 65%). Flowers had moderate digestibility (64%), but contained twice as much protein (18%) as the leaves. The digestibility data suggest that nutritional qualities of meadow hawkweed leaves may be utilized by ruminants. However, questions related to palatability and utilization remain to be answered.

Biology and Ecology

New plants begin as seedlings or as leaves sprouting from stolons, rhizomes (Thomas and Dale 1975), or roots (Peterson 1979). Juvenile plants consist of low-lying rosettes that develop into flowering plants as erect, slender stems arise from the rosette center. Yeung and Peterson (1972) showed that flowering in the closely related *H. floribundum* is dependent on photoperiod, and that flowering occurs only after exposure to a specific amount and quality of light. Hawkweeds have a relatively short life cycle. Seedlings planted in March can produce flowers by mid-June and set seeds by early August (L. Wilson, unpublished data).

All of the weedy, non-indigenous hawkweeds are polyploid and reproduce asexually by apomixis. Apomictic reproduction includes both vegetative reproduction and asexual seed production. Seeds are occasionally produced sexually by periodic pollination and outcrossing. Vegetative reproduction is accomplished in three ways: stolons, rhizomes and, in the case of meadow hawkweed, adventitious root buds. These reproductive strategies ensure a high degree of genetic variability within and between populations (Bishop and Davy 1994).

Mature seeds do not have an after-ripening period (Thomas and Dale 1974), and can germinate as soon as they are released from the plant (Stergios 1976). Studies have shown that seeds are viable in the soil for up to seven years (Panebianco and Willemsen 1976). Seeds are less important than vegetative spread to the population dynamics of non-indigenous hawkweeds in North America. In a study of the seed dynamics in an established population of the closely related *H. floribundum*, Thomas and Dale (1974) found that only 1% of new plants in a population were derived from seedlings. Moreover, most seeds (80%) are dispersed within the plant colony, and less than 1% are found farther than 30 feet (10 m) from the patch (Thomas and Dale 1974). Most of the seeds that germinated outside the colony succumbed to summer drought or winterkill (Johnson and Thomas 1978). However, most surviving seedlings were those that germinated the following spring (Panebianco and Willemsen 1976).

Several studies have shown that density-dependent phenomena are important in regulating hawkweed populations. The timing and rate of flowering, the number and viability of seeds, and stolon production are all regulated to some degree by the density of plants in a colony (Thomas and Dale 1975). Most of the flowering plants in a population are located near the periphery of the population. Plants in the center of the patch, where density is highest, have a lower rate of flowering (Thomas and Dale 1974). Stergios (1976) showed that only about 10% of the plants in the middle of the patch flowered.

Although most new hawkweed infestations are probably started by seeds, most expansion of established populations is vegetative. Orange hawkweed sends from three to eight long, slender stolons out along the soil surface. Meadow hawkweed also produces long, slender stolons, but some of its vegetative structures grow as shallow, underground rhizomes (L. Wilson, unpubl data). Stolons and rhizomes, initiated from axillary buds at the base of rosette leaves, begin to grow when the plant initiates flowering. Once established, vigorous stolon growth quickly expands the colony, forming dense patches that can have as many as 3,200 plants per square yard (3,500 plants per square meter). New plants develop after stolons elongate through the summer and form daughter rosettes at their tips (Thomas and Dale 1974, L. Wilson unpubl. data). In meadow hawkweed, new rosettes also develop from plants that sprout from swollen, adventitious root buds located on the fibrous roots (L. Wilson, unpublished data).

Management

• *Prevention and Early Detection*

Hawkweed prevention is possible where sound land management prevents new infestations from establishing. Small, newly-formed infestations should be controlled before the patch spreads vegetatively. Large infestations are more difficult to control. Early detection of hawkweeds is difficult because they infest mountainous, forested habitats where satellite infestations are well concealed. Conventional ground-based surveys to determine the location and extent of hawkweed infestations are not practical where infestations cover remote, mountainous terrain. High-resolution, multispectral, digital images may offer a better alternative to visual surveys over large areas. In a study conducted in northern Idaho, Carson et al. (1995) used multispectral digital imaging to detect flowering hawkweed, and produced images with 1-meter resolution where hawkweed cover exceeded 60%. Less dense patches resulted in images with less resolution, but meadow hawkweed can be detected with sufficient resolution across a range of phenological stages (Lass and Callihan 1997).

• *Mechanical Control*

Mechanical control of hawkweed has had limited success. Digging the plants or otherwise disturbing the stolons, rhizomes, or roots only serves to spread the weed, since plants can grow from buds on small root, stolon, and rhizome fragments. Disturbance by machinery will only spread the weeds across the field. Local disturbances caused by grazing livestock, ungulates, and rodents enhance the rate of spread of hawkweed.

In lawns, mowing does not kill the weeds because the low-lying stolons and rhizomes are missed by the mower blades. Although mowing prevents seed production by removing flowering stems, repeated mowing encourages faster vegetative spread.

• *Chemical Control*

Meadow and orange hawkweeds are effectively controlled by phenoxy-type herbicides, including 2,4-D, clopyralid, and picloram (Noel et al. 1979, Lass and Callihan 1992b). Applications of 2,4-D at 1.5 to 2.0 pounds of active ingredient per acre (1.68 to 2.24 kilograms active ingredient per hectare) should be made early

243

during the growing season. Plants should be treated with herbicides in the rosette stage to prevent flowering and seed production. A surfactant should be included in all herbicide tank mixtures, to ensure adherence of the material to the hairy leaf and stem surfaces. Studies conducted at the University of Idaho showed that over 50% control was achieved for six years following treatment with clopyralid, applied at the rate of 0.5 lb ai/ac (0.56 kg ai/ha) (Lass and Callihan 1992b). Similar results were achieved using picloram at the rate of 0.25 to 0.5 lb ai/ac (0.42 to 0.56 kg ai/ha). Other herbicides either failed to control meadow hawkweed or suppression was for fewer than three years (Lass and Callihan 1992b, Miller et al. 1987). Additional herbicide trials are currently under way at the University of Idaho. Herbicide recommendations vary by region and site. For specific recommendations, consult a state Extension weed specialist or refer to the *Pacific Northwest Weed Control Handbook* (William et al. 1998).

• *Cultural Control*

Where perennial grasses, legumes, and other beneficial forbs are present in the plant community, fertilizers can help control hawkweed by increasing the competitive ability of the more desirable species. This may be particularly important in rangeland and pastures, because these lands are generally not priority areas for supplemental fertilization, and soil nitrogen levels may be inadequate for optimal grass health. Marked reductions of hawkweed density and vigor have been obtained by fertilizer treatments in the United States (Reader and Watt 1981), Canada (Hay and Ouellette 1959), and New Zealand (Scott et al. 1990). However, Reader and Watt (1981) found that repeated fertilizer treatments had no effect on dense patches of hawkweed that contained few grasses or other forbs. Depending on soil productivity and grass condition, a single nitrogen application may be sufficient for grasses competitively to suppress hawkweed growth for three to five years (Reader 1990). Good grazing management will likely extend this period. Thus, it appears that a combination of herbicides and fertilizers will contain the spread of hawkweed once it becomes established in a field. Hawkweed does not persist in cultivation because crops can outcompete hawkweed, especially where herbicides are used in the cropping system.

• *Biological Control*

Biological control is another weed management strategy being developed for non-indigenous hawkweeds in the United States. In 1995, the USDA Agricultural Research Service in Bozeman, MT, and in France, began to investigate the feasibility of a biological control program for meadow and orange hawkweed. As of 1998, the International Institute of Biological Control (IIBC) in Delémont, Switzerland, is testing the host-specificity of insects attacking mouse-ear hawkweed, which is a serious problem in New Zealand. IIBC is screening four insect species for hawkweed (Syrett et al. 1996). In addition to the insects, two species of fungi are being evaluated for potential biological control possibilities in New Zealand (Morin and Syrett 1996) and in North America (Hasan and Fornasari 1996). The work of IIBC and USDA in Europe will likely be expanded to include insects and pathogens that attack meadow and orange hawkweed in North America.

The University of Idaho is contributing to the biological control program against hawkweed by 1) surveying the indigenous insects associated with the native hawkweed species, and determining which of these species have transferred to the non-indigenous species, and 2) conducting pre-release studies to determine how hawkweed compensates for damage, and what structures are most vulnerable to damage. These studies will serve to curtail the costly testing that is required to implement a biological control program.

• *Education*

Increased awareness of and growing public concern about the rapid expansion of hawkweeds in northern Idaho culminated in 1994 with the formation of the Hawkweed Action Committee, based in St. Maries, Idaho. Comprised of local weed control personnel, private landowners, state and federal land managers, Native American tribe land managers, and representatives of the timber industry, the Hawkweed Action Committee is a grassroots, non-profit organization whose aim is to increase the awareness of the hawkweed problem, and then promote the need for an aggressive, well-coordinated, broad-based management program against these weeds. In working closely with the University of Idaho and the Idaho Department of Agriculture, this grassroots organization has established cooperation between public agencies and other organizations throughout the inland Northwest, and has taken steps to increase public awareness of hawkweeds. In addition, the Hawkweed Action Committee funds publication of the newsletter *Hawkweed News*, a publication of the University of Idaho.

• *Integrated Weed Management*

Meadow and orange hawkweeds are well suited to an Integrated Weed Management (IWM) program. Strategies for sustainable, long-term management include prevention, early detection, minimizing disturbance, maintaining soil health, controlling grazing, judicious use of herbicides, periodic applications of fertilizer, and cooperation between private and public land managers. Small, isolated infestations in remote areas should be recorded, mapped and immediately controlled with judicious use of herbicides. Larger infestations will require additional measures to prevent or minimize further spread of the infestation. Broad-based land management will be the key to successful long-term hawkweed management. A combination of biotic stresses (biological control agents, grazing management, and competition from perennial grasses and desirable forbs) and abiotic stresses (weather, drought, severe cold, and soil productivity) will be involved in an integrated approach to hawkweed management.

Literature Cited

Abrams, L., and R.S. Ferris. 1960. Illustrated Flora of the Pacific States: Vol. IV. Stanford University Press, CA.

Bishop, G.F., and A.J. Davy. 1994. *Hieracium pilosella* L. (*Pilosella officinarum* F. Schultz & Schultz-Bip.). J. Ecol. 82(1):192-210.

Britton, N.L., and A. Brown. 1970. An Illustrated Flora of the Northern United States and Canada: Vol. III. Dover Pubs, NY.

Callihan, R.H., C.W. Hunt, and B.B. Marsh. 1995. Yellow hawkweed forage quality. Res. Prog. Rep. West. Soc. Weed Sci.

Callihan, R.H., L.M. Wilson, J.P. McCaffrey, and T.W. Miller. 1997. Hawkweeds. PNW Bulletin 499.

Carson, H.W., L.W. Lass, and R.H. Callihan. 1995. Detection of yellow hawkweed with high resolution digital images. Weed Technol. 9: 477-83.

Dawes, D.S., and N.C. Maravolo. 1973. Isolation and characteristics of a possible allelopathic factor supporting the dominant role of *Hieracium aurantiacum* in the bracken-grasslands of northern Wisconsin. Trans. Wis. Acad. Sci. Arts. Lett. 61:235-51.

Grundy, T.P. 1989. An economic evaluation of biological control of *Hieracium*. Research report number 202, Agribusiness and Economics Research Unit, Lincoln College, New Zealand.

Guppy, G.A. 1976. The hawkweeds of British Columbia. Davidsonia 7:13-17.

Hasan, S. and L. Fornasari. 1996. *Entyloma hieracii* and *Puccinai hieracii*, two promising pathogens for the biological control of *Hieracium* spp. (Asteraceae, hawkweeds) in North America. *In*: V.C. Moran and J.H. Hoffmann (eds.). Proc. IX Int. Symp. Biol. Cont. Weeds, 19-26 January 1996, Stellenbosch, South Africa. Univ. Cape Town.

Hay, J.R., and G.J. Ouellette. 1959. The role of fertilizers and 2,4-D in the control of pasture weeds. Can. J. Plant Sci. 39:278-83.

Hulten, E., and M. Fries. 1986. Atlas of North European Vascular Plants North of the Tropic of Cancer. Koeltz Scientific Books, Germany.

Johnson, M.F. 1977. The genus *Hieracium* L. (Cichorieae-Asteraceae) in Virginia. The Vir. J. of Sci. 28(4):151-56.

Johnson, C.D., and Thomas, A.G. 1978. Recruitment and survival of seedlings of perennial *Hieracium* species in a patchy environment. Can. J. Bot. 56:572-80.

Lass, L.W., and R.H. Callihan. 1992a. Response of yellow hawkweed to range herbicides in a non-crop site. Res. Prog. Rep., West. Soc. Weed Sci.

Lass, L.W., and R.H. Callihan. 1992b. Response of yellow hawkweed to sulfonylurea and pyridine herbicides. Res. Prog. Rep., West. Soc. Weed Sci.

Lass, L.W., and R.H. Callihan. 1997. The effect of phenological stage on detectability of yellow (*Hieracium pratense*) and oxeye daisy (*Chrysanthemum leucanthemum*) with remote multispectral digital imagery. Weed Technol. 11:248-56.

Makepeace, W. 1976. Allelopathy of mouse-ear hawkweed (*Hieracium pilosella*). Proc. New Zealand Weed Pest Control Conference. 29th:106-109.

Maycock, P.F., and M. Guzikowa. 1984. Flora and vegetation of an old-field community at Erindale, southern Ontario. Can. J. Bot. 62(11):2193-207.

Miller, T.W., L. Lass, R.H. Callihan, and D.C. Thill. 1987. Response of meadow hawkweed to sulfonylurea and pyridine herbicides. Res. Prog. Rep, West. Soc. Weed Sci.

Morin, L., and P. Syrett. 1996. Prospects for biological control of *Hieracium pilosella* with the rust *Puccinia hieracii* var. *piloselloidarum* in New Zealand. *In:* V.C. Moran and J.H. Hoffmann (eds.). Proc. IX Int. Symp. Biol. Cont. Weeds, 19-26 January 1996, Stellenbosch, South Africa. Univ. Cape Town.

Noel, W.O., W.S. Belles, D.W. Wattenbarger, and G.A. Lee. 1979. Chemical control of orange hawkweed on rangeland. Proc. West. Soc. Weed. Sci. 32:77.

Panebianco, R., and R.W. Willemsen. 1976. Seed germination of *Hieracium pratense*, a successional perennial. Bot. Gaz. 137(3):255-61.

Peck, M. E. 1941. A Manual of the Higher Plants of Oregon. Binford and Mort, Portland, OR.

Peterson, R. I. 1979. Root buds in *Hieracium florentinum*: effects of nitrogen and observations on bud growth. Bot. Gaz. 140(4):407-13. University of Chicago Press.

Reader, R.J. 1990. Competition constrained by low nutrient supply: an example involving *Hieracium floribundum* Wimm. & Grab. (Compositae). Funct. Ecol. 4(4):573-77.

Reader, R.J., and W.H. Watt. 1981. Response of hawkweed (*Hieracium floribundum*) patches to NPK fertilizer in an abandoned pasture. Can. J. Bot. 59:1944-49.

Rickett, H.W. 1973. Wild flowers of the United States. Vols. I-VI. New York Botanical Garden.

Scott, D., J.S. Robertson, and W.J. Archie. 1990. Plant dynamics of New Zealand tussock grassland infested with *Hieracium pilosella*. I. Effects of seasonal grazing, fertilizer, and overdrilling. J. Appl. Ecol. 27:224-34.

Sell, P.D. 1974. *Hieracium. In:* Tutin et al. (eds.). Flora Europaea, Vol. 4. Cambridge Univ. Press, London.

Skalinska, M. 1967. Cytological analysis of some *Hieracium* species, subg. *Pilosella*, from mountains of southern Poland. Acta. Biol. Cracov. Ser. Bot. 10:128-41.

Skalinska, M. 1976. Cytological diversity in the progeny of octaploid facultative apomicts of *Hieracium aurantiacum*. Acta. Biol. Cracov. Ser. Bot. 19(1):39-46.

Stergios, B.G. 1976. Achene production, dispersal, seed germination and seedling establishment of *Hieracium aurantiacum* in an abandoned field community. Can. J. Bot. 54:1189-97.

Suzuki, S., and T. Narayama. 1977. Orange hawkweed (*Hieracium aurantiacum* L.) as an alien pasture weed in Hokkaido. Hokkaido Nat. Agric. Exp. Sta. Res. Bull. 117:45-55.

Syrett, P., H.M. Harman, G. Grosskopf, and L.A. Smith. 1996. Insects for biological control of *Hieracium* in New Zealand: a progress report. *In:* V.C. Moran and J.H. Hoffmann (eds.). Proc. IX Int. Symp. Biol. Cont. Weeds, 19-26 January 1996, Stellenbosch, South Africa. Univ. Cape Town.

Thomas, A.G., and H.M. Dale. 1974. Zonation and regulation of old pasture populations of *Hieracium floribundum*. Can. J. Bot. 52:1451-58.

Thomas, A.G., and H.M. Dale. 1975. The role of seed production in the dynamics of established populations of *Hieracium floribundum* and a comparison with that of vegetative reproduction. Can. J. Bot. 53:3022-31.

Voss, E.G., and M.W. Bohlke. 1978. The status of certain hawkweeds (*Hieracium* subgenus *Pilosella*) in Michigan. The Michigan Bot. 17(2):35-47.

Weber, W.A. 1990. Colorado Flora: Eastern Slope. University Press of Colorado, Niwot, CO.

William, R.D., D. Ball, T.L. Miller, R. Parker, J.P. Yenish, R.H. Callihan, E. Eberlein, G.A. Lee, and D.W. Morishita (eds.). 1998. Pacific Northwest Weed Control Handbook. Oregon St. Univ., Coop. Ext. Serv., Corvallis, OR.

Wilson, L.M., J.P. McCaffrey, P.C. Quimby, and J.L. Birdsall. 1997. Hawkweeds in the Northwestern United States. Rangelands 19(4):18-23.

Yeung, E.C., and Peterson, R.L. 1972. Studies on the rosette plant *Hieracium floribundum*: I. Observations related to flowering and axillary bud development. Can. J. Bot. 50:73-78.

Leafy Spurge

Sherry Lajeunesse, Roger Sheley, Celestine Duncan, and Rodney Lym

Leafy spurge (*Euphorbia esula*) infests close to 3 million acres (1.2 million ha) of land in 29 states. This invasive perennial weed overruns and destroys grazing lands for cattle and horses, degrades wildlife habitat and wildlife-associated recreation, decreases rangeland plant diversity, threatens native plants, and reduces land values. The legal responsibility for managing this noxious weed is an additional liability. Intensive, long-term, integrated management is necessary to reduce leafy spurge infestations.

Identification

Leafy spurge is a long-lived, deep-rooted perennial that reproduces vegetatively and by seeds. Stems are hairless and pale green or blue-green. They grow 16 to 32 inches (40 to 81 cm) in dense patches. The narrow, hairless leaves are alternate on the stem. Stems and leaves contain a milky latex. The small flowers are green and inconspicuous, but are surrounded by a pair of yellow-green, heart-shaped leaves (bracts) that are often mistaken for flowers.

Origin, History, and Distribution

In its native Eurasia, leafy spurge is widespread and adapted to many sites and habitat types ranging from riparian to dry hillsides (Smith and Tutin 1968). Infestations are usually small and widely scattered, probably due, in part, to naturally occurring biological factors and intensive grazing by sheep and goats, conditions under which leafy spurge evolved.

The weed was brought to northeastern North America as an ornamental in 1829. It soon spread from yards and gardens, and by the 1900s infestations had reached the west coast of North America (Best et al. 1980). The weed is increasingly widespread throughout the United States and Canada (Lajeunesse et al. 1994).

Potential Invasion

Because of its high genetic variability, the species easily adapts to local growing conditions. It can be found in dry to subhumid habitats, in subtropical to subarctic habitats, on sites ranging from flood plains and river banks to grasslands, ridges, and mountain slopes (Hanson and Rudd 1933, Bakke 1936, Best et al. 1980). Leafy spurge often dominates bottomlands, and is less frequent on topslope, summit, and shoulder slopes (Lym 1994). It is primarily found in pastures, rangelands, roadsides, waste areas, abandoned cropland, and increasingly in areas disturbed by development. The species will thrive in many soil types, especially after soil disturbance.

Impacts

A highly competitive plant, leafy spurge displaces native vegetation, forming pure stands in many cases. As of 1997, it infested more than 2.7 million acres (1.1 million hectares), mainly in the northern Great Plains of the United States and the prairie provinces of Canada. Loss of plant diversity and loss of wildlife forage and habitat are additional environmental impacts of leafy spurge.

Economic losses in North Dakota alone exceed $14.4 million annually due to reduced forage production and use, and control costs (Lym 1994). Because cattle will not graze in areas with a 10% to 20% leafy spurge cover, the weed reduces carrying capacity of infested rangeland to near zero (Hanson and Rudd 1933, Selleck et al. 1962, Best et al. 1980). The value of heavily infested land can also be reduced because of legal responsibilities, expense, and difficulty associated with management. However, leafy spurge has been shown to provide good forage for sheep and goats (Parker 1991, Olsen and Lacey 1994).

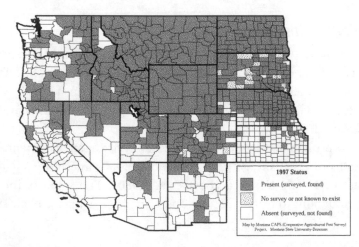

Distribution of leafy spurge by county in the western United States. Data were collected by surveying various weed authorities in each state.

Biology and Ecology

Several biological characteristics make leafy spurge highly competitive and extremely difficult to manage:

• *Seed Production and Dispersal*

Seeds develop in a three-sided capsule that "explodes" at maturity, projecting seeds up to 15 feet (4.6 m) (Bakke 1936, Dersheid et al. 1960). Each flowering stem produces an average of 140 seeds. Seed production, which ranges from 25 to 4,000 pounds per acre (28 to 4,480 kg per hectare) depending on plant density and site productivity, is usually completed by mid-August in northern climates (Bakke 1936, Best et al. 1980).

Leafy spurge seeds float on water, often resulting in new infestations along rivers and in areas that are periodically flooded. Seeds are also spread in mud on equipment, motorbikes, or vehicles, or on feet or hair of animals, including sheep (Lym 1994). They are ingested by sheep, goats, rodents, birds, and to a lesser extent by whitetail deer, then deposited in dung (Selleck et al. 1962, Blockenstein et al. 1987, Pemberton 1988). Crop seed, feed grain, and hay containing leafy spurge seeds spread the weed over long distances, as do recreational and other vehicles and equipment that have been used in infested areas.

• *Seed Viability, Dormancy, and Germination*

Leafy spurge seeds can remain viable eight years or longer; the deeper they are buried, the longer they tend to remain viable and dormant. Dormancy may be broken at any time, with most germination occurring in the first two years. Germination can occur in a wide range of temperatures, any time adequate moisture is available. The optimum range is 68° to 86° F (20° to 30° C). Germination and seedling emergence peak in late May and early June in northern climates (Hanson and Rudd 1933, Selleck et al. 1962, Best et al. 1980).

• *Vegetative Reproduction*

Large numbers of buds are found on each root to depths of 10 feet (3 m) or more. Each bud is capable of producing a new, independent plant. This is a major factor in the spread and persistence of the weed. Shoots that arise from vegetative root buds are immediately aggressive and highly competitive because of the deep, established root reserves. Shoots usually emerge in late April in northern climates. By the time surrounding plants begin growth, leafy spurge can already be dominant. When older spurge shoots are removed or when the crown is damaged, for example by hand-pulling, mowing, or burning, new shoots simply develop from root buds. A second flush of growth occurs in the fall after most other species are dormant. The early- and late-season growth results in monopolization of moisture, light, and nutrients.

• Seedlings

New plants originating from germinating seeds can emerge through 4 inches (10 cm) of soil, but optimum soil depths for emergence are from $^1/_2$ to 2 inches (1.3 to 5 cm) (Selleck et al. 1962). Although seedling mortality is about 80%, survivors grow rapidly, becoming highly competitive within four months (Raju et al. 1963, Morrow 1979), even in pristine areas or on rangeland in excellent condition. Seedlings rarely produce flowers the first year unless they are free from competing vegetation.

• Adults

Yellow-green bracts, often mistaken for flowers, develop on stems in May or early June in northern climates. However, the small, green true flowers develop about two weeks later—an important distinction when applying herbicides. Flowering is usually completed by mid-July, seeds develop for 20 to 30 days, and seed dispersal begins (Bakke 1936, Best et al. 1980). In late summer leafy spurge becomes dormant, but growth resumes in the fall.

• Roots

The competitive, rapidly growing, and extensive root system makes leafy spurge very difficult to manage. An understanding of the root system can help land managers develop the most effective strategies.

In the absence of competition, seedling roots can penetrate to a depth of 3 feet (2.7 m) and spread laterally 40 inches (102 cm) in four months. At this point, the seedlings can outcompete most range plants. Leafy spurge is most vulnerable to management efforts and competition from desirable plants before this stage of development. Seedlings develop root buds a short time after emergence and new shoots quickly develop from these buds if the main shoot is injured or removed (Raju et al. 1963, Morrow 1979).

The mature root system comprises vertical and horizontal roots. Main vertical roots can penetrate 26 feet (7.9 m) deep. Horizontal roots, which grow nearer the soil surface, can extend outward from the parent plant 15 feet (4.6 m) per year. Thick, corky bark and depth of roots enable leafy spurge to survive drought, grazing stress, or herbicide applications. The roots sometimes exude translocated herbicides into the soil. Large carbohydrate reserves in the roots can sustain the plant for years, and can provide energy needed for regrowth after chemical or physical treatments (Raju et al. 1963, Morrow 1979). Mature root systems are extremely difficult and expensive to eradicate.

Management

High genetic variability, even within the same plant, enables leafy spurge to adapt quickly to localized conditions, sometimes making control inconsistent. Early- and late-season growth, when many other plants are dormant, and deep root systems allow the weed to avoid direct competition. Leafy spurge is particularly successful in areas with few desirable species that can consistently compete for surface moisture, especially if those species are managed poorly. Grazing programs that allow consistent selection of desirable species and avoidance of leafy spurge also enhance success of the weed. As population and human activity increase, soil disturbances and transport of weed seeds into uninfested areas provide more opportunities for successful establishment.

Five methods are used to manage leafy spurge: prevention, plant competition, physical control, biological control, and chemical control. The only effective management programs developed as of the 1990s incorporate several or all of these methods. Leafy spurge control must be considered a long-term management program.

• Prevention

All new infestations of leafy spurge originate from seeds or vegetative buds on root pieces which can be brought into uninfested areas by activities such as outdoor recreation, agriculture, and construction. Seeds can be transported in mud on motorbikes, construction and farming equipment, tires, vehicles, and on the feet of livestock. Check or clean equipment when moving from infested to uninfested areas. Landscaping soil and fill dirt can also contain seeds and root pieces; avoid importing such materials from weedy areas.

Livestock feed and crop seeds can contain leafy spurge seeds. Certified weed-free crop seeds and feed, including feed for pack animals, are available. Before moving livestock from infested areas, hold cattle 6 days and sheep 11 days in corrals or small pastures, to give seeds time to pass through the digestive tract. Monitor and treat the holding areas regularly for seedling establishment.

Prevent the formation of seeds, which can be carried into uninfested areas, by using herbicides or sheep grazing.

• Grazing

Plant communities can be manipulated to favor particular plant groups (for example, grasses) over others (such as leafy spurge) by implementing a careful grazing program and by reseeding. It may take 3 to 10 years, or even longer, to shift a plant community from leafy spurge to grasses. The use of grazing animals must be considered a long-term suppression (as opposed to eradication) program.

Sheep can be used to graze leafy spurge often enough and at sufficient stocking rates to weaken the plants over time. Grazing rotations must be timed to minimize stress to grasses when they are growing stems and flowering. Generally, sheep graze forbs (including leafy spurge) more heavily in the spring and summer, when plants are young and succulent, and will select grasses more heavily in fall and winter. (Dietary overlap between sheep and cattle ranges from about 30% during spring and summer to 70% during fall and winter [Van Dyne et al. 1980, Squire 1982, Kirby et al. 1988].) If sheep are leased, a contract of at least five years might be necessary for the project to be feasible for the sheep producer. An intensive grazing system that includes a minimum of two grazing periods, each followed by a rest period to allow grass regrowth, is recommended, rather than season-long grazing by sheep. Begin grazing leafy spurge in early spring when plants are 2 to 6 inches (5 to 15 cm) tall. Schedule rotations to prevent leafy spurge from producing seeds. When "the yellow is gone" (the yellow-green bracts), the sheep should be moved to other patches. A goal of grazing management should be prevention of seed production. However, sheep that graze leafy spurge plants with mature seeds should be corralled for at least 11 days to pass seeds from their digestive tracts before being moved to uninfested areas.

Stocking rates vary with leafy spurge density, terrain, and rainfall. Approximately 3 to 6 animals per acre (7 to 15 animals per hectare) of leafy spurge per month, or 1 to 2 ewes with lambs per acre (2 to 5 ewes with lambs per hectare) of leafy spurge per summer may be an initial guideline (Lym and Zollinger 1995). Adjust the stocking rate as needed to prevent blooming and suppress regrowth. Sheep that have not grazed leafy spurge may require a 5- to 10-day adjustment period (Parker 1991, Olsen and Lacey 1994).

Sheep have been used in conjunction with insect biological control agents in the genus *Aphthona* (the flea beetles) with positive results when sheep are grazed once in spring and again in the fall. This schedule avoids disturbing adult beetles when they are laying eggs and feeding.

Goats also will consume leafy spurge (along with other woody species) and are particularly useful where terrain is too rough for sheep, or where burrs and other plant species problematic for sheep are plentiful. Goats can be used initially to clear the site of problematic plants, to be followed by a sheep-grazing program. Although goats are sometimes better suited to a particular area, they require more management than sheep, and markets may be more limited.

Stocking rates for goats vary with terrain, leafy spurge density, and rainfall, but 12 to 16 Angora goats per acre (30 to 40 Angora goats per hectare) of leafy spurge per month, or 3 to 4 goats per acre (7 to 10 goats per hectare) of leafy spurge for four months may be sufficient (Lym and Zollinger 1995).

Fall grazing by goats followed by herbicide application (picloram at one quart + 2,4-D at one quart product per acre) in mid-September has provided good control in northern climates. Allow 3 to 4 inches (7.5 to 10 cm) regrowth after grazing before herbicide application (Lym et al. 1994). When using goats, monitor desirable species regularly, especially woody species such as young trees and desirable shrubs, to prevent overgrazing.

Cattle avoid leafy spurge and will not control it. Letting cattle graze grasses during stem elongation and flowering will favor leafy spurge growth. However, well-timed grazing by cattle at proper stocking rates can increase the competitiveness of grasses by removing excess plant material that will smother new grass growth. The milky latex found in leafy spurge is toxic to cattle when ingested, resulting in lesions around the eyes and mouth.

• *Physical Control*

Cultivation is used to control leafy spurge on cropland where few options are available. Two types of tillage programs—intensive throughout the growing season, and fall-only cultivation—have proven useful.

The most successful intensive programs begin in the spring, two to four weeks after leafy spurge emerges. Duckfoot cultivators are used to till 4 inches (10 cm) deep, repeating every three weeks until the soil freezes in the fall, for one to two growing seasons (Dersheid et al. 1960, 1963). The cultivation schedule should not be interrupted, because leafy spurge recovers quickly from rootstock, and pieces of roots as small as $^1/_2$-inch (1.2 cm) long and $^1/_{10}$-inch (0.25 cm) diameter can produce new shoots. Care should be taken not to transport root pieces caught on machinery into uninfested portions of the field or other areas.

Fall-only cultivation can be helpful. When fall regrowth of spurge is 3 to 6 inches (7 to 15 cm) tall, cultivating one or two times post-harvest for three years has been effective (Lym and Messersmith 1993). Compared to season-long cultivation, fall-only cultivation allows crops to be grown during the season, limits organic matter degradation, and reduces soil erosion. Integrating cultivation with herbicides applied at crop label rates at least seven days before the first fall cultivation can enhance leafy spurge control.

Mowing and burning are ineffective for reducing leafy spurge infestations (Maxwell et al. 1984, Ferrell and Alley 1985, Scholes and Clay 1994). However, burning prior to herbicide application can increase visibility of the weed, improve spray coverage by eliminating old stems and ground litter (Winter 1992, Wolters et al. 1994), and provide uniform regrowth for more effective herbicide treatment. Allowing at least five weeks of regrowth before herbicide application appears most effective (Lym and Messersmith 1986). Burning combined with herbicide applications does not increase control other than improving spray coverage, but may reduce leafy spurge seed viability (Bjugstad 1986).

Mowing will reduce seed production if repeated every two to four weeks during the growing season, but will provide little long-term control. Mowing and/or herbicide application combined with applications of nitrogen fertilizer to stimulate grass growth does not control leafy spurge better than herbicides applied alone (Madsen and Miller 1988, Scholes and Clay 1994).

Pulling leafy spurge by hand is ineffective, even for small isolated patches, because of the depth of the root system and numerous root buds. However, it may be possible to hand-pull a few plants that are in their first year of growth. Seedlings only a few weeks old already have vegetative root buds.

• Biological Control

Insects, pathogens, and animals that consume or otherwise use leafy spurge are considered biological controls. With careful management, biological controls can be integrated with other management methods. Biological control agents will not eradicate a weed, but in some cases will reduce weed populations over a period of years.

Insects and pathogens that attack leafy spurge in its area of origin, Eurasia, are carefully studied and screened for their appropriateness for importation into the United States. As of 1996, 13 species of insect biological control agents of leafy spurge had been approved by the U.S. Department of Agriculture for introduction into the United States (Rees et al. 1996). The most effective and well-established agents appear to be six species of root- and foliage-feeding beetles in the genus *Aphthona*, and a stem- and root-boring beetle, *Oberea erythrocephala*. All of the biological control species stress the plant to some extent. Most species are widely available through the USDA Animal and Plant Health Inspection Service (APHIS), or from local insectaries.

Plant diseases can also be used to manage weeds. Several promising bacteria and fungi that attack leafy spurge roots, crowns, and seedlings have been identified and are being investigated. The integration of pathogens with other control methods is being investigated.

Current information about the availability and effectiveness of biological control agents is available from state Departments of Agriculture, the U.S. Department of Agriculture, the Cooperative Extension Service, and regional colleges and universities.

• Chemical Control

Annual applications of herbicides are one component of an effective management strategy. When leafy spurge patches are small and isolated, it is possible to eradicate them with herbicides. Treat an extra 10 to 15 feet (3 to 4.5 m) around leafy spurge patches to control spreading roots and seedlings. Neither biological control nor grazing should be used to manage small patches because these methods work slowly, will not prevent seed production (in the case of biological control), and will not eradicate the weed. Eradicating small patches of leafy spurge should be the highest priority of a management program.

Picloram is the most effective herbicide for controlling leafy spurge. For small, isolated patches where annual retreatment is not feasible, picloram at one gallon of product per acre will give 90% or more leafy spurge control the first year after treatment. Control will decline gradually to approximately 70% three years after treatment and more rapidly thereafter. This application rate should not be used on infestations larger than half an acre.

Research at North Dakota State University has shown that a cost-effective treatment for leafy spurge control is a tank mix of picloram at one pint per acre + 2,4-D at one quart per acre (based on a concentrate with four pounds active ingredient per gallon), applied during true flowering (approximately two weeks after development of the yellow bracts) and repeated annually. This treatment has provided 85% leafy spurge control after four annual applications, and was the most cost-effective herbicide treatment for leafy spurge (Lym et al. 1993). Fall application of this treatment does not effectively control leafy spurge. Picloram at one quart per acre + 2,4-D at one quart per acre provides greater control, does not require retreatment as frequently, but is more expensive. This treatment can be applied either during true flowering or in the fall. An economical management option is to apply picloram at one quart per acre + 2,4-D at one quart per acre the first year, followed by picloram at one pint per acre + 2,4-D at one quart per acre for three consecutive years.

Dicamba will provide an intermediate level of leafy spurge control in pastures and rangeland, but annual treatments with dicamba are not cost-effective.

Glyphosate + 2,4-D will provide season-long leafy spurge control, but there is a risk of grass injury, especially when fall-applied. A single treatment of one pint glyphosate plus one pint 2,4-D per acre applied during early seed-set growth stage will provide at least 75% leafy spurge control with 0% to 10% grass injury. This treatment should not be applied to the same area for two consecutive years, because grass injury could be severe. Applications of glyphosate + 2,4-D should be followed by picloram + 2,4-D during the flowering stage the second year.

For short-term management of leafy spurge topgrowth while increasing forage production, 2,4-D applied annually in the spring at one quart per acre (based on a formula with four pounds active ingredient per gallon) has shown good results. Either the amine or the ester formulations can be used. Higher rates of 2,4-D do not provide significantly better control. If 2,4-D is applied to grass stands twice a year, the hay may be removed for feed, and seed production in leafy spurge may be prevented.

Amine formulations of 2,4-D at one quart per acre may be used for leafy spurge control among trees. Avoid contacting tree foliage and green bark of young trees with either direct spray or drift. Glyphosate at one quart per acre applied from mid-July to mid-September when grasses are dormant will give 80% to 90% leafy spurge control. Control drops sharply the second year, to about 10%, if it is not retreated. Treatments made prior to mid-July have not given good leafy spurge control. Glyphosate is non-selective and will kill grasses and other vegetation; avoid contacting tree foliage and green bark of young trees with either spray or drift. Glyphosate + 2,4-D (one pint each) can be used under trees to control leafy spurge and decrease the amount of grass and vegetation injury caused by glyphosate alone. Glyphosate + 2,4-D can be applied two to four weeks earlier than glyphosate alone.

• *Revegetation*

Initial reseeding with grasses followed by eventual revegetation with forbs and shrubs may contribute to long-term suppression of leafy spurge. Weed seeds will germinate in areas exposed by the removal or suppression of weeds by biological control or herbicides. To provide early competition for leafy spurge seedlings, reseed with a mixture of species to compete effectively with the weed. Several grass species have been evaluated for establishment and production in leafy spurge-infested areas. The most competitive grasses tested (Ferrell et al. 1993, Christianson et al. 1994, Biesboer et al. 1993) were Russian wildrye (*Psathyrostachys juncea*), pubescent wheatgrass (*Agropyron intermedium* var. *trichophorum*), smooth brome (*Bromus inermis*), western wheatgrass (*Pascopyrum smithii*), pubescent wheatgrass, Dahurian wildrye (*Elymus dahuricas*), and little bluestem (*Schizachyrium scoparium*). Effectiveness of species varies by region. Local agronomists or Natural Resources Conservation Service (NRCS) Plant Materials Centers can suggest species best adapted to a particular location.

After grasses are well established, in perhaps three to five years, desirable forbs and shrubs may be introduced. If grazing animals such as sheep and goats will be used as part of the long-term management program, select plant species that are not preferred by the animals, so the plants retain their competitiveness with the weeds.

For reseeding and revegetation, select a mixture of competitive, well-adapted species with deep, shallow, and intermediate root systems to maximize competition for water, nutrients, and space. Maximize competition with mature spurge plants as well as with seedlings. Also consider the time of season at which each species actively grows, and include some that grow actively in early spring (Morrow 1979), some in midseason, and others in fall. When one species matures or enters its dormant phase, another should be starting its active growing phase to provide continuing competition for moisture and nutrients. Where the roots of one species stop in the soil profile, the roots of another, deeper-rooted desirable species can provide competition for deeper soil moisture and nutrients. Winter and summer annuals and shallow-rooted perennials may help provide season-long competition against germinating leafy surge seedlings. Using these mixtures can maximize competition through the entire soil profile, during the entire year, especially when paired with good grazing management.

• *Long-Term Management*

A long-term, sustainable program must be adjusted and readjusted as plant populations and conditions change and better management information becomes available. Success with leafy spurge is possible, but only with persistence. Because infestation size and economic and environmental variables change, annual evaluation and re-mapping are suggested. This enables the land manager to evaluate effectiveness of the management strategies by indicating changes in the infestation. This information is necessary to determine the best combination of strategies for each site and to make adjustments as conditions change.

Literature Cited

Bakke, A.L. 1936. Leafy spurge, *Euphorbia esula* L. Iowa Agric. Exp. Sta. Res. Bull. 189:209-45.

Best, K.F., G.G. Bowes, A.G. Thomas, and M.G. Maw. 1980. The biology of Canadian weeds. 39. *Euphorbia esula* L. Can. J. Plant Sci. 60:651-63.

Biesboer, D.D., W.L. Koukkari, and B. Darveaux. 1993. Controlling leafy spurge in Minnesota with competitive species and combined management practices. Proc. Great Plains Ag. Council 14. Leafy spurge control in the Great Plains. Silver Creek, CO.

Bjugstad, A.J. 1986. Herbicide and heat to reduce seed germination of leafy spurge. Proc. Great Plains Ag. Council 14. Leafy spurge control in the Great Plains. Riverton, WY, 7-9.

Blockenstein, D.E., B.D. Maxwell, and P.K. Fay. 1987. Dispersal of leafy spurge seed (*Euphorbia esula*) by mourning doves (*Zenaida macroura*). Weed Sci. 35:160-66.

Christianson, K.M., R.G. Lym, and C.G. Messersmith. 1994. Herbicides and grass competition for leafy spurge control. Proc. Great Plains Ag. Council. 14. Leafy spurge control in the Great Plains. Bozeman, MT, 8.

Dersheid, L.A., K.E. Wallace, and R.L. Nash. 1960. Leafy spurge control with cultivation, cropping and chemicals. Weeds 8:115-27.

Dersheid, L.A., G.A. Wicks, and W.H. Wallace. 1963. Cropping, cultivation, and herbicides to eliminate leafy spurge and prevent reinfestation. Weeds 11:105-11.

Ferrell, M.A., and H.P. Alley. 1985. Evaluation of mowing as a setup treatment prior to herbicide treatment for leafy spurge shoot control. Res. Prog. Rep., West. Soc. Weed Sci., 30.

Ferrell, M.A., T.D. Whitson, D.W. Koch, and A.E. Gade. 1993. Integrated control of leafy spurge (*Euphorbia esula*) with Bozoisky Russian wildrye (*Psathyrostachys juncea*) and Luna pubescent wheatgrass (*Agropyron intermedium* var. *trichophorum*) 1993. Proc. West. Soc. Weed Sci. 46:36-38.

Hanson, H.C., and V.E. Rudd. 1933. Leafy spurge life history and habits. N. Dak. Agric. Coll., Agric. Exp. Sta. Bull. 266.

Kirby, D., M. Parman, M. Pessin, and M. Humann. 1988. Dietary overlap of cattle and sheep on rotationally grazed rangeland. SID Res. J. 4:6-11.

Lajeunesse, S., R. Sheley, R. Lym, D. Cooksey, C. Duncan, J. Lacey, N. Rees, and M. Ferrell. 1994. Leafy spurge: Biology, ecology, and management. Ext. Bull. EB34, Montana State Univ., Bozeman, MT.

Lym, R.G. 1994. Ecology, economic impact, and control of leafy spurge. Proc. Western Section, Amer. Soc. of Animal Sci. Vol. 45.

Lym, R.G., and C.G. Messersmith. 1986. Mowing as a pretreatment for leafy spurge control with herbicides. Res. Prog. Rep. West. Soc. Weed Sci., 9-10.

Lym, R.G., and C.G. Messersmith. 1993. Fall cultivation and fertilization to reduce winterhardiness of leafy spurge (*Euphorbia esula*). Weed Sci. 41:441-46.

Lym, R.G., and R.K. Zollinger. 1995. Integrated management of leafy spurge. N. Dak. State Univ. Ext. Serv. Pub. W-866.

Lym, R.G., C.G. Messersmith, and R. Zollinger. 1993. Leafy spurge identification and control. N. Dak. State Univ. Ext. Serv. Pub. W-765.

Lym, R.G., D.R. Kirby, and K.K. Sedivec. 1994. Personal communication.

Madsen, K.A., and S.D. Miller. 1988. Mow/fertilization treatments and their effect on leafy spurge (*Euphorbia esula*) control with herbicides. Proc. West. Soc. Weed Sci. 41:62-67.

259

Maxwell, B.D., P.K. Fay, and M.E. Foley. 1984. Leafy spurge research update, 1984. Proc. Great Plains Ag. Council 14. Leafy spurge control in the Great Plains. Riverton, WY, 35-37.

Morrow, L.A. 1979. Studies on the reproductive biology of leafy spurge (*Euphorbia esula*). Weed Sci. 27:106-109.

Olsen, B.E., and J.R. Lacey. 1994. Sheep: a method for controlling rangeland weeds. Sheep Res. J., Special Issue, 105-12.

Parker, J.W. 1991. Multispecies grazing systems. Sheep Forage Production Systems Symposium, Sheep Industry Dev. Progr. Englewood, CO.

Pemberton, R.W. 1988. Myrmecochory in the introduced range weed, leafy spurge (*Euphorbia esula* L.). Amer. Midland Naturalist 119:431-35.

Raju, M.V.S., T.A. Steeves, and R.T. Coupland. 1963. Developmental studies on *Euphorbia esula* L. Morphology and root systems. Can. J. Bot. 41:579-89.

Rees, N., P.C. Quimby, Jr., G.L. Piper, E.M. Coombs, C.E.Turner, N.R. Spencer, and L.V. Knutson (eds.). 1996. Biological Control of Weeds in the West. W. Soc. Weed Sci., Bozeman, MT.

Scholes, C.M., and S.A. Clay. 1994. Evaluation of season-long mechanical and low herbicide input treatments for leafy spurge suppression. Proc. Great Plains Ag. Council 14. Leafy spurge control in the Great Plains. Bozeman, MT, 33.

Selleck, G.W., R.T. Coupland, and C. Franklin. 1962. Leafy spurge in Saskatchewan. Ecol. Monogr. 32:1-29.

Smith, A.R., and T.G. Tutin. 1968. *In:* T.G. Tutin (ed.). Flora Europaea, Vol. 2, 213-226. Cambridge Univ. Press.

Squire, V.R. 1982. Dietary overlap between sheep, cattle, and goats when grazing in common. J. Range Manage. 35:116-19.

Van Dyne, G.M., N.R. Brockington, A. Sozoss, J. Duek, and C.A. Ribic. 1980. Large herbivore subsystems. *In:* Grassland, Systems Analysis and Man. Int. Biol. Programme 19. Cambridge Univ. Press, 269-535.

USDA-ARS. 1995. Purge Spurge: Leafy spurge database CD-ROM, v. 2.5. USDA Agric. Res. Serv. and Montana State Univ., Sidney, MT 59270.

Winter, B. 1992. Leafy spurge control in a tallgrass prairie natural area. Proc. Great Plains Ag. Council 14. Leafy spurge control in the Great Plains. Lincoln, NE, 2-31.

Wolters, G.L., C. Hull Sieg, A.J. Bjugstad, and F.R. Gartner. 1994. Herbicide and fire effects on leafy spurge density and seed germination. Proc. Great Plains Ag. Council 14. Leafy spurge control in the Great Plains. Bozeman, MT, 44.

Mediterranean Sage

Cindy Talbott Roché and Linda M. Wilson

Like the silver candelabra atop the piano, Mediterranean sage (*Salvia aethiopis*) stands majestically on western rangeland, belying its intruder status. Native to the northern and eastern Mediterranean region, it has invaded shrub steppe rangelands (shrub overstory with bunchgrass understory) in six western states. In the hundred years after its introduction around 1892, it spread to more than 1.3 million acres (500,000 ha). Dense infestations of Mediterranean sage, unpalatable to grazing animals, reduce forage production on rangeland and pasture. Its felty covering of white hairs require adjuvants for herbicidal control; tumbleweed-mobility frustrates containment efforts. Establishment of competitive grasses, coupled with proper grazing management and a crown-mining weevil, contribute to long-term reduction of Mediterranean sage populations.

Identification

Mediterranean sage, a member of the mint family (Lamiaceae), has erect, sturdy, squarish stems up to 3 feet tall (ca. 1 m), with opposite leaves and a stout taproot. Leaves are densely woolly with white hairs, especially when young. As plants age, the upper sides of the leaves lose some of the felty covering of hairs, revealing prominent veins and a wrinkled surface.

The seedling's two oval cotyledons are notched at the tips. The first true leaves develop a distinctive mat of tangled white hairs. Pre-reproductive plants form a ground-hugging rosette. Rosette leaves are indented or shallowly toothed and have a stalk $1^1/_2$ to $3^1/_2$ inches (4 to 9 cm) long. Second-year rosettes are very leafy, almost succulent, and are usually 7 to 10 inches (18 to 25 cm) in diameter, although they can grow to 4 feet (ca. 1 m) across. Dense, silvery-white hairs make leaves appear light- to gray-green. As rosettes, Mediterranean sage could be confused with common mullein (*Verbascum thapsus*) which also has felty leaves. However, mullein leaves tend toward yellow-green, in contrast to the gray- or blue-green cast of Mediterranean sage leaves. Mullein leaves lack petioles, and are not toothed along the margin. In addition, Mediterranean sage emits a pungent, sage-like odor when crushed.

Mature plants have upright stems with clasping leaves that become progressively smaller up the stem. The uppermost leaves are merely bracts, purple-tinged and tapering to long points. The branched panicle that resembles a candelabra bears numerous flowers in woolly clusters. Four to six whitish flowers are clustered in whorls, each encircled by silvery-haired bracts with pointed tips. Each flower is about $^1/_2$ inch (13 mm) long, shaped like a mint flower, with the upper lip resembling a hooked beak. The pale yellow lower lip divides into three lobes, with the central lobe smaller than the outer lobes. Each flower produces four seeds. Seeds are about $^1/_8$ inch (3 mm) in diameter, somewhat egg-shaped, with darker brown veins forming an irregular pattern on the smooth brown surface.

References containing additional photographs, line drawings, or descriptions of Mediterranean sage include Robbins et al. (1970), Gaines and Swan (1972), Hitchcock and Cronquist (1973), Dennis (1980), Hawkes et al. (1985), Polunin (1987), Roché (1991), and Whitson et al. (1991).

Origin, History, and Distribution

Mediterranean sage is native to southern and southeastern Europe, as far north in central Europe as former Czechoslovakia and to 51° N latitude in southcentral Russia, including Crimea, and east through Turkey into Iran (Tutin et al. 1972, Davis 1975).

Probably introduced into the United States with alfalfa (*Medicago sativa*) seeds (Dennis 1980), Mediterranean sage has also been planted as a garden flower (Bailey 1935). The earliest U.S. record is from roadsides in Susanville, Lassen County, California, in July 1892 (Howell 1941) where it "figuratively stood still by the edge of the road for about 60 years" until widely transported by new highway construction (Bellue 1950). Spreading to Plumas County by 1919 (Howell 1941), it was widespread in Modoc County by the 1940s (Bellue 1950). In 1994 Mediterranean sage infested an estimated 7,000 acres (2,800 ha) in Siskiyou, Modoc, Lassen and Plumas counties, California (Barbe 1990, Andres et al. 1995).

In Oregon, Mediterranean sage was present in Lake County by the 1920s, and had established in Klamath County by 1949 (Bellue 1950). At that time, the largest infestation of Mediterranean sage in Lake County was about 5 miles (8 km) wide, extending 25 miles (40 km) north of Lakeview, with scattered small infestations to the north, south and east. Additional widely separated but larger populations between Lakeview and Klamath Falls contributed to the overall infestation of about 42,240 acres (17,000 ha) (Bellue 1950). By 1954 the estimated size of the infestation had grown to more than 100,000 acres (40,500 ha) (White 1955). The largest infestations still occupy southern Lake County, with smaller scattered populations in Baker, Grant, Harney, Klamath, Malheur and Wheeler counties. The overall infested area is estimated at 1,300,000 acres (526,000 ha) (Andres et al. 1995).

The earliest Idaho record was collected from Payette in 1967 (Herbarium, Weed Diagnostics Lab., Univ. Idaho). In 1976, Mediterranean sage grew along the Salmon River and on dry hillsides in the vicinity of Lucille, Idaho County (Herbarium specimens, Univ. Idaho and Wash. State Univ.). As of 1994, most of the Mediterranean sage in Idaho (4,000 acres or 1,600 ha) infested Idaho County, with additional small populations reported near Council, Adams County (G. Keetch 1994, personal communication), and near Orofino, Clearwater County.

The first record in Washington was from pastureland and flood plain areas of the Touchet River in Columbia County in 1951 (Gaines and Swan 1972, Wash. State Univ. Herbarium). Infestations in Washington have been contained in Columbia County, comprising about 400 acres (160 ha) (F. Gritman 1994, personal communication.

Mediterranean sage was first reported in Colorado in 1947 in a pasture near Longmont (W. Weber 1994, personal communication). After remaining a small, stable colony near Boulder for many years, it began spreading rapidly in the late 1980s along the Foothills Highway north of Boulder (Weber 1990). In 1994 infestations were scattered over an area of approximately 4 square miles (1,000 ha) (W. Weber 1994, personal communication).

In Arizona, Mediterranean sage was reported in 1951 on the south rim of the Grand Canyon (Coconino County), and in Prescott, Yarnell, and Peeples Valley (Yavapai County), having spread rapidly on overgrazed rangeland in the Peeples Valley during the previous 12 years (Kearney and Peebles 1951). Since then it has been collected on Lake Mary Road near Flagstaff and at Lee's Ferry on the Colorado River (Coconino County) (T. Ayers 1994, personal communication).

As of 1998, Mediterranean sage had not been found in Montana (MSU Herbarium and H. Stepper 1994, personal communication), Utah (BYU Herbarium and G. Rasmussen 1994, personal communication), or Wyoming (T. Whitson 1994, personal communication).

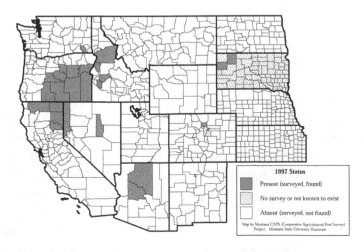

Distribution of Mediterranean sage by county in the western United States. Data were collected by surveying various weed authorities in each state.

Potential Invasion

In its native range, Mediterranean sage is usually associated with early to mid-seral habitats, never reaching densities considered a problem (Bogavac and Mitic-Muzina 1971). These include ruderal habitats with dry soils, such as roadside cutbanks, pastures, abandoned fields, and other areas of disturbance. In Serbia and Macedonia, Mediterranean sage is most frequently found on alluvial deposits of sand and clay, on limestone and in chernozemic soils (Bogavac 1972). These soils are generally unsuitable for agriculture (Bogavac 1972). It rarely grows as a dominant member in plant communities, nor is it associated extensively with weedy vegetation. It only occasionally grows in crops (Bogavac 1972).

In Oregon, major populations are found in the sagebrush steppe of central southeastern counties. Vegetation types include Wyoming and big sagebrush (*Artemisia tridentata* subsp. *wyomingensis* and subsp. *tridentata*) with bluebunch wheatgrass (*Agropyron spicatum*), Thurber needlegrass (*Stipa thurberiana*) and Idaho fescue (*Festuca idahoensis*), juniper/sagebrush/bunchgrass and ponderosa pine/bitterbrush (*Pinus ponderosa/Purshia tridentata*) or bunchgrass. Mediterranean sage has been observed in greasewood (*Sarcobatus vermiculatus*) or shadscale (*Atriplex confertifolia*) types only on slopes and it has not invaded the shallow, saline/alkaline sites of internally drained basins. Mediterranean sage grows on moderate (14 to16 inches, 35 to 40 cm) or deeper soils with good drainage. It invades disturbed or degraded sites more rapidly, and may attain understory dominance in sagebrush/cheatgrass communities, but invasion also occurs in climax sites where rosettes establish between the grass bunches (B. Bolton 1998, personal communication).

In Idaho, Mediterranean sage grows in the Canyon Grasslands and extends up into the adjacent ponderosa pine woodlands. Mediterranean sage flourishes where native vegetation was replaced by exotic annual grasses and weedy forbs following overgrazing, excessive trampling, and soil erosion. It also grows in depleted riparian areas and dry pastures.

In Washington, Mediterranean sage grows in openings in ponderosa pine associated with snowberry (*Symphoricarpos albus*), ninebark (*Physocarpus malvaceus*), and bluebunch wheatgrass, and in Douglas hawthorn (*Crataegus douglasii*) floodplains currently dominated by Kentucky bluegrass (*Poa pratensis*). Some of the ponderosa pine habitat has had the overstory removed, some is grazed by cattle, and some is relatively undisturbed. All of the infested floodplain and riparian areas are grazed and highly disturbed (R. Schirman 1994, personal communication).

Based on current infestations, the steppe, shrub steppe and ponderosa pine zones in west-central Idaho and eastern Oregon and Washington are susceptible to invasion by Mediterranean sage. This large region of steppe and shrub steppe communities includes big sagebrush/bluebunch wheatgrass, and the Canyon Grasslands of bluebunch wheatgrass/Sandberg bluegrass (*Poa sandbergii*) bordering the Snake, Salmon, and Columbia rivers, extending through the *Agropyron/Festuca* zone into the Ponderosa pine/shrub types surrounding the Blue Mountains (Daubenmire 1970,

Franklin and Dyrness 1984, Tisdale 1985, Ferguson et al. 1987). Andres et al. (1995) suggested that much of the Salmon and Snake River watersheds, Great Basin, and northern California is susceptible to Mediterranean sage invasion. The climate in this region is similar to that in the native range of Mediterranean sage, which appears to grade between Mediterranean and Continental climates (Davis 1965, Polunin 1987).

Once established, Mediterranean sage is able to spread into non-disturbed land, but is not normally found in pristine habitats. Disturbances such as livestock grazing and trampling, off-road vehicles, and logging increase a site's susceptibility to invasion.

Impacts

Mediterranean sage is closely related to culinary sage (*Salvia officinalis*) and has a variety of secondary plant metabolites (Ulubelen and Uygur 1976, Rodriguez et al. 1984). In Serbia, Mediterranean sage is considered a medicinal herb; the leaves are used as a wound dressing (Bogavac 1972). Volatile oils, predominantly terpenes, are exuded as aromatics from the epidermal hairs (Lovett and Weerakoon 1983) and from the roots (Rodriguez et al. 1984). It has been suggested that secondary metabolites contribute to the allelopathic properties of several *Salvia* species (Lovett and Weerakoon 1983). However, it has not been demonstrated that allelopathy is a factor mediating Mediterranean sage expansion in the western United States. The aromatic chemicals combined with dense pubescence are believed to deter attack by many plant-feeding insects (Strong et al. 1984). Mediterranean sage does not harbor crop insect or disease pests (Andres et al. 1995).

Although Mediterranean sage is not toxic to livestock (Andres 1966), it is unpalatable to most grazing animals (Bogavac and Mitic-Muzina 1971) and its presence reduces the value of grazing lands.

Biology and Ecology

Morphological and physiological characteristics typical of xerophytic plants make Mediterranean sage well-adapted to warm, dry environments. These characteristics include hairiness, a wrinkled leaf surface, a thick cuticle, mucilaginous seeds, and dormancy to avoid summer drought.

Seeds germinate in the spring or fall, depending on moisture, and develop into flat rosettes the first growing season. Young seedlings quickly establish a taproot. During the first winter, the rosettes shed most of their leaves, which become mulch to protect the crown. In order to flower, plants appear to need vernalization (a period of cold temperatures and short photoperiods), a characteristic typical of temperate plants with a biennial or monocarpic perennial life cycle. Rosettes resume growth in the spring, produce new leaves, and may or may not flower the second year. Reproductive plants bolt by late May and reach full height around the middle of

265

June. Flowering begins in early June and peaks around early July. By late July, reproductive plants begin to senesce, and most die by autumn.

Like many other biennials, Mediterranean sage does not adhere to a strict two-year life cycle. Rosettes may persist in the vegetative stage for two or more years. At two sites in northern Idaho only 54% of second-year rosettes flowered (Wilson 1992). All other rosettes remained in the vegetative stage. Some researchers (Werner 1975, Klinkhamer and DeJong 1987, Thompson and Stout 1991) have suggested that timing of flowering in biennial plants is largely dependent on the size, not age, of the rosette. During the hottest part of summer, non-flowering rosettes, including two-year-old plants, enter dormancy; the outer leaves wither and the central leaves stop growing. Growth resumes with cooler temperatures and rain in the fall. Thus, rosettes often undergo both summer and winter dormancy.

Mediterranean sage reproduces solely by seeds. Plants produce between 50 and 100,000 seeds (White 1955). Seeds mature in late July and August, but are not usually dispersed until September or October. Fresh seeds are dormant through an after-ripening period. Seeds collected from plants in late July and August did not germinate in the laboratory until late September (L. Wilson, unpubl. data). Under natural conditions, this prevents early germination following summer showers and subsequent high mortality before reliable autumn rains.

Mediterranean sage is a tumbleweed. Stout, candelabra-shaped seed stalks become stiff, brittle, and lightweight when mature, and break off at a natural abscission line on the stem 4 to 6 inches (10 to 15 cm) above the soil surface. Seeds scatter as plants tumble in the wind and new infestations often establish where plants lodge in fence rows, in thickets, or along creek bottoms. Seeds are carried longer distances by human activities, including movement of vehicles and products, livestock, hay, and gravel.

Seeds fall on the soil surface and may roll into crevices. Upon wetting, seeds imbibe water and within five minutes develop a mucilaginous cover which protects them from the drying effects of wind and sun (Young et al. 1970, Young and Evans 1973).

Management

Containment and control of Mediterranean sage in the United States has been achieved with a number of methods. Containment includes preventing seed movement and eradicating small, scattered infestations. In addition to scattering by wind-tumbled plants, seeds may move with contaminated soil, hay, agricultural equipment, livestock, wildlife (including birds), and vehicles. The methods discussed below must be repeated for years to deplete seed reserves, requiring persistence and continuity in a weed management program. Management of the grazing resources to favor the conservation of forage species in competition with the weed is necessary for long-term management success.

• Manual and Mechanical

To eradicate scattered or outlying infestations, dig out individual plants with a shovel. Cutting the taproot 2 to 3 inches (5 to 8 cm) below the crown when plants are starting to bolt prevents most resprouting (Roché 1991). Cultural methods such as tillage are effective for accessible pastures and abandoned fields. Mowing can prevent or reduce seed production only if repeated several times during the growing season, because plants regrow and resume flowering after cutting. Rosettes remain undamaged because they are too low to be cut. Mowing done too late in the season may spread seeds.

• Chemical

Several herbicides effectively control Mediterranean sage, particularly when applied with a surfactant to plants in the rosette stage. Aerial applications are an option for steep, rugged, or inaccessible rangeland infestations. Selective herbicides are especially useful in containment programs for roadsides and other rights-of-way, as well as for reseeding programs. Picloram at a rate between 0.375 and 0.5 lb active ingredient per acre (ai/acre) (0.42 to 0.56 kg ai/ha) applied before bolting kills existing plants and seedlings from seeds that germinate later. Spring application appears to have a longer residual effect on new seedlings, while fall application releases desirable vegetation from competition during earliest spring growth (R. Schirman 1994, personal communication). Clopyralid (0.5 lb ai/acre, 0.56 kg ai/ha) and 2,4-D (1.5 to 2 lb ae/acre, 1.68 to 2.24 kg ae/ha) kill existing plants, but lack residual activity for future seedlings. Specific recommendations vary by region; consult the state Extension weed specialist, and in Idaho, Oregon or Washington consult the current *Pacific Northwest Weed Control Handbook* (William et al. 1998).

• Biological Control

A biological control program for Mediterranean sage was initiated with the introduction of a root-feeding weevil, *Phrydiuchus tau*, in 1969. It was followed by *P. spilmani* in 1971, which did not establish (Rees and Coombs 1996). *Phrydiuchus tau*, introduced from Turkey and released in Oregon and Idaho, has established in most Mediterranean sage populations in Idaho, Oregon, and California.

Phrydiuchus tau larvae damage plants by feeding inside the root crown, destroying vegetative buds and meristematic tissue (Wilson and McCaffrey 1993, Wilson et al. 1994). In some instances, flowering is either prevented or delayed because of damage caused to the root crown (Wilson and McCaffrey 1993). Damage by the weevil in combination with competitive vegetation is credited with reducing the density of some Mediterranean sage populations in Oregon (E. Coombs 1998, personal communication) and in Idaho County, Idaho (C. Crabtree 1998, personal communication). Predation by ants on weevil larvae is a problem in some Oregon populations (E. Coombs 1998, personal communication). Long-term studies are needed to determine the impact of the weevil on Mediterranean sage populations and why it is more effective in some infestations than others.

• *Integrated Management*

Successful management of Mediterranean sage will ultimately depend on the interactions between land management applications, biotic factors (biological control, competition from other plants in the community, grazing effects) and abiotic factors (weather and drought factors, soil disturbance).

Weed density in Mediterranean sage populations appears to fluctuate in natural cycles. Between 1990 and 1992, density in some populations in Oregon and northern Idaho declined severely, apparently through a combination of drought-induced seedling mortality, winterkill of young rosettes, and damage by crown-feeding weevils. This was followed by high seedling recruitment in 1993, a year with above-average moisture (L. Wilson, personal observation).

In the absence of competitive grasses, or when grazing is poorly managed, the biological control weevil cannot be expected to provide long-term weed reduction. Grazing prescriptions should consider timing and level of utilization for the resource objectives. In Oregon, fall/winter or dormant-season grazing has been used for bunchgrasses. Level of use is regulated to maintain grass vigor, because the associated Mediterranean sage is ungrazed. In contrast, greater utilization of cheatgrass (*Bromus tectorum*) in spring grazing may remove annual grass residue prior to applying a selective herbicide to kill the Mediterranean sage rosettes and seeding perennial grass. The success rate of seeding is improved by reducing the weed seedbank and by incorporating the desired seed in the surface soil. Incorporation has been done by rangeland drill, broadcasting in ash after prescribed fire or wildfire, and by livestock trampling. Disturbances such as fire and trampling often stimulate a flush of Mediterranean sage rosettes, which can be killed by a selective herbicide, thus reducing the seedbank.

Integrated strategies for large, established infestations will attempt to prevent spread to other locations, preserve existing desirable vegetation or replace weedy vegetation by reseeding, prescribe grazing practices that maintain vigor of desired grasses, and establish *Phrydiuchus tau* weevils to reduce the vigor of remaining Mediterranean sage plants.

Acknowledgments

The authors thank the following individuals who contributed to this chapter: Tina Ayers and Abril Perez, Deaver Herbarium, Northern Arizona University; Bob Bolton, BLM, Lakeview, Oregon; Eric Coombs, Oregon State Dept. Agric.; Carl Crabtree, Idaho County Weed Coordinator; Fred Gritman, Columbia County Weed Coordinator; Gordon Keetch, Adams County Extension Agricultural Agent; Jeff Knight, Nevada Dept. Agric.; G.A. Rasmussen, Utah Extension Range Specialist; Roland Schirman, Columbia County Extension Agent; Harold Stepper, Weed Coordinator, Montana Dept. Agric.; Tom Whitson, Wyoming Extension Weed Specialist; and W.A. Weber, Univ. Colorado Herbarium.

Literature Cited

Andres, L.A. 1966. Host specificity studies of *Phrydiuchus toparius* and *Phrydiuchus* sp. J. Econ. Entomol. 59:69-76.

Andres, L.A., E. Coombs, and J P. McCaffrey. 1995. Mediterranean sage. *In:* Nechols, J.R., L.A. Andres, J.W. Beardsly, R.D. Goeden, and C.G. Jackson (eds.). Biological Control in the Western United States: Accomplishments and Benefits of Regional Research Project W-84. Berkeley, Univ. Calif., Division of Agric. and Nat. Res. Pub. No. 3361, 296-98.

Bailey, L.H. 1935. The Standard Cyclopedia of Horticulture. Vol. III. The Macmillan Co., New York, 3060.

Barbe, D. 1990. Noxious Weeds of California I. Distribution Maps. State of Calif. Dept. of Food and Agric., Sacramento, CA.

Bellue, M.K. 1950. Mediterranean sage moves. Calif. Dept. Agric. Bull. 39:43-46.

Bogavac, M. 1972. Distribution and natural enemies of the weeds Dalmatian toadflax (*Linaria dalmatica* Mill.) and Mediterranean sage (*Salvia aethiopis* L.) in Yugoslavia. Final Technical Report. Institute for Plant Protection, Belgrade.

Bogavac, M., and Mitic-Muzina. 1971. *Phrydiuchus tau* R.E. Warner (Col. Curculionidae) as reducing agent of Mediterranean sage. Zast. Bilja. 22(114):233-46.

Daubenmire, R. 1970. Steppe Vegetation of Washington. Tech. Bull. 62. Wash. Agric. Expt. Sta., Pullman, WA.

Davis, P.H. (ed.). 1965. Flora of Turkey and the East Aegean Islands, Vol. I. Edinburgh Univ. Press, Edinburgh, 1-7.

Davis, P.H. (ed.). 1975. Flora of Turkey and the East Aegean Islands, Vol. VII. Edinburgh Univ. Press, Edinburgh, 440.

Dennis, L.R.J. 1980. Gilkey's Weeds of the Pacific Northwest. Oregon State Univ. Press, Corvallis, OR, 245-46.

Ferguson, D.E., P. Morgan, and F. Johnson. 1987. Proc. Land Classification based on vegetation: applications for resource management. Gen. Tech. Rep. INT-257, USDA-FS Intermountain Res. Sta., Ogden, UT.

Franklin, J.F., and C.T. Dyrness. 1984. Natural Vegetation of Oregon and Washington. Oregon State Univ. Press, Corvallis, OR.

Gaines, X.M., and D.G. Swan. 1972. Weeds of Eastern Washington and Adjacent Areas. Camp Na-Bor-Lee Assoc., Inc., Davenport, WA, 236-37.

Hawkes, R.B., T.D. Whitson, and L.J. Dennis. 1985. A Guide to Selected Weeds of Oregon. Oregon State Dept. Agric., Salem, OR, 57.

Hitchcock, C.L., and A. Cronquist. 1973. Flora of the Pacific Northwest. Univ. Wash. Press, Seattle, WA, 269.

Howell, J.T. 1941. Plants new to California. Leafl. West. Bot. 3:79-80.

Kearney, T.H., and R.H. Peebles. 1951. Arizona Flora. Univ. Calif. Press, Berkeley, CA, 743.

Klinkhamer, P.G., and T. DeJong. 1987. Plant size and seed production in the monocarpic perennial *Cynoglossum officinale* L. New Phytol. 106:773-83.

Lovett, J.V., and W.L. Weerakoon. 1983. Weed characteristics of the Labiatae, with special reference to allelopathy. Biol. Agric. and Hort. 1:145-58.

Polunin, O. 1987. Flowers of Greece and the Balkans. Oxford Univ. Press, Oxford, 400.

Rees, N.E., and E.M. Coombs. 1996. Mediterranean Sage. *In:* Rees, N.E., P.C. Quimby, G.L. Piper, C.E. Turner, E.M. Coombs, N.R. Spencer, and L.V. Knutson (eds.). Biological control of weeds in the West. Western Soc. Weed Sci., USDA Agric. Res. Serv., Montana Dept. Agric., and Montana State Univ., Bozeman, MT.

Robbins, W.W., M.K. Bellue, and W.S. Ball. 1970. Weeds of California. Calif. Dept. of Agric., Sacramento, CA.

Roché, C. 1991. Mediterranean Sage (*Salvia aethiopis* L.). Coop. Ext. Bull. PNW 381, Wash. State Univ., Pullman, WA.

Rodriguez, B., F. Fernandez-gadea, and G. Savona. 1984. A rearranged abietane diterpenoid from the root of *Salvia aethiopis*. Phytochem. 23(8):1805-06.

Strong, D.R., J.H. Lawton, and R. Southwood. 1984. Insects on plants: Community patterns and mechanisms. Harvard University Press, Cambridge, MA.

Thompson, D.J., and D.G. Stout. 1991. Duration of the juvenile period in diffuse knapweed (*Centaurea diffusa*). Can. J. Bot. 69:368-71.

Tisdale, E.W. 1985. Canyon grasslands and associated shrublands of west-central Idaho and adjacent areas. Forest, Wildlife and Range Expt. Sta. Bull. No. 40, Univ. Idaho, Moscow, ID.

Tutin, T.G., V.H. Heywood, N.A. Burges, D.M. Moore, D.H. Valentine, S.M. Walters, and D.A. Webb (eds.). 1972. Flora Europaea, Vol. 3. Cambridge Univ. Press, Cambridge.

Ulubelen, A., and I. Uygur. 1976. Flavonoidal and other compounds of *Salvia aethiopis*. Planta Medica 29:318-20.

Weber, W.A. 1990. Colorado Flora: Eastern Slope. Univ. Press of Colorado. Niwot, CO, 221.

Werner, P.A. 1975. Predictions of fate from rosette size in teasel (*Dipsacus fullonum* L.). Oecologia 20: 197-201.

White, L. 1955. Mediterranean Sage: what it is and how to recognize it. Modoc County, Calif. Dept. Agric.

Whitson, T.D., L.C. Burrill, S.A. Dewey, D.W. Cudney, B.E. Nelson, R.D. Lee, and R. Parker. 1991. Weeds of the West. Western Soc. Weed Sci. and West. U.S. Land Grant Universities Coop. Ext. Serv., 368-69.

William, R.D., D. Ball, T.L. Miller, R. Parker, J.P. Yenish, R.H. Callihan, C. Eberlein, G.A. Lee, and D.W. Morishita, (eds.). 1998. Pacific Northwest Weed Control Handbook. Oregon St. Univ. Coop. Ext. Serv., Corvallis, OR.

Wilson, L.M. 1992. Bionomics and bioacoustics of *Phrydiuchus tau* (Coleoptera: Curculionidae) in Idaho. M.S. Thesis, University of Idaho, Moscow, ID.

Wilson, L.M., and J.P. McCaffrey. 1993. Bionomics of *Phrydiuchus tau* (Coleoptera: Curculionidae) associated with Mediterranean sage in Idaho. Environ. Ent. 22(3):704-708.

Wilson, L.M., J.P. McCaffrey, and E.M. Coombs. 1994. Biological control of Mediterranean sage. PNW Ext. Pub. 473, Univ. of Idaho, Moscow, ID.

Young, J.A., and R.A. Evans. 1973. Mucilaginous seed coats. Weed Sci. 21:52-54.

Young, J.A., R.A. Evans, and P.C. Martinelli. 1970. Mucilaginous coatings on weed seeds. Proc. West. Soc. Weed Sci. 23:35.

Medusahead

Heather C. Miller, David Clausnitzer, and Michael M. Borman

Medusahead (*Taeniatherum caput-medusae* subsp. *asperum*) is a member of the Triticeae tribe of the grass family (Young 1992). Medusahead is a serious threat to intermountain rangelands, especially those sites that have high clay content soils. It competes aggressively with other plants and is a low-value forage species for both livestock and wildlife. Medusahead-dominated rangelands and pastures have very low diversity and low value for wildlife habitat and watershed function. Medusahead can be found on thousands of acres of rangeland in California, Oregon, Washington, and Idaho, and continues to expand its influence in these states as well as Nevada and Utah.

Identification

During the seedling stage in late fall or early spring, medusahead may be recognized by its slender and delicate-looking leaves and its bright green color. When a seedhead starts to develop in the late spring, it is wrapped in leaves and has visible awn tips. As the plant continues to mature, the seed head becomes totally visible and the awns stick straight up from the seed. The stems of the mature plant are wiry and slender and contain a few short, narrow leaves. It is not until the plant starts to dry out, turning from a purplish color to a tan color, that the awns start to take on the twisted appearance by which the plant is customarily identified (Turner et al. 1963).

Medusahead has the capability to produce tillers. Miller (1996) observed one plant to produce up to 11 tillers. Plant height can range from 8 to 20 inches (20 to 50 cm). Medusahead contains two or sometimes three spikelets at each node of a spike. Each spikelet contains one seed. The typical number of seeds per spike is between 8 and 15, but up to 45 have been observed (Turner et al. 1963, Clausnitzer 1996, Miller 1996). Most plants produce single spikes, but large individuals can have multiple spikes (Pyke 1993). Medusahead has two kinds of awns. Both awns are flat, and the longer of the two contains barbs that point upward, and can be felt by rubbing in the opposite direction. The shorter of the two awns is attached to a seed that is approximately $1/4$ inch (6 mm) long. The longer of the two awns ranges from $1/4$ to 1 inch (6 to 25 mm) in length, and arises below the seeds at the nodes of the central

271

axis of the spike. It remains attached after the seed (with the attached long awn) shatters (Turner et al. 1963).

Origin, History, and Distribution

Medusahead is an annual grass native to Eurasia. Three distinct subspecies have been described by Frederiksen (1986). *Taeniatherum caput-medusae* subsp. *caput-medusae* is found in Spain, Portugal, southern France, Morocco, and Algeria; *Taeniatherum caput-medusae* subsp. *crinitum* occupies the region from Greece and the former Yugoslavia eastward to Afghanistan; and *Taeniatherum caput-medusae* subsp. *asperum* overlaps the range of the other two subspecies (Young 1992). There has been repeated conjecture about which taxon was introduced into the United States. Young (1992) suggested that *Taeniatherum caput-medusae* subsp. *asperum* (Simk.) Melderis is the correct taxon for medusahead in the western United States.

Medusahead was first collected in the United States in the Umpqua Valley, southwestern Oregon, in 1884 (Turner et al. 1963). Medusahead was abundant in the upper Willamette Valley of Oregon by 1915. In Oregon, the southwestern interior valleys have the most extensive distribution of medusahead, but it has also expanded and occupies extensive areas in central and eastern Oregon.

In 1901, G.R. Vasey discovered medusahead near Steptoe, Washington, where it started spreading rapidly around 1914 from Steptoe Butte (Turner et al. 1963, Young 1992). Medusahead was discovered near Mountain Home, Idaho, by 1930. Although it was discovered in southern Idaho, medusahead was first collected in Idaho in 1944 near Payette, Washington County, Idaho. Ranchers reported that it occurred in this area as early as 1942 (Sharp and Tisdale 1952, Young 1992). By 1995 it was estimated that medusahead occupied close to 1 million acres (400,000 ha) (Turner et al. 1963, Young 1992, Callahan 1994).

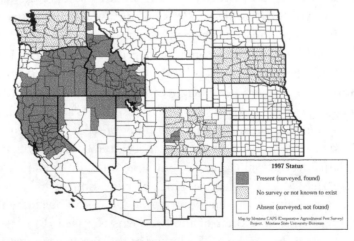

	1997 Status
■	Present (surveyed, found)
▨	No survey or not known to exist
□	Absent (surveyed, not found)

Map by Montana CAPS (Cooperative Agricultural Pest Survey) Project, Montana State University-Bozeman

Distribution of medusahead by county in the western United States. Data were collected by surveying various weed authorities in each state.

Medusahead spread south through the mountain valleys of western Oregon and reached the upper Sacramento Valley of California by 1900 (Murphy and Turner 1959, Young and Evans 1970). By 1959 it had traveled 600 miles (965 km) south of the Oregon border and was found existing in Ventura County (Murphy and Turner 1959). Medusahead has spread throughout the annual-dominated ranges of northern and central California (Young and Evans 1970). Current estimates for medusahead distribution in the Great Basin of northeastern California, extending across Modoc and Lassen Counties, are approximately 5 million acres (2 million ha). Medusahead may have invaded all suitable sites in California (Young 1994).

As of 1994, Nevada had fewer than 100,000 acres (40,500 ha) occupied by medusahead with most of the concentration occurring within Elko County in northeastern Nevada. Most of Nevada seems to be able to avoid extreme problems with medusahead because of its salt deserts and coniferous forests (Young 1994).

As of 1994, medusahead occupied approximately 2,000 acres (810 ha) in northern Utah within Cache and Box Elder counties, which lie along the Idaho border. Box Elder County also borders Nevada. A small amount of medusahead has been found in Weber County which lies south of Cache County (Dewey 1994).

As of 1994, Montana, Wyoming, New Mexico and Arizona were not reporting medusahead infestations.

Potential Invasion

Medusahead is likely to invade areas in which the native vegetation has been weakened by overgrazing, intense fires, or cultivation. It can also take over from previously established weeds such as cheatgrass (*Bromus tectorum*). Medusahead is capable of establishing in highly disturbed communities as described above, but it is also capable of establishing and maintaining itself in diverse communities of native perennial plants (Miller 1996). On those sites in which it becomes established, the implications for additional expansion are ominous when normal disturbances such as fire or ant and rodent activity occur.

Medusahead occurs in areas with annual precipitation of 10 to 40 inches (25 to 100 cm) occurring during fall, winter, and spring (Major et al. 1960). Seasonal distribution of precipitation is more important than total amount of precipitation in favoring medusahead (Parish 1956). Lack of summer precipitation favors medusahead over the later starting and longer growing perennial grass seedlings (Harris 1977). The upper precipitation limit for medusahead growth seems to be about 50 inches (127 cm) (Major et al. 1960).

Infestations primarily occur in former sagebrush-grass or bunchgrass communities that receive 10 to 20 inches (25 to 50 cm) of precipitation (Sharp and Tisdale 1952). In these drier areas, medusahead is at a competitive advantage where extra moisture collects due to topography, east or north exposures, decreased evaporation, or where high soil clay content within 10 to 12 inches (25 to 30 cm) of the surface provides longer water-holding capacity (Dahl and Tisdale 1975). It does

273

well in clay soils that shrink, swell, and crack. Soils that stay moist through summer do not seem to support medusahead (Turner et al. 1963). Thus, a climate with a summer precipitation pattern would probably limit medusahead invasion.

Well-drained, coarser-textured soils with poorly developed profiles, areas above 4,500 feet (1,370 m) elevation, and sites receiving less than 9 to 12 inches (23 to 30 cm) of annual precipitation may be less susceptible to invasion (Horton 1991). Current observations indicate potential for medusahead invasion onto loamy soils (Miller 1996).

Impacts

Medusahead crowds out other annuals and generally outcompetes perennial grass seedlings. Where it is allowed a foothold, it tends to form exclusive stands that are reinvaded by other vegetation very slowly if at all. However, there is evidence that squirreltail (*Sitanion hystrix*) can reinvade medusahead stands (Hironaka and Tisdale 1963, Hironaka and Sindelar 1973, 1975).

The dense, long-lasting litter layer formed by medusahead can burn readily. Frequent fires destroy the shrub component of the community and potentially part of the bunchgrass community without destroying significant amounts of medusahead seed. The shrub subcanopy mounds and microphytic crusts (soil-covering crusts composed of algae, lichens, and/or mosses) characteristic of the native community disappear (Young 1992).

The persistent root system in soil associated with perennials is lost once medusahead dominates the system. Although the litter layer may be of some value in protecting the soil from wind and water erosion (Turner et al. 1963), the short-lived roots of medusahead will not hold the soil as well as the root network of an established perennial community (Hironaka 1965).

Medusahead is almost worthless as forage for cattle and sheep (Turner 1965). Animals will graze it for a short time early in spring during the pre-seedhead stage, especially if there is not a heavy standing litter layer (Hironaka 1965). Chemical analysis reveals that the composition of medusahead is comparable to many desirable forages in moisture content, crude protein, crude fat, crude fiber, and lignin (Bovey et al. 1961), but coarseness due to the high silica content makes it unpalatable to livestock. In addition, awns can cause injury to eyes and mouths of animals (Hilken and Miller 1980). It has been estimated that grazing capacity can be reduced 50% to 80% after a few years of medusahead infestation (Hironaka 1961). Where medusahead has replaced cheatgrass stands, grazing capacity has been reduced to between 50% and 75% of its former extent (Harris and Goebel 1976). Wild birds eat very little medusahead seed (Goebel and Berry 1976). Captive chukar partridges ate medusahead seeds readily, but the seeds appeared to be largely indigestible (Savage et al. 1969).

Biology and Ecology

Medusahead germinates in fall, winter, or spring. Root growth can proceed through the winter, when little above-ground growth may be apparent. Winter root growth is mostly a downward extension of the primary root, with greater lateral development in spring (Hironaka 1961). Medusahead roots have been measured as deep as 40 inches (100 cm) (Hironaka 1961). Growth accelerates in the spring, with maturity generally occurring late June to early July, a few weeks later than most other annuals. Seeds tend to remain on heads until late summer or early fall (Young 1992, Sharp and Tisdale 1952, Hironaka and Tisdale 1958). Medusahead is primarily self-pollinating (Young 1992).

Most medusahead germination occurs at 50° to 60° F (10° to 15° C). Germination drops off considerably at higher temperatures until after an afterripening period of about 180 days (Young et al. 1968). Medusahead seed thus can avoid premature germination and wait for cool, wet conditions in the autumn. Nelson and Wilson (1969) observed greater than 90% germination of viable three-year-old medusahead seed, although the percentage of viable seeds decreases annually (Sharp et al. 1957, Kay 1965).

Small silica barbs (Young 1992) enable the seed to cling to animals, clothing, and machinery for dispersal. Dispersal can also occur through animal droppings or by wind and water (Furbish 1953, Turner et al. 1963).

Medusahead litter is slow to decompose because of its high silica content (Young 1992), causing buildups of litter 2 to 4 inches (5 to 10 cm) thick. This litter layer may inhibit seedlings of more desirable species by shading, or by keeping their seeds from gaining contact with the soil. However, medusahead seeds can germinate within the litter layer. Humidity and temperature conditions within the litter can stimulate medusahead germination. If the initial root dries out and dies, a new root can later develop when moisture conditions improve (Young et al. 1971, Young 1992). The litter buildup also provides fine fuels that can result in intense fires which kill or injure its competitors.

Medusahead possesses several additional characteristics that make it a successful competitor. Medusahead exhibits plastic, prolific seed production. Uncrowded medusahead can produce six or more seedheads per plant, while dense stands will produce at least one seedhead per plant (Murphy and Turner 1959). A seedhead may contain more than 20 seeds (Clausnitzer 1996) and in ideal situations can produce 4,000 to 10,000 seeds per square meter (Harris and Goebel 1976). Medusahead germination rates tend to be faster and more complete than competitors. Germination has been observed 8 to 10 hours after moistening, at low temperatures (50° F [10° C]) and droughty conditions (-11.4 bars) (Harris 1977). Germination rates are often over 90% (Sharp et al. 1957, Clausnitzer 1996).

Autumn germination followed by fast, deep root growth through the winter provides a competitive advantage. After the roots have gone relatively deep, medusahead produces branching roots (Hironaka 1961). By the time preferred plants initiate growth in the spring, medusahead has a well-developed, deep root system.

Medusahead roots have thicker cell walls than do cheatgrass roots, which allows it to conduct water from deep sources through dry upper horizons (Harris 1977). This allows medusahead to extend its growth two to three weeks later into the year relative to cheatgrass. Thus, medusahead has a greater competitive edge than cheatgrass over establishing perennial grass seedlings, because it can deplete deep as well as shallow soil moisture.

Finally, medusahead's low palatability, due to the high silica content of tissues, confers upon it an additional competitive advantage.

Management

A combination of management treatments has generally been necessary to reduce established, dense stands of medusahead. Reseeding is likely to be necessary to keep medusahead from reclaiming a treated site if too few preferred plants are in place to occupy the site quickly. To be successful, a seeded species must be adapted to the site, easily established, and aggressive (Vallentine 1989).

• *Mechanical*

Spring plowing after most of the medusahead has germinated has given some control, with results of up to 95% reduction in medusahead. Disking has also shown results (50% control reported), and has been used as a follow-up to plowing (Erickson et al. 1956, Higgins and Torell 1960, Harris and Goebel 1976). These measures may not be practical under certain conditions of terrain or soil. The results from mechanical efforts are much greater if combined with burning or especially herbicide treatments.

• *Prescribed Fire*

Burning medusahead stands can destroy significant amounts of seeds if the seedhead has not disseminated, reducing the stand by 60% to 95% in the next growing season (Harwood 1960, Murphy and Lusk 1961, McKell et al. 1962). Slow fires that burn downslope or against the wind are most effective (Murphy and Turner 1959). Prescribed burning should be conducted during the soft dough stage of seed development when the seeds exude a milky substance. High moisture content in the seed accentuates the effects of burning (McKell et al. 1962).

Burning also removes the medusahead litter layer. This decrease in litter and seeds helps prepare the seedbed for more desirable species and improves the effectiveness of mechanical and/or chemical treatment (Harris and Goebel 1976, Torell et al. 1961).

An effective management strategy practiced by the USDI Bureau of Land Management in central Oregon is burning medusahead in late spring or early summer before the seeds have dropped off the plant. This was followed the next spring with a herbicide treatment of glyphosate (0.375 lbs/ac, 0.41 kg/ha) after remaining seeds have germinated. Since glyphosate is a broad-spectrum herbicide, this is recommended only when reseeding is required, owing to a shortage of preferred plants. Finally, the area was reseeded with desirable grass and shrub species because of the medusahead seed that remained in the seedbank (Crockett 1995, Zalunardo 1995).

• Chemical

[This discussion of chemical treatments is based on research conducted over several years. Over time, labelling of the chemicals discussed may have changed. It is critical to determine current labelling for each chemical before planning to use it as a management tool.]

Glyphosate has had good results when applied at 0.375 lbs/ac (0.41 kg/ha) in early spring while the plants are still small, and before the seedhead is produced. The rate is low enough to avoid damaging the desired competitors to medusahead, such as squirreltail (Crockett 1995). This application is most effective after the area has been burned. The herbicide treatment is also effective if done on two occasions, one month apart.

Good results have been obtained by the pre-emergence application of soil-active herbicides such as atrazine (see special state labels to determine approved areas), bromacil, and siduron. Atrazine at 1 lb ai/ac (1.1 kg/ha) in late fall, and bromacil at 0.5 lb/acre (0.55 kg/ha) in early spring or fall has selectively controlled medusahead in stands of perennial grasses. Atrazine injures and kills Sandberg bluegrass (*Poa secunda*); bromacil apparently does not (Turner 1965). Sandberg bluegrass is an important species in resisting reinvasion by annuals. Pre-emergence applications of 3 lbs/ac (3.3 kg/ha) siduron plus 0.3 lb/ac (0.33 kg/ha) picloram have been effective (Young and Evans 1970).

Foliar applications of dalapon have been effective when applied during the vegetative stage, usually mid-April to early May. Rates typically used have been 2 or 3 lbs/acre (2.2 to 3.3 kg/ha) resulting in 96% to 100% control (Higgins and Torell 1960, Kay 1963, Torell and Erickson 1967).

Combining herbicides with mechanical treatment or burning has shown very good results. Burning followed by fall application of glyphosate (1.2 liters/hectare) before minimum-till drill seeding has allowed establishment of crested wheatgrass (*Agropyron cristatum*) (Horton 1991).

• Biological

Research by the USDA Agricultural Research Service and Montana State University indicates that crown rot (*Fusarium culmorum*), a common pathogen found on wheat, causes severe disease on medusahead, while having a less severe impact on squirreltail and western wheatgrass (*Agropyron smithii*), making the fungus a possible biological control agent (Grey 1994). Further research has been proposed to examine other root pathogens of medusahead found in the Great Basin region.

• Grazing

Forcing livestock to graze medusahead stands heavily before seedset may reduce the seed crop appreciably, if done over several years (Horton 1991). Sheep will eat green medusahead when confined to a small area, and possibly will eat some medusahead even after it has dried (Lusk et al. 1961). However, since medusahead is unpalatable relative to most desired species, grazing is not likely to be an effective control treatment.

• Revegetation

Revegetation and grazing management must follow any control effort to prevent medusahead from reestablishing dominance (Major et al. 1960). Broadcast-seeding perennial grasses into stands of medusahead without some prior control of the weed has been very unsuccessful. Good seeding results have been obtained with prior control, primarily on sites that are suitable for tillage. Shallow, steep, or rocky sites of low potential are much more difficult to revegetate (Turner 1965).

Combined treatments of two tillages, or tillage combined with herbicide or burning is most effective (Torell and Erickson 1967). Treatment of two successive crops of annual weeds enhances survival of desirable seedlings by reducing the weed seed reserve.

If herbicide use precedes reseeding, it is important to wait until residual herbicide activity subsides, to prevent damage to the seeded species. Several weeks are required with dalapon, whereas atrazine takes about a year to break down sufficiently (Turner et al. 1963). Strip tilling or some other method of removing the herbicide from the drill row might preclude this problem, and allow seedling establishment of preferred species while the herbicide actively suppresses medusahead between rows.

Squirreltail, bluebunch wheatgrass (*Agropyron spicatum*), crested wheatgrass, intermediate wheatgrass (*A. intermedium*), orchardgrass (*Dactylis glomerata*), Thurber needlegrass (*Stipa thurberiana*), needle-and-thread grass (*Stipa comata*), Indian ricegrass (*Oryzopsis hymenoides*), Sandberg bluegrass, and sheep fescue (*Festuca ovina*), because of their high vigor and competitiveness, are good choices on sites to which they are adapted for reseeding medusahead stands (Torell and Erickson 1967, Barrington 1995, Washington 1995, Miller 1996). Sowing in late autumn or early spring using a rangeland drill is the usual practice (Turner 1965, Robocker and Schirman 1976). Newly established wheatgrass stands will suppress but not eliminate medusahead.

• Integrated Strategies for Sustainable, Long-Term Management

Medusahead competes most severely when desirable species are overgrazed (Higgins and Torell 1960). Grazing management and control of new infestations while they are still small are the best strategies for long-term management (Christen et al. 1974).

Since medusahead is unpalatable compared with most desired species, grazing management must be prescribed to facilitate establishment of desired species following treatment of medusahead. Grazing prior to establishment of the desired seeded species will likely result in stand failure and recolonization by medusahead. Once a reseeded pasture has been established, timing and duration of grazing must be managed to maintain the desired species. Winter and early spring grazing are likely to be most conducive to maintaining desirable species, since sufficient soil moisture probably remains to allow completion of the plants' growth cycle. Winter and early spring grazing may also result in some use of medusahead because it is a winter-growing annual. Late spring, summer, and fall grazing are likely to provide a competitive advantage to medusahead, since grazing preference will focus on desired species. Monitoring is essential to determine whether or not a grazing prescription is maintaining desired species.

Acknowledgments

Miller and Clausnitzer were previously Graduate Research Assistants, Dept. of Rangeland Resources, Oregon State University, supported by the Vegetation Diversity Project, USGS Biological Resources Division, Forest & Rangeland Ecosystem Science Center, OSU, Corvallis, OR 97331. Support for preparation of this chapter was provided by FRESC and by the Dept. of Rangeland Resources at Oregon State University.

Literature Cited

Barrington, M.R. 1995. Identification of restoration potential on southwestern Oregon rangelands. Ph.D. Thesis. Oregon State University, Corvallis, OR.

Bovey, R.W., D. Le Tourneau, and L.C. Erickson. 1961. The chemical composition of medusahead and downy brome. Weeds 9:307-11.

Callahan, R. 1994. Personal communication. Extension Weed Specialist, University of Idaho, Moscow, ID.

Christen, M.D., J.A. Young, and R.A. Evans. 1974. Control of annual grasses and revegetation in ponderosa pine woodlands. J. Range Manage. 27:143-45.

Clausnitzer, D. 1996. Field study of competition between medusahead (*Taeniatherum caput-medusae* subsp. *asperum*) [Simk.] Melderis) and squirreltail (*Elymus elymoides* [Raf.] Swezey). M.S. Thesis, Dept. of Rangeland Resources, Oregon State Univ., Corvallis, OR.

Crockett, R.P. 1995. Personal communication. Product Development Associate, Monsanto, Vancouver, WA.

Dahl, B.E., and E.W. Tisdale. 1975. Environmental factors related to medusahead distribution. J. Range Manage. 28:463-68.

Dewey, S. 1994. Personal communication. Dept. of Plant, Soils and Biometeorology, Utah State University, Logan.

Erickson, L.C., C. Lambert, and R.S. Parrish. 1956. Chemical and cultural treatments for the control of medusahead rye. Res. Prog. Rep. WWCC, 20-21.

Evans, R.A., and J. A. Young. 1970. Plant litter and establishment of alien annual weed species in rangeland communities. Weed Science 18:697-703.

Frederiksen, S. 1986. Revision of *Taeniatherum* (Poaceae). Nordic J. of Botany 6:389-97.

Furbish, P. 1953. Control of medusahead on California ranges. J. Forest. 51:118-21.

Goebel, C.J., and G. Berry. 1976. Selectivity of range grass seeds by local birds. J. Range Manage. 29:393-95.

Grey, W. 1994. Personal communication. Asst. Research Professor, Dept. of Plant Pathology, Montana State University, Bozeman.

Harris, G.A. 1977. Root phenology as a factor of competition among grass seedlings. J. Range Manage. 30:172-77.

Harris, G.A., and C.J. Goebel. 1976. Factors in plant competition in seeding Pacific Northwest ranges. Washington State Univ. Agr. Exp. Sta., Bull. 820.

Harwood, L. 1960. Programs to control medusahead. Proc. Calif. Sec., Soc. Range Manage. Fresno, CA, 45-49. (Mimeo release.)

Higgins, R.E., and P.J. Torell. 1960. Medusahead: range menace. Univ. of Idaho Agr. Exp. Sta., Ext. Bull. 331.

Hilken, T.O., and R.F. Miller. 1980. Medusahead (*Taeniatherum asperum* Nevski): A review and annotated bibliography. Oregon State Univ. Agric. Expt. Bulletin 644.

Hironaka, M. 1961. The relative rate of root development of cheatgrass and medusahead. J. Range Manage. 14:263-67.

Hironaka, M. 1965. The medusahead problem. *In*: Proc. of the Cheatgrass Symposium, Vale, Oregon. (Portland) Bureau of Land Management, 62-65.

Hironaka, M., and B.W. Sindelar. 1973. Reproductive success of squirreltail in medusahead infested ranges. J. Range Manage. 26:219-21.

Hironaka, M., and B.W. Sindelar. 1975. Growth characteristics of squirreltail seedlings in competition with medusahead. J. Range Manage. 28:283-85.

Hironaka, M., and E.W. Tisdale. 1958. Relative rate of root development of medusahead and cheatgrass. Res. Prog. Rep. WWCC, 28.

Hironaka, M., and E.W. Tisdale. 1963. Secondary succession in annual vegetation in southern Idaho. Ecology 4:810-12.

Horton, W.H. 1991. Medusahead: importance, distribution, and control *In*: James, Evans, Ralphs, and Childs (eds.), Noxious range weeds. Westview Press, Boulder, CO.

Kay, B.L. 1963. Effects of dalapon on a medusahead community. Weeds 11:207-209.

Kay, B.L. 1965. The medusahead problem in California—what progress is research making? *In*: Proc. of the Cheatgrass Symposium, Vale, Oregon. Bureau of Land Management, Portland, 74-80.

Lusk, W.C., M.B. Jones, P.J. Torell, and C.M. McKell. 1961. Medusahead palatability. J. Range Manage. 14:248-51.

Major, J., C.M. McKell, and L.J. Berry. 1960. Improvement of medusahead infested rangeland. Calif. Agr. Exp. Sta., Ext. Serv. Leaf. 123.

McKell, C.M., A.M. Wilson, and B.L. Kay. 1962. Effective burning of rangelands infested with medusahead. Weeds. 10:125-31.

Miller, H. C. 1996. Demography of medusahead on two soil types: potential for invasion into intact native communities. M.S. Thesis. Department of Rangeland Resources. Oregon State University, Corvallis, OR.

Murphy, A.H., and W.C. Lusk. 1961. Timing medusahead burns to destroy more seed—save good grasses. Calif. Agric. 15:6-7.

Murphy, A.H., and D. Turner. 1959. A study on the germination of medusahead seed. Calif. Dept. of Agric. Bull., 48(1):6-10.

Nelson, J.R., and A.M. Wilson. 1969. Influence of age and awn removal and dormancy of medusahead seeds. J. Range Manage. 22:289-290.

Parish, R.L. 1956. A study of medusahead rye, *Elymus caput-medusae* L., including some of the morphological and physiological factors influencing its growth and distribution, and determining some possible methods for its control on Idaho ranges. M.S. Thesis. Univ. of Idaho, Moscow, ID.

Pyke, D.A. 1993, 1994. Personal communications. U.S. Biological Survey, Biol. Res. Div., and Department of Rangeland Resources Professor, Oregon State University, Corvallis, OR.

Robocker, W.C., and R.D. Schirman. 1976. Re-seeding trials on Columbia Basin rangelands dominated by winter annual grasses. J. Range Manage. 29:492-97.

Savage, D.E., J.A. Young, and R.A. Evans. 1969. Utilization of medusahead and downy brome caryopses by chukar partridges. J. Wildl. Manage. 33:975-78.

Sharp, L.A., and E.W. Tisdale. 1952. Medusahead, a problem on some Idaho ranges, a preliminary study. Research Notes. Forest, Wildlife and Range Expt. Station. Univ. of Idaho. Research Note No. 3.

Sharp, L.A., M. Hironaka, and E.W. Tisdale. 1957. Viability of medusahead seed collected in Idaho. J. Range Manage. 10:123-26.

Torell, P.J., and L.C. Erickson. 1967. Reseeding medusahead-infested ranges. Univ. of Idaho, Agr. Exp. Sta., College of Agr. Bull. 489.

Torell, P.J., L.C. Erickson, and R.H. Hass. 1961. The medusahead problem in Idaho. Weeds 9:124-31.

Turner, R.B. 1965. Medusahead control and management studies in Oregon. *In*: Proc. of the Cheatgrass Symposium, Vale, Oregon. Bureau of Land Management, Portland, OR 70-73.

Turner, R.B., C.E. Poulton, and W.L. Gould. 1963. Medusahead—a threat to Oregon rangeland. Oregon State Univ., Agr. Exp. Sta., Spec. Rep. 149.

Vallentine, J.F. 1989. Range Developments and Improvements (3rd ed.). Academic Press, Inc. San Diego, CA.

Washington, J. 1995. Personal communication. Graduate Student Oregon State University, Corvallis, OR.

Young, J.A. 1992. Ecology and management of medusahead (*Taeniatherum caput-medusae* subsp. *asperum* [SIMK.] *melderis*) Great Basin Naturalist 52:245-53.

Young, J.A. 1994. Personal communication. USDA Agricultural Research Service, Reno, NV.

Young, J.A., and R.A. Evans. 1970. Invasion of medusahead into the Great Basin. Weed Sci. 18:89-97.

Young, J.A., R.A. Evans, and R.E. Eckert, Jr. 1968. Germination of medusahead in response to temperature and afterripening. Weed Sci. 16:92-95.

Young, J.A., R.A. Evans, and B.L. Kay. 1971. Germination of caryopses of annual grasses in simulated litter. Agron. J. 63:551-55.

Zalunardo, D. 1995. Personal communication. Range Conservationist, Prineville BLM, Prineville, OR.

Oxeye Daisy

Bret E. Olson and Roseann T. Wallander

Oxeye daisy (*Chrysanthemum leucanthemum*) is a forb that has been introduced from Europe and has naturalized in the northeastern United States. It is often spread intentionally as an ornamental, or incidentally as a contaminant in hay and grain seed. It is an aggressive competitor and often forms dense patches, especially in areas grazed by cattle. Heavy infestations of oxeye daisy reduce hay or forage production.

Oxeye daisy is not a problem with annual crops, because cultivation easily destroys its shallow root system. However, cultivating is not possible in many pastures, rangelands, and roadside areas. Mowing oxeye daisy as the flowerheads open may eliminate seed production if the growing season is too short for new flower stems to develop. However, mowing may stimulate the production of rosettes along the rootstalks.

Herbicides may control oxeye daisy in pastures for two to three years until new plants establish from seeds in the seedbank (B. Roché 1992, P. Fay 1993, personal communication). However, herbicides may harm native perennial forbs on mountain rangelands and along roadsides. In addition, herbicides may not be cost-effective in many areas.

Identification

Oxeye daisy is a perennial herb with oblique, shallow, branched rhizomes and strong adventitious roots (Howarth and Williams 1968). Basal stems are prostrate and will root, the other stems are erect (12 to 30 inches [30 to 80 cm]), and simple or slightly branched. Stems are glabrous to slightly pubescent. Basal leaves are on long stalks, spatulate to round, and dentate. Stem leaves are spiral, sessile, and narrow lanceolate or ligulate coarsely toothed, often with lobes at the base. Flower heads are mostly solitary on long terminal peduncles, 1 to $2^{3}/_{16}$ inches (2.5 to 5.5 cm) in diameter.

The cotyledons of seedlings open above the soil surface. The first true leaf is lobed. The cotyledons wither soon after the first leaves develop and the stem (hypocotyl) does not elongate above the ground. A rosette of leaves is considered a juvenile plant (Howarth and Williams 1968). It is not known whether oxeye daisy can flower and produce seeds in its first year.

Origin, History, and Distribution

Fruits of oxeye daisy have been identified from the Iron Age and from the Roman period. The plant appeared in Britain during the post-glacial period along with other weeds. Oxeye daisy is distributed throughout the British Isles, as well as Europe to northern Scandinavia, Lapland, and central and Russian Asia (Howarth and Williams 1968). Italians use oxeye daisy in salads; it was used more extensively in the past as a food item.

Oxeye daisy was carried to North America and New Zealand as a contaminant in seed. It was also introduced as an ornamental. In the northeastern United States, this plant has escaped cultivation and has naturalized. Because it is showy, many landowners will not mow oxeye daisy plants in their lawns. Oxeye daisy is locally abundant in the Great Plains (Great Plains Flora Association 1986). It grows along roadsides, in waste places, and in pastures in western and southcentral Montana (Dorn 1984). It is the most common roadside weed in the Pacific Northwest (Taylor 1990). However, its general distribution in the United States has not been described.

Potential Invasion

Oxeye daisy occurs in meadows, native grasslands, pastures seeded to introduced species, waste grounds, and along railway embankments and roadsides. It grows well on relatively nutrient-poor to nutrient-rich soils; it has a wide edaphic tolerance. Oxeye daisy is found in every state, but is less common in the southern United States. Cattle grazing of pastures encourages its spread because cattle avoid the plant. It is seldom noticed in ungrazed or lightly grazed grasslands, possibly because it has a low shade tolerance.

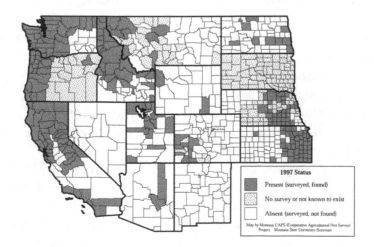

Distribution of oxeye daisy by county in the western United States. Data were collected by surveying various weed authorities in each state.

Impacts

The ecological, environmental, economic, or sociological impacts of oxeye daisy have not been well documented. It frequently invades fields where it competes aggressively, especially in pastures grazed by cattle, and forms dense populations. In turn, this reduces plant species diversity. Bare soil is more prominent in areas with high densities of oxeye daisy (personal observation), so the potential for soil erosion is increased. Oxeye daisy has a relatively small taproot compared with the extensive fibrous root systems of associated grasses. Thus, a heavy infestation of oxeye daisy may reduce the amount of organic matter contributed belowground annually, and in turn slow the rate of nutrient cycling.

Because cattle avoid grazing oxeye daisy, carrying capacity of heavily infested pastures is reduced when cattle are the primary grazers. Oxeye daisy is continuing to expand its range.

Biology and Ecology

Oxeye daisy is commonly found on basic or neutral soils, whereas it is less common on acid soils (Howarth and Williams 1968). Ferdinandsen (1918) found oxeye daisy growing optimally on soils with pHs ranging from 6.5 to 7.0.

Oxeye daisy is indifferent to water and soil friability, but has a moderate requirement for nitrogen (Ellenburg 1950). Its requirements are very similar to English plantain (*Plantago lanceolata*), common plantain (*P. major*), and Canada thistle (*Cirsium arvense*), with which it often grows. When grown under low nutrient levels, oxeye daisy allocates more biomass to the root system at the expense of the flower heads (Boutin and Morisset 1988). Low nutrient levels had little effect on allocation to leaf material. In general, it is more prolific on poorer soils (Sanders 1993). This may not necessarily indicate a preference for or tolerance of soils with low fertility, but that it cannot compete with established vegetation on more fertile soils.

Allocation to reproductive effort is higher under low light levels than under high light levels, indicating a strategy of maximizing seed production when shaded by other plants. Reproductive effort was unaffected by nutrient level.

Oxeye daisy is unaffected by light frost and tolerates drought well, although it is usually found in more moist areas. It is a pioneer species in several habitats exposed to soil drying. During periods of water stress, deeper rooting species (e.g., *Taraxacum officinale*) wilt before oxeye daisy (Howarth and Williams 1968).

Besides reproducing vegetatively along a rhizome, oxeye daisy is a prolific seed (achene) producer. A vigorous daisy plant growing in a grass field produced about 26,000 seeds; smaller plants at the same site produced from 1,300 to 4,000 seeds per plant (Dorph-Peterson 1925). Salisbury (1942) noted that an oxeye daisy plant may produce 2,688 offspring per year. Seeds become viable 10 days after flowers open (Georgia 1914). Usually the seeds are dispersed by wind close to the parent plant because it lacks a pappus, but it may also be carried by animals. In the past, oxeye daisy seeds have contaminated grass seed sold in the United States (Georgia 1914, Gilkey 1957).

Ripening of the achene is not followed by a period of dormancy unless enforced by environmental conditions. Germination is insensitive to light, nitrates, chilling, and sulphuric acid treatments. Thus, oxeye daisy seeds will germinate throughout the growing season, but most seedlings become established in spring (Howarth and Williams 1968). Oxeye seedlings are considered to be drought tolerant. Seeds that do not germinate in the spring or summer may remain viable for a long time. In a buried seed trial, 82% of the seeds were still viable after six years, 1% of the seeds were viable after 39 years (Toole and Brown 1946).

At about the six-leaf stage, the primary root starts to be replaced by a well-developed system of laterals which are relatively shallow. As the rhizome system develops, the main root system becomes less important. The plant is characterized as a hemicryptophyte, whereby the perennating buds at the soil surface are protected from environmental extremes by snow or organic debris.

An individual oxeye daisy plant may consist of one to many rosettes on the soil surface. Each rosette usually produces only one flowering stem. A population of oxeye daisy can form a dense mat of rosettes. As the plants bloom, a field infested with oxeye daisy may appear "white as snow."

Horses, sheep, and goats graze oxeye daisy, but cows and pigs tend to refuse it because it is acrid (Howarth and Williams 1968). When oxeye daisy plants are not grazed, they gain an advantage over more desirable forage plants in pastures (Gilkey 1957). Oxeye daisy is not common on grasslands that are lightly grazed in the British Isles (Howarth and Williams 1968). Its abundance is partly related to the intensity of cutting or grazing of associated species. This suggests that oxeye daisy requires reduced competition from existing vegetation by grazing, or possibly a disturbance, to establish. Kydd (1964) found that canopy coverage of oxeye daisy was highest with "undergrazing" and "unadjusted" cattle grazing treatments compared with "overgrazing," rotational grazing," or a "haying-with-the-aftermath-grazed-twice" treatment. The "unadjusted" pastures were grazed frequently; cattle numbers were not adjusted to the herbage base, which resulted in close autumn grazing and light spring grazing.

Where oxeye daisy was a dominant member of the community (20% canopy coverage), it increased greatly with continuous cattle grazing (Norman 1957). Increases were much smaller with close rotational grazing by cattle, and close rotational and continuous grazing by sheep. The canopy coverage of oxeye was essentially unchanged under lenient cattle grazing.

The effects of intensive cattle grazing on oxeye daisy have been assessed in southwestern Montana (Olson et al. 1997). Although cattle generally do not graze forbs such as oxeye daisy, the landowner wanted to see whether high cattle densities for short periods of time could overcome the tendency of cattle to avoid oxeye daisy.

Two years of intensive grazing reduced densities of oxeye seedlings and rosettes, but had no effect on densities of adult plants compared with densities in adjacent, ungrazed exclosures. Nonetheless, by reducing densities of these recruitment-age classes, the seedlings and rosettes, densities of adult plants would have decreased in subsequent years, if the study could have been continued. Intensive grazing had minimal impact on the associated perennial grasses. The cattle tended to pull up many of the oxeye daisy adult stems as if they had initially intended to graze the plant but changed their mind, possibly because of its acridity. However, overall "use" on the daisy was similar to use on the other vegetation, so cattle did not completely avoid the plant. Based on European studies, sheep or goats would probably have had a more significant impact than cattle on oxeye daisy.

When the inflorescence is removed, many lateral stems develop rapidly. When cows consume ripe seedheads, less than 40% of the seeds passing through the cow are viable (Howarth and Williams 1968). This is a considerable reduction in seed numbers, but an oxeye daisy plant may produce 1,300 to 4,000 fruits (Dorph-Petersen 1925). Thus, many seeds will survive the gastrointestinal tract of the ruminant.

Horse manure may contain seeds of oxeye daisy (personal observation), and it is likely that other large ungulates may intentionally or incidentally ingest oxeye daisy, and then pass seeds in their feces. Animals may also pick up seeds in their fleeces or coats. Hay from pastures infested with oxeye daisy may contain oxeye daisy seeds. In southwestern Montana the first cutting of hay often coincides with the beginning of seed set of this species.

Management

Oxeye daisy is a showy plant, and is often encouraged as an ornamental. It has been included in popular seed mixes, such as "Meadow in a Can," thus consumers should read the label carefully to ensure that this species is not a component of these mixes. Also, many landowners intentionally avoid mowing oxeye daisy on their lawns, allowing seed development and spread. Making the public aware of its potential to displace native species could help reduce its intentional spread.

As with any weed, it is important to minimize the amount of bare soil exposed by farming or haying practices, or livestock grazing. The seeds of oxeye daisy are quite small and have a difficult time germinating and establishing through a relatively thick layer of litter, whereas they can readily germinate on bare soil. In addition, maintaining an adequate plant canopy of associated plant species ensures their competitiveness and minimizes light penetrating to the lower canopy levels, and thus to seedlings or basal rosettes of oxeye daisy.

• Mechanical

Oxeye daisy is generally not found in intensively cultivated areas because its shallow root system is easily killed. In pastures, plants should be mowed as soon as the first flowers open to eliminate seed production (Georgia 1914); however, mowing may stimulate shoot production and subsequent flowering if the growing season is long enough. Oxeye daisy became the dominant plant in a field 14 months after herbicides were used to kill existing plants, followed by plowing and disking (Marks and Mohler 1985). This apparent contradiction could reflect that the plowing and disking stimulated abundant seed germination, and/or that the one-time plowing and disking simply fragmented oxeye daisy plants and vegetative reproduction was enhanced.

• *Chemical*

Oxeye daisy is moderately resistant to some 2,4-D-based herbicides, except at high rates: 5 lb/acre 2,4-D (Howarth and Williams 1968). In the early 1970s, Roché (unpublished data) compared 2,4-D at 2 lbs acid equivalent (ae) per acre with picloram at 2 oz active ingredient (ai) per acre for their effectiveness in controlling oxeye daisy on a mountain meadow in eastern Washington. Across these herbicide treatments, he applied nitrogen fertilizer at four different rates (0, 40, 80, 160 lbs as N, using ammonium nitrate-sulfate) beginning in 1972. Some plots were refertilized in 1973, 1975, and 1976; others were not refertilized, to assess residual effects of the initial application. Another set of plots was fertilized at the same rates but was not treated with either herbicide. Both herbicides were effective at reducing canopy cover of oxeye daisy, but fertilizer alone was almost as effective as the herbicides. Eighty pounds of N was the most cost-effective treatment after seven years. Grass yields increased 500% with high levels of N. Forage production in 1981, five years after the last fertilization treatment, was still 2.5 times greater than the control. On a mountain pasture in southwestern Montana, Fay (unpublished data) applied 1.5 pt of picloram with 1 qt 2,4-D per acre on a heavily infested oxeye daisy site in 1990. There was 100% control for two years, although he does not recommend this treatment as a long-term control method for oxeye daisy.

Roché (1992, personal communication) found that herbicides can control oxeye daisy; however, rodent burrows create small areas of bare soil, exposing seeds from the seedbank to mineral soil and minimal competition. Given long viability in the seed state, seedbanks can potentially reinfest a site for many years.

• *Biological Control*

As of 1998, effective biological controls have not been developed for this weed. Unfortunately, an introduced plant can invade thousands of hectares during the time required to introduce biological control agents.

• *Grazing*

Using appropriate livestock species may be a potential solution for controlling oxeye daisy. Even appropriate livestock species will seldom eradicate a weed, but at least they can minimize spread and potentially the competitiveness of the weed by reducing seed production.

287

Sheep or goats would be the most likely class of livestock to control this species, because they readily graze it. However, many infested areas have fencing and handling facilities appropriate only for cattle. In addition, sheep or goat grazing on mountain rangelands where oxeye daisy is often found may be uneconomical because of predation by coyotes, bears, and mountain lions.

• Integrated Management

Attempts at integrated weed management have been limited to combining herbicides with fertilizers, with considerable success. The herbicide kills at least the top of the plant, while the fertilizer stimulates growth of the associated grasses. Apparently, either the fibrous-rooted grasses are more adept than the tap-rooted oxeye daisy at taking up the additional nitrogen, or increased grass canopies shade the relatively smaller oxeye daisy, or possibly both are important factors. Potentially, combining sheep or goat grazing with fertilizers may have a similar effect without the costs associated with herbicides, as long as the animals preferentially graze the oxeye daisy and other forbs, and do not graze too much of the grass.

If an area will continue to be grazed by cattle only, herbicides will keep the oxeye daisy under control, but at significant cost. Some producers have had considerable success with fertilizing areas grazed by cattle. Rotational or intensive grazing may also keep the oxeye daisy under control, as long as utilization does not have a negative effect on the physiology of the associated grasses. Most importantly, a significant grass canopy needs to be maintained to shade oxeye daisy.

Conclusion

Oxeye daisy is an aggressive competitor and often forms dense patches, especially in areas grazed by cattle. Heavy infestations of oxeye daisy reduce hay or forage production. It continues to spread, partly because it is such an attractive plant, and the public is not aware of the plant's ability to displace native species. Herbicides may control oxeye daisy in pastures for two to three years, until new plants establish from seeds in the seedbank, but generally herbicides have to be reapplied every three to five years, to prevent new plants from establishing from seeds in the seedbank. Combining herbicides with fertilizers has been effective in some areas. On pastures grazed by cattle, rotational or intensive grazing may reduce densities of oxeye daisy. Over the long term, sheep or goat grazing has the greatest potential to reduce oxeye daisy's importance in plant communities.

Literature Cited

Boutin, C., and P. Morisset. 1988. Etude de la plasticite phenotypique chez le *Chrysanthemum leucanthemum*. I. Croissance, allocation de la biomasse et reproduction. Can. J. Bot. 66:2285-98.

Dorn, R.D. 1984. Vascular Plants of Montana. Mountain West Publishing, Cheyenne, WY.

Dorph-Petersen, K. 1925. Examinations of the occurrence and vitality of various weed seed species under different conditions, made at the Danish State Seed Testing Station during the years 1896-1923. Rep. 4th Int. Seed Test. Congr. 4:128-38.

Ellenburg, H. 1950. Unkrautgemeinschaften als Zeiger fur Klima und Boden. Stuttgart.

Ferdinandsen, C. 1918. Undersogeber over danske Unkrudsformationes paa Mineraljorder. Tidssk. Pl. Avl. 28:629.

Georgia, A. 1914. A manual of weeds. MacMillan, NY.

Gilkey, H.M. 1957. Weeds of the Pacific Northwest. Oregon State College.

Great Plains Flora Association. 1986. Flora of the Great Plains. University Press of Kansas. Lawrence, KS.

Howarth, S.E., and J.T. Williams. 1968. Biological flora of the British Isles. J. Ecol. 56:585-95.

Kydd, D.D. 1964. The effect of different systems of cattle grazing on the botanical composition of permanent downland pasture. J. Ecol. 52:139-49.

Marks, P.L., and C.L. Mohler. 1985. Succession after elimination of buried seed from a recently plowed field. Bull. Torrey Bot. Club. 112:376-82.

Norman, M.J.T. 1957. The influence of various grazing treatments upon the botanical composition of a downland permanent pasture. J. Br. Grassl. Soc. 12:246-56.

Olson, B.E., R.T. Wallander, and P.K. Fay. 1997. Intensive cattle grazing of oxeye daisy (*Chrysanthemum leucanthemum*). Weed Technol. 11:176-81.

Salisbury, E.J. 1942. The Reproductive Capacity of Plants. G. Bell and Sons Ltd, London.

Sanders, J. 1993. Hedgemaids and Fairy Candles. The Lives and Lore of North American Wildflowers. Ragged Mountain Press, Camden, ME.

Taylor, R.J. 1990. Northwest Weeds: The Ugly and Beautiful Villains of Fields, Gardens, and Roadsides. Mountain Press Publishing. Missoula, MT.

Toole, E.H., and E. Brown. 1946. Final results of the Duvel buried seed experiment. J. Agri. Res. 72:201-10.

Poison-Hemlock

Joseph M. DiTomaso

Poison-hemlock (*Conium maculatum*) is a member of the family Apiaceae (=Umbelliferae), which also contains a few important vegetable crops, including carrots and celery, and herbs such as parsley, anise, dill, caraway, and coriander. It is an erect biennial, or sometimes perennial, which generally grows 4 to 6 feet (1.2 to 1.8 m) in height, but can reach 10 feet (3 m) high in fertile soils (Tucker et al. 1964). In the first year of growth the plants form a basal rosette of leaves. Erect stems and flowers are produced in the second year. Poison-hemlock has long been recognized for its toxic properties, and has played an important role in human history.

Identification

Cotyledons are narrow at the base, elliptical at the tip, with the undersurface being prominently veined. The first true leaves are smooth, pale green, triangular, and two to three times deeply dissected. Seedlings have a parsnip-like odor when crushed. Subsequent leaves are alternate, dark glossy-green, sheathed at the base, at least 2 feet long, and several times pinnately divided. Leaf veins terminate at the tips of the teeth. Poison-hemlock usually remains vegetative during the first year of growth, producing a large rosette of leaves. During the spring of its second year, the plant develops branching, erect, leafy stems. Stems are ridged, stout, and hollow, except at the nodes. The lower portions of the stems are mottled with purple spots. After senescence, stems can persist long into the winter. Stems and leaves lack hairs and exude an unpleasant odor when crushed. The fleshy white taproot is long and sometimes branched. Small white flowers are clustered on the end of flat to slightly convex compound umbels. The fruits (schizocarps) are small, grayish-brown with conspicuous, wavy, longitudinal ribs. Two seeds are enclosed within each fruit. The foliage of poison-hemlock can resemble wild carrot, but poison-hemlock lacks hairs on the leaves and stems.

Origin, History, and Distribution

Of the six species of *Conium* found worldwide, none is native to North America. Poison-hemlock is the only species of the genus now established in North America. It was introduced from Europe as an ornamental, probably during the 1800s (Parish 1920, Kingsbury 1964). Since then, it has spread to become naturalized in nearly every state in the United States and in southern Canada. In both its native habitat and in North America, poison-hemlock is commonly found at lower elevations along roadsides, ditch and stream banks, creek beds, fence-lines, waste places, and in or on the edge of cultivated fields where moisture is sufficient (Hutchinson 1973, Cheeke and Shull 1985, Baskin and Baskin 1990). It can also invade native plant communities in riparian woodlands and flood plains of natural aquatic systems (Goeden and Ricker 1982). Although it tends to be more competitive under wetter soil conditions, it can also survive in dry sites (Tucker et al. 1964). Poison-hemlock was also introduced into New Zealand, Australia, and parts of South America, where it has also become a significant problem (Goeden and Ricker 1982).

Impacts

Poison-hemlock can invade perennial crops. In alfalfa, it only poses a significant problem in the first cutting (Jeffery and Robison 1990). Subsequent regrowth of alfalfa can suppress regrowth of poison-hemlock. It can also be found in grain fields, where it can contaminate the harvested seed (Panter and Keeler 1990). Poison-hemlock more commonly invades grazing areas, particularly pastures and meadows. It crowds out more desirable forage species, and its toxicity causes serious livestock losses when animals feed on fresh forage, harvested silage or, to a lesser degree, hay contaminated with poison-hemlock (Panter and Keeler 1988).

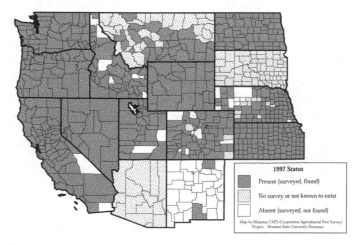

Distribution of poison-hemlock by county in the western United States. Data were collected by surveying various weed authorities in each state.

291

Toxicity

Poison-hemlock is toxic to humans. It was commonly used to kill political prisoners in ancient Greece, and is best known for the death of Socrates in Athens in 329 B.C. Native Americans poisoned the tips of their arrows by dipping them in a mixture of crushed poison-hemlock seeds and decomposed deer liver (Smith et al. 1982). Presently, cases of human poisoning are generally associated with children using the hollow stems as flutes, or adults confusing poison-hemlock for parsley, parsnip, or anise (Smith et al. 1982).

Human poisonings from the plant are much less common than livestock poisonings. Although animals tend to avoid poison-hemlock when other forage is available, they will feed on the plant when little other green vegetation is present, or when greenchop, silage, or hay is contaminated with the toxic plant (Panter and Keeler 1988, Jeffery and Robison 1990).

Although all parts of the plant are considered poisonous, the roots are the least toxic (Smith et al. 1982). There are eight known alkaloids in poison-hemlock, but the principal toxicants are the piperidine alkaloids coniine and γ-coniceine (Panter and Keeler 1988). Coniine is more common in the seed and in mature plants (Panter et al. 1988a), whereas γ-coniceine makes up 98% of the total alkaloids in the early vegetative stage (Keeler 1978, Panter et al. 1988a). γ-coniceine is eight times more toxic than coniine. Environmental conditions such as soil moisture, soil type, temperature, and the season of growth can alter the plant's alkaloid composition and concentration (Panter and Keeler 1988).

Coniine and γ-coniceine can affect the central nervous system and the reproductive cycle of an animal. All classes of livestock and wildlife are sensitive to poison-hemlock toxicity, including cattle, horses, pigs, goats, elk, and turkeys. Of the domesticated animals, cattle, goats, and horses are most sensitive. A lethal dose in horses and cattle is as low as 0.25 to 0.5% (fresh plant weight) of the animal's weight (Leininger et al. 1977, Smith et al. 1982). Sheep and pigs are less susceptible (Panter and Keeler 1990). A 10-fold difference in sheep susceptibility compared with cattle is due to increased liver metabolism of the toxic agents (Cheeke and Shull 1985).

Symptoms of poisoning include nervousness, trembling, knuckling at the fetlock joints, ataxia, dilation of the pupils, a weak and slow heartbeat, coma, and eventually death from respiratory paralysis (James 1984, Panter et al. 1988b, Panter and Keeler 1990). These symptoms can occur within 30 to 40 minutes in horses and 1.5 to 2 hours in cows and sheep (Cheeke and Shull 1985).

In the western United States, poison-hemlock produces teratogenic effects similar to those reported for *Lupinus* and *Nicotiana* (Cheeke and Shull 1985). Fetal deformity (crooked calf disease) is caused by coniine and γ-coniceine. It occurs when pregnant cattle, pigs, and goats consume poison-hemlock during a specific interval in the gestation period (Keeler 1978, Panter and Keeler 1988, Panter et al. 1990). Sheep and horses are less susceptible to the plant's teratogenic effects (Cheeke and Shull 1985).

Central nervous system toxicity in livestock usually occurs in the spring, when poison-hemlock is among the first green plants to emerge (James 1984, Panter et al. 1988a). This is also when concentrations of γ-coniceine are at their highest. In the fall, regrowth or newly germinated poison-hemlock may be the last green forage available (Panter and Keeler 1988). Since ingestion at this time coincides with the critical period of gestation in many animals, fall poisonings are usually associated with teratogenic toxicity. Winter poisonings are also common when harvested hay is fed to cattle (Keeler 1978).

The toxic alkaloids in poison-hemlock are volatile, and can cause toxicity following long-term inhalation (Cheeke and Shull 1985). However, rapid volatilization accounts for the loss in toxicity upon drying (Panter and Keeler 1988).

Like the quinolizidine alkaloids from *Lupinus*, piperidine alkaloids have the ability to pass into milk when animals feed on sublethal amounts of poison-hemlock (Panter and Keeler 1990). These alkaloids can harm the developing fetus in pregnant animals, and affect the flavor and safety of milk used for human consumption (Jeffery and Robison 1990).

Biology and Ecology

Although few studies have focused on the competitive interactions between poison-hemlock and crop or pasture plant species, many studies have examined its reproductive biology.

• Seed Dispersal

Poison-hemlock usually behaves as a biennial that reproduces solely by seed. Despite its prolific seed production, it does not have a well-developed mechanism for long-distance seed dispersal. While some seeds may be spread by water, birds or rodents, most simply drop close to the parent plant (Panter and Keeler 1988). Seeds are dispersed over a considerable time period, beginning in September and ending in late February (Baskin and Baskin 1990).

• Dormancy and Seed Longevity

The extended period in which poison-hemlock disperses its seed is of great importance to its long-term survival in a particular area. Approximately 85% of the seeds mature by mid-July, prior to dispersal (Baskin and Baskin 1990). Once dispersed, these seeds can germinate almost immediately if environmental conditions (i.e., moisture and temperature) required for germination are met. The remaining 15% of the seeds are dormant. These seeds require high summer and/or low winter temperatures before dormancy can be broken (Baskin and Baskin 1990). In addition, many nondormant seeds that remain attached to the inflorescence during winter can become dormant.

The long dispersal period and the transition of nondormant seeds to a dormant state is of particular importance to the population biology of poison-hemlock. It

293

insures that seed germination will not be confined to a single month or season. For example, seeds dispersed in late fall would give rise to seedlings in late winter and the following fall, whereas those dispersed in late winter would germinate in the spring, fall, or the following year (Baskin and Baskin 1990). In effect, a longer delay in seed dispersal increases poison-hemlock seed longevity in the soil. This strategy does not provide a large soil seed reserve for poison-hemlock. Unlike the long-lived seed banks of many weed species, the seed reserve of poison-hemlock only persists for about three years.

• *Germination Requirements*

Once the seeds are no longer dormant, they can germinate from late summer to early spring, as long as moisture and temperature conditions are suitable (Baskin and Baskin 1990). Rapid germination occurred in seeds exposed to maximum daily temperatures between 49° and 83° F (9.4° and 28.3° C) and minimum daily temperatures between 36° and 63° F (2.2° and 17.2° C) (Baskin and Baskin 1990). These temperatures are characteristic of the cooler periods of the year. Unlike many other weed species, poison-hemlock does not require light to germinate.

Management

Most management strategies are designed to reduce the incidence of poisoning in livestock. When poison-hemlock infestations are present, toxicity can be minimized by preventing grazing at times when the plant is the only available forage, or by removing pregnant livestock at the most susceptible period in their gestation (Panter and Keeler 1988).

It is important to prevent a small-scale infestation of poison-hemlock from becoming a more significant problem. This can be accomplished by periodic scouting and identification of newly invading plants. Once identified, individual plants should be removed by hand-pulling, hoeing, or spot herbicide application. It is essential to prevent isolated plants or a small cluster of plants from producing seed.

• *Mechanical*

Plowing or repeated cultivation will prevent poison-hemlock establishment (Panter and Keeler 1988). In areas where cultivation is not practical or possible, mowing can be used once the plants have bolted (Smith et al. 1982). However, a single mowing will not control poison-hemlock. Repeated mowing can reduce its competitive ability, deplete carbohydrate energy reserves in the taproot, and prevent seed production (Smith et al. 1982). Close mowing has the additional advantage of reducing the amount of leaf material available for livestock grazing (Tucker et al. 1964).

• Biological Control

Before 1982, it was thought that few insects attacked poison-hemlock (Goeden and Ricker 1982). Although no serious efforts have been made to introduce biological control organisms into the United States to control poison-hemlock, the European palearctic moth, *Agonopterix alstroemeriana* (Clerck) (Oecophoridae), has recently proven to be very effective. It is not known how the moth was initially introduced into the United States. It was first discovered in Tompkins County, New York, in 1973 (Berenbaum and Passoa 1983). By 1983, it was established in less arid regions of California, Oregon, and Utah, and by 1987 it was found in Colorado, Idaho, and Washington (Berenbaum and Harrison 1994). It has colonized the western United States so rapidly that it is difficult to establish from what direction(s) it spread (Powell 1991).

As in its native European range, the moth's only known host plant is poison-hemlock (Berenbaum and Harrison 1994). The larvae live in conspicuous leaf rolls, and feed on foliage, buds, and flowers in spring and early summer (Powell 1991). The adult moths emerge in summer, and can be found from June until March of the following year. In 1996, the *Pacific Northwest Weed Control Handbook* (William et al. 1996) listed *Agonopterix* as a widespread biological control agent in Idaho, Oregon, and Washington. Its infestation levels were considered to be high, and the level of poison-hemlock control was good to excellent (William et al. 1996).

• Cultural Control

Although burning has not been tried, it is probably not a practical method for managing poison-hemlock. In areas where poison-hemlock is the dominant plant, sufficient dried material would not be available as fuel to control poison-hemlock prior to fruit maturation.

• Chemical Control

Tebuthiuron is a herbicide that provides excellent preemergence control of poison-hemlock (*Crop Protection Chemical Reference* 1995). Among the sulfonylurea herbicides, chlorsulfuron and a combination of chlorsulfuron and metsulfuron not only give excellent preemergence control, but can also provide some foliar activity (William et al. 1996). Other preemergence photosynthetic inhibitors, such as hexazinone (1.5 lb ai/acre [1.68 kg ai/ha]), metribuzin (1.0 lb ai/acre [1.12 kg ai/ha]), and terbacil (1.0 lb ai/acre [1.12 kg ai/ha]) give excellent control of poison-hemlock (Jeffery and Robison 1990). However, in alfalfa, herbicides should be applied when the forage crop is dormant.

Postemergence application of phenoxy herbicides or glyphosate can also provide effective poison-hemlock control. Phenoxy herbicides such as 2,4-D, 2,4-DB, and MCPA are most effective when applied in early spring (Jeffery and Robison 1990). Both the amine and ester formulations of 2,4-D are effective at 1.0 to 2.0 lb ae/acre (1.12 to 2.24 kg ae/ha) combined with a good wetting agent (Smith et al. 1982,

295

Jeffery and Robison 1990). 2,4-D may make poison-hemlock more attractive to livestock, but does not change its toxicity (Smith et al. 1982). Glyphosate at 1.0 lb ae/acre (1.12 kg ae/ha) also provides excellent postemergence control when applied to plants in the rosette stage (Jeffery and Robison 1990). However, cooler temperatures can reduce the efficacy of glyphosate.

Treatment of poison-hemlock with herbicides may require repeated applications for a couple of years until the seedbank has been significantly depleted (Panter and Keeler 1988). Once it is under control, maintenance of desirable forage species with proper pasture management, fertilization, irrigation, or drainage will help prevent reinfestation (Tucker et al. 1964).

• *Integrated Management*
Few integrated approaches have been attempted for the control of poison-hemlock. It is possible, however, that widespread establishment of the European palearctic moth may provide sustainable control when used in combination with chemical treatment. Such approaches have yet to be investigated, but may offer a long-term management strategy for poison-hemlock in North America.

Literature Cited

Baskin, J.M., and C.C. Baskin. 1990. Seed germination ecology of poison-hemlock. Can. J. Bot. 68:2018-24.

Berenbaum, M.R., and T.L. Harrison. 1994. *Agonopterix alstroemeriana* (Oecophoridae) and other Lepidopteran associates of poison-hemlock (*Conium maculatum*) in east central Illinois. Great Lakes Entomol. 27:1-5.

Berenbaum, M.R., and S. Passoa. 1983. Notes on the biology of *Agonopterix alstroemeriana* (Clerck), with descriptions of the immature stages (Oecophoridae). J. Lepidopt. Soc 37:38-45.

Cheeke, P.R., and L.R. Shull. 1985. Natural Toxicants in Feeds and Poisonous Plants. AVI Publ. Co., Westport, CT.

Crop Protection Chemical Reference. 1995. 11th ed., Pharmaceutical Press, Inc., NY.

Goeden, R.D. and D.W. Ricker. 1982. Poison-hemlock, *Conium maculatum*, in southern California—an alien weed attacked by few insects. Ann. Entomol. Soc. Am. 75:173-76.

Hutchinson, J. 1973. British Wild Flowers. 1st American ed. Vol 2. Fairleigh Dickinson Univ. Press, Rutherford, NJ.

James, L.F. 1984. Poisonous plants: the hemlocks. Rangelands 6:128-29.

Jeffery, L.S., and L.R. Robison. 1990. Poison-hemlock (*Conium maculatum*) control in alfalfa (*Medicago sativa*). Weed Technol. 4:585-87.

Keeler, R.F. 1978. Alkaloid teratogens from *Lupinus, Conium, Veratrum*, and related genera. *In:* R.F. Keeler, K.R. Van Kampen, and L.F. James (eds.), Effects of Poisonous Plants on Livestock, Academic Press, NY, 401-414.

Kingsbury, J.M. 1964. Poisonous Plants of the United States and Canada. Prentice-Hall, Inc., Englewood Cliffs, NJ.

Leininger, W.C., J.E. Taylor, and C.L. Wambolt. 1977. Poisonous Range Plants of Montana. Montana State Univ. Bull. No. 632. Bozeman, MT.

Panter, K.E., and R.F. Keeler. 1988. The hemlocks: poison-hemlock (*Conium maculatum*) and waterhemlock (*Cicuta* spp.). *In:* L.F. James, M.H. Ralphs, and D.B. Nielsen (eds.), The Ecology and Economic Impact of Poisonous Plants on Livestock Production. Westview Press, Boulder, CO, 207-25.

Panter, K.E., and R.F. Keeler. 1990. *Conium, Lupinus,* and *Nicotiana* alkaloids: Fetal effects and the potential for residues in milk. Vet. Hum. Toxicol. 32 (suppl.):89-94.

Panter, K.E., T.D. Bunch, and R.F. Keeler. 1988a. Maternal and fetal toxicity of poison-hemlock (*Conium maculatum*) in sheep. Am. J. Vet. Res. 49:281-83.

Panter, K.E., R.F. Keeler, and D.C. Baker. 1988b. Toxicoses in livestock from the hemlocks (*Conium* and *Cicuta* spp.). J. Anim. Sci. 66:2407-13.

Panter, K.E., R.F. Keeler, W.B. Buck, and R.J. Callan. 1990. Congenital skeletal malformations and cleft palate induced by ingestion of *Lupinus, Conium* and *Nicotiana* species. Toxicon. 28:1377-85.

Parish, S.B. 1920. The immigrant plants of southern California. Bull. South. Calif. Acad. Sci. 19: 3-30.

Powell, J.A. 1991. Rapid colonization of the western United States by the Palearctic moth, *Agonopterix alstroemeriana* (Oecophoridae). J. Lepidopt. Soc. 45:234-36.

Smith, L.J., D.C. Thill, R.H. Callihan, and J.M. Lish. 1982. Poison-hemlock. A threat to man and livestock. Univ. of Idaho. Curr. Inf. Ser. No. 632. Moscow, ID.

Tucker, J.M., M.E. Fowler, W.A. Harvey, and L.J. Berry. 1964. Poisonous hemlocks. Their identification and control. Univ. of California Circ. No. 530. Berkeley, CA.

William, R.D., D. Ball, T. L. Miller, R. Parker, K. Al-Khatib, R.H. Callihan, C. Eberlein, and D.W. Morishita. 1996. Pacific Northwest Weed Control Handbook. Washington State Univ, Coop. Ext. Serv. Pullman, WA.

Purple Loosestrife

Barbra Mullin

Purple loosestrife (*Lythrum salicaria*) is a member of the loosestrife (Lythraceae) family. It is usually associated with wetland, marshy, or riparian sites. Ornamental plantings can readily move into aquatic sites, crowding out native vegetation and wildlife habitat. It is found across northern U.S. states and Canadian provinces.

Identification

Purple loosestrife is a stout, erect perennial herb that sends up multiple stems from a strongly developed root system. The plants range in height from 18 inches to more than 10 feet (0.5 to 3.5 m). The stem is four- to eight-sided and can be either smooth or hairy. These erect stems are tough, often appearing to be woody at the base. Leaves are lance-shaped and notched at the base, attached to the stalk without stems in an alternate, opposite, or whorled pattern.

The flowers are arranged on a spike that is from 2 inches to 3 feet (0.5 cm to 0.9 m) long. Individual flowers have five to seven petals, with 8 to 10 stamens of various lengths. Petals are typically purple, but can range from white to pink to deep purple or even red. Flowers open from July through September or October. The fruit is a capsule containing many small seeds. Mature capsules are brown, $^1/_8$ to $^3/_{16}$ inch (0.32 to 0.48 cm) long, and persist through the winter on the plant stalk. Tiny (1 mm) seeds are angular and light tan. Purple loosestrife is a prolific seed producer, with each spike capable of producing up to 120,000 seeds. Seeds germinate when exposed on bare soil. Seedlings are extremely small, with oval cotyledons. Young plants generally have oval leaves with an opposite or whorled stem attachment (Mal et al. 1992).

Lythrum virgatum, a closely related species that often hybridizes with *L. salicaria*, is similar but is smooth, with narrower leaves that are sharply pointed rather than heart-shaped at the base (Gleason 1952).

298

Origin, History, and Distribution

Purple loosestrife comes from Europe and Asia, but its exact geographical origins are unknown (Thompson et al. 1987). It was introduced into North America from Europe in the early 1800s as horticultural stock and as a contaminant of ship ballast (Stuckey 1980). It comes from similar habitats in Europe and Asia, and is well-adapted to many North American habitats (Hulten 1971).

Purple loosestrife was well established along the New England seaboard by the 1830s, and spread into vast stretches of interior drainage basins. It was so well-established by the 1830s that Torrey and Gray (1840) referred to it as "probably native." As agricultural settlements moved west, wetlands, watersheds, and forests were cleared for cropland and pasture. This provided the disturbance and stress to native wetlands that allowed invasion by purple loosestrife. Development of early canals (such as the Erie Canal, Delaware Canal, and feeder segments) also provided disturbance and habitat for establishment of purple loosestrife. The spread of purple loosestrife was closely related to canal traffic moving inland from northeastern shipping estuaries. All evidence suggests that the early phase of spread into the interior of North America was by waterborne commerce into recently disturbed habitats (Thompson et al. 1987).

Between 1881 and 1900, canal traffic declined and railroads took over priority shipments and much of the bulk cargo shipments. However, very little spread can be attributed to railroads. Coastal sites in the maritime states and provinces continued to be colonized by purple loosestrife from 1901 to 1940. During this time, the first establishments were reported from marine estuaries in the Pacific Northwest, suggesting that again, marine commerce was the principal mode of spread. Transcontinental railroad routes and the construction of the first state and federal highway networks seemed to have little effect on the spread of purple loosestrife (Thompson et al. 1987).

The range of purple loosestrife has greatly expanded since 1941 (Balogh and Bookhout 1989). The northern Midwest is almost completely colonized, with infestations occurring in western Minnesota, the upper Red River Valley in North Dakota, into Manitoba, and the wetlands of lakes Winnipeg and Manitoba. The most dramatic expansion, however, has been in the arid West, including the states of California, Idaho, Washington, Montana, and Wyoming (Thompson et al. 1987).

Purposeful introduction of purple loosestrife as an ornamental occurred very early and has become an increasing problem. It was recommended in early herbal medicinals and was found in many early herb gardens. Purple loosestrife was noted in early literature as a "plant of great beauty" to be used along banks of water as an ornamental. Many horticultural varieties of *Lythrum* are advertised as sterile hybrids. Plantings of supposedly "sterile" ornamental hybrids of purple loosestrife can cross with wild types and spread. Studies in Minnesota, Montana, and Manitoba have shown that sterile hybrids readily cross with common varieties, producing viable pollen and seed (Wright and Fay 1992, Anderson and Asher 1993, Lindgren and Clay 1993, Ottenbreit and Staniforth 1994).

Potential Invasion

Purple loosestrife can escape from ornamental settings into aquatic sites along streambanks or shorelines of shallow ponds. Infestations readily spread along open and flowing water. Much of the recent expansion in the West may be tied to the number of irrigation projects in the West. Recent purple loosestrife infestations in California, Idaho, Washington, and Wyoming are all found in irrigation areas. There is limited movement along highways unless they are near watercourses.

Remote sensing using small-format aerial photography can be successfully applied to discriminate flowering loosestrife plants from other background wetland vegetation. Colorslide transparencies taken at 1:5000 or larger scale can show flowers in sufficient detail for detection and mapping of isolated plants. Purple loosestrife is most difficult to detect on sites with few loosestrife plants or with large amounts of dead plant material below the flowerheads.

Impacts

Purple loosestrife reduces the diversity of native wetland ecosystems. Infestations severely degrade wildlife habitats and reduce species diversity. This plant crowds out wildlife-supporting native vegetation, such as cattails and bulrushes. Songbirds do not eat the small seeds of purple loosestrife. Muskrats cannot use it for food or shelter (Rawinski and Malecki 1984). Waterfowl are negatively affected when dense, impenetrable stands of purple loosestrife eliminate nesting sites and open water (Thompson et al. 1987).

It spreads effectively along waterways. Its thick, matted root system can rapidly fill irrigation ditches, decreasing water flow and increasing maintenance.

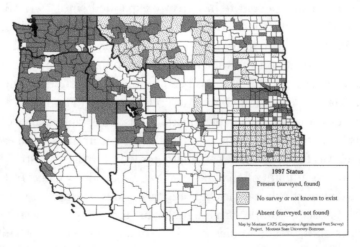

Distribution of purple loosestrife by county in the western United States. Data were collected by surveying various weed authorities in each state.

Biology and Ecology

Purple loosestrife is an invasive aquatic and wetland plant capable of growing in habitats from wetlands to moist, wet soils on upland sites.

Seeds are dispersed mainly by water, but they can also be transported on the feet and bodies of waterfowl and other birds, as well as by numerous wetland animals. Most seeds sink, then rise to the surface upon germination. These cotyledon-stage seedlings, as well as other plant parts, are buoyant and can be transported by water to root in other locations (Balogh 1986).

The major roots become thick and woody in mature plants. The aerial shoots die in the fall, but these dead stalks may remain for one to two years, making stands of purple loosestrife very dense. New shoots arise the following spring from buds at the top of the root crown (Rawinski 1982).

Infestations of purple loosestrife appear to follow a pattern of establishment, maintenance at low numbers, and then dramatic population increases when conditions are optimal. Purple loosestrife first establishes in wetland habitats that have been disturbed or degraded from draining, natural drawdown in dry years, or siltation. Once established, it can spread to other wetland sites. Seeds are usually present in such high numbers and germinate in such high densities that seedlings of native plants are suppressed. Purple loosestrife crowds out native vegetation and eventually creates a monoculture (Rawinski 1982).

From a distance, purple loosestrife may be confused with a number of other plants, including fireweed (*Epilobium*), blue verbena (*Vervain*), dotted gayfeather (*Liatris*), germander (*Teucrium*), smartweed (*Polygonum*), dame's violet (*Hesperis*), woodland salvia (*Salvia*), or foxglove (*Digitalis*). Upon close examination, purple loosestrife is readily distinguished from these plants by its multiple-sided stems and spike flower arrangements (Thompson et al. 1987).

Management

Effective management of purple loosestrife along waterways and in riparian areas requires integrating management strategies to prevent further introductions, detecting and eradicating new infestations, and containing and controlling large-scale infestations.

• *Education*

Educating the public about the purple loosestrife problem should be a major part of any weed management strategy. Prevention is always the best method of weed control. Since purple loosestrife is a popular ornamental plant commonly referred to as "lythrum," the public needs to be very much aware of and involved in any purple loosestrife control program. They need to know that they should not cultivate purple loosestrife or any of the horticultural varieties of *Lythrum*. Establishing an embargo on the importation of purple loosestrife seed and plant parts and listing it

301

on the state noxious weed list aids in implementing a long-term management plan. The embargo should include *L. salicaria*, *L. virgatum*, and all hybrids.

• Eradication

Eradication of purple loosestrife may be feasible in small infestations. All seed production must be eliminated, with a concurrent depletion of viable seeds and plant parts from the soil. To do this, infestations should first be treated with a herbicide to eliminate the majority of the plants before seed production (late June through July). These areas should receive follow-up spot treatments the same year to prevent escaped plants from producing seeds. In succeeding years, the infested sites must be revisited to find and eliminate plants that originate from seeds and roots in the soil.

Persistence and dedication to a long-term monitoring program on each infested site is the key to eradication. These sites must be resurveyed and treated yearly, until no viable seeds or rootstocks remain in the soil. There are no data on how long seeds remain viable in the soil, but an eradication program should continue until no seedlings or plant growth have been found for at least three years.

When resources or terrain limit the area that can be treated, divide large infestations into units geographically separated from other populations in the infested area. As long as seeds are being produced, wildlife may pick up seeds and spread them to other wetlands.

Land managers should adopt a strict management program for horticultural plantings. These include: 1) remove all purple loosestrife plants from horticultural planting, where possible; 2) clip and bag flowerheads from all purple loosestrife plants growing more than 500 feet from a waterway or wetland before seed production begins; 3) remove all purple loosestrife plants growing within 500 feet of a waterway or wetland; and 4) prevent all new plantings or transplanting of purple loosestrife in the area.

• Control Measures

Several management practices help control purple loosestrife. These include chemical, physical, and biological techniques. Each infested site should be evaluated to determine the appropriate control measure. Factors to be considered include the proximity and type of vegetation on the site, the nature of the water involved (flowing or still), and the utilization of the site and the water (domestic, irrigation, recreation, or scenic value).

• Chemical Control

The following chemicals are labeled for use in sites where purple loosestrife is known to occur. Extensive research on the chemical susceptibility of purple loosestrife has not yet been conducted. The rates shown in this section have demonstrated some control of purple loosestrife at some sites. Adequate and accurate coverage is critical for effective chemical control. As weed densities increase, spray coverage should also increase.

Refer to the current label for recommended application rates, approved sites, and application restrictions and precautions. The label may also recommend additional adjuvants and compatible colorants.

Glyphosate (Rodeo®)

1. Selectivity: Glyphosate kills both broadleaf and grassy plants.

2. Rate: Apply 4 pints per acre as a broadcast spray or as a 1% solution using hand-held equipment. Rodeo® is registered for use in and around water; do not apply within $1/2$ mile upstream of a potable water intake in flowing water or within a $1/2$ mile of a potable water intake in a standing body of water.

3. Surfactant: Use 2 or more quarts of a nonionic surfactant per 100 gal. of spray solution. Use a nonionic surfactant labeled for use in water. The surfactant must contain 50% or more active ingredient.

4. Plant growth stage at application: Apply to actively growing plants; during early to late bloom appears to be the best time. Although best results are achieved during late bloom, the results are only slightly less effective at early bloom. Since glyphosate takes one to two weeks to affect the plants, late applications can allow some production of seeds. It is important, therefore, to begin early, or plan to clip and bag the seedheads prior to late applications.

5. Time of year for application: Mid-July to early September is recommended for best results.

6. Special considerations: Use selective spot treatment applications to avoid damaging adjacent nontarget plants, which are essential to revegetate the area left bare by the killed loosestrife.

2,4-D

1. Selectivity: 2,4-D selectively kills broadleaf plants. Damage to grasses and grass-like plants can occur under conditions of high temperatures or over-application. Careful spot spraying is recommended for suppression of purple loosestrife.

2. Rate: Effective rates have been 1 to 2 quarts or a $1/2$-to-1% solution in 100 to 200 gallons per acre. There are several 2,4-D labels registered for use around water. Contact your state Department of Agriculture for registrations specific to your state. Do not contaminate domestic or irrigation water.

3. Surfactant: Use of a surfactant can improve the effectiveness of 2,4-D. Follow the recommendations on the 2,4-D and surfactant labels.

4. Plant growth stage at application: Apply when the plants are actively growing or until the seeds mature. Early-bud to early-bloom appears most effective.

5. Time of year for application: Apply whenever the plants are actively growing. Fall application must be done before a killing frost.

6. Special considerations: Application should be on a spray-to-wet basis, with spray volume increasing as the weed density increases. Refer to the label for all precautions.

Triclopyr (Garlon®)

1. Selectivity: Triclopyr selectively controls broadleaf plants and is a very effective brush killer. Damage to grasses and grass-like plants can occur under some conditions. Spot spraying is recommended.

2. Rate: Recommended rates are from $1/2$ to 2 gallons of Garlon® in 20 to 200 gallons of spray mixture per acre. Control of larger plants will require the greater volume of water.

3. Surfactant: Surfactants can improve effectiveness. Follow the label for recommended rates.

4. Plant growth stage at application: Plants should be at bud to mid-bloom and actively growing.

5. Time of year for application: Recommended time for application is from mid-July to mid-August.

6. Special considerations: Triclopyr is labeled for use on non-irrigation ditch banks, but cannot be used in aquatic settings. Seasonally dry wetlands can be treated. Do not contaminate water.

Application Equipment and Methods. All equipment used to apply herbicides must be clean, maintained, and calibrated to assure that the equipment is functioning properly and is applying the designated amount of herbicide.

A variety of sprayers is available, including backpack sprayers and truck- or boat-mounted sprayers. Experience has shown that wick applications can be effective in applying a chemical to control purple loosestrife. Wick application is very labor-intensive, and the higher concentration of herbicide (33% glyphosate) requires that care be used.

It is important to leave as much of the surrounding beneficial vegetation as possible to fill in where the loosestrife is killed. This is accomplished by careful spot-spraying with low pressures, large droplets, and narrow patterns. Drift reduction agents can be used in some situations to increase the droplet size and lower the potential for drift. Dyes and colorants can indicate uniform application without skips and overlaps.

• *Physical Control*

When removing purple loosestrife by hand (pulling or digging), it is difficult to get all the roots and stems, so choose sites carefully. Small infestations can be controlled by this method, but it is seldom effective for older plants or large infestations. Pulling is most effective on 1- to 2-year-old plants, because they have immature root systems. Carefully remove as much of the root and stems as possible, since all pieces can sprout and form new plants. Pulling is easiest when the water level is at or slightly above the ground surface.

Try to minimize soil disturbance. Bare or disturbed soil will favor loosestrife seedlings.

Disposal of plants and roots is best accomplished by piling, drying, and burning away from water. Take care to prevent spread of any seed from the transported plants.

Cutting can spread purple loosestrife plants because cut portions not properly disposed of can re-sprout. Cutting is not effective and requires a return to the site year after year.

Burning seems to favor loosestrife rather than native plants.

Changing water levels (flooding) may enhance spread by increasing the sites where purple loosestrife seeds can germinate, grow, and produce more seeds. Flooding is not an effective approach to limiting or preventing establishment of purple loosestrife seedlings.

If revegetation is necessary because few beneficial species are left in the site, use plants that are adapted to aquatic/moist conditions, such as native grasses, cattails, or rushes, whenever possible. This allows the infested area to be treated with selective herbicides that will control purple loosestrife seedlings but not harm the seeded vegetation (Malecki & Rawinski 1985). Studies have shown that seeding Japanese millet (*Echinochola crusgalli* var. *frumentacea*) on exposed, moist soil sites following a drawdown resulted in successful control of loosestrife on those areas, as well as providing for production of a desirable waterfowl food (Malecki and Rawinski 1985).

• *Biological Control*

Biological control agents do not eradicate or contain the target pest, but they may suppress the weed population to a non-detrimental level (Rees et al. 1996). Five species of insects have been identified with a potential for control of purple loosestrife. All of these species have been screened by the International Institute of Biological Control (IIBC) in Switzerland, and were approved for field release in the United States in 1992. Field research is currently being conducted on these insects, and redistribution is limited. (Malecki et al. 1993)

Hylobius transversovittatus (loosestrife root weevil) larvae live in the roots of purple loosestrife, while the adults feed on plant foliage. Small roots can be destroyed within two years, and larger roots may die after several consecutive years of infestation (Blossey et al. 1991a). *Galerucella pusilla* (golden loosestrife beetle) and *G. calmariensis* (black margined loosestrife beetle) feed on buds and foliage, resulting in stunted plants and reduced seed production (Blossey et al. 1991b). Two flower-feeding weevils, (*Nanophyes marmoratus* [loosestrife seed weevil] and *N. brevis* [blunt loosestrife seed weevil]), feed on flower buds, shoot tips, and immature seed capsules. Studies have shown seed reductions of up to 50% (Blossey and Schroeder 1995).

The use of biological control agents should be monitored, with prudent site selection. If eradication of purple loosestrife is the goal for a management area, biological control agents should not be used alone, or should only be released in areas that are inaccessible to other control measures.

• *Integrated Weed Management*

Purple loosestrife is a perennial weed that is difficult to control, through its ability to reproduce and the habitat in which it thrives. It is found in marshy, wetland, and riparian sites. Its use as an ornamental planting has increased spread in the northern

305

United States and Canadian provinces. Glyphosate, 2,4-D, and triclopyr can provide effective control, but are limited in wetland sites. Biological control and pulling may also provide control. Integrated management systems have proven to be long-term and cost-effective. Each management plan works best when designed for a specific area. Weed managers in the states of Montana and North Dakota have adopted long-term management plans that are designed to integrate all management techniques to stop the spread of purple loosestrife in those states.

Literature Cited

Anderson, N.O., and P.D. Ascher. 1993. Male and female fertility of loosestrife (*Lythrum*) cultivars. J. Amer. Soc. Hort. Sci. 118(6):851-58.

Balogh, G.R. 1986. Distribution and seedling ecology of purple loosestrife in Ohio's Lake Erie marshes. Ohio J. Sci. 86(2):51.

Balogh, G.R., and T.A. Bookhout. 1989. Purple loosestrife *Lythrum-salicaria* in Ohio's Lake Erie Marshes USA. Ohio J. Sci. 89(3):62-64.

Blossey, B., and D. Schroeder. 1995. Host specificity of three potential biological control agents attacking flowers and seed of *Lythrum salicaria* (purple loosestrife). Biol. Control 5:47-53.

Blossey, B., D. Schroeder, S.D. Hight, and R.A. Malecki. 1991a. Host specificity and environmental impact on the weevil *Hylobius transversovittatus*, a biological control agent of purple loosestrife (*Lythrum salicaria*). Weed Sci. 42(1):128-33.

Blossey, B., D. Schroeder, S.D. Hight, and R.A. Malecki. 1991b. Host specificity and environmental impact of two leaf beetles (*Galerucella calmariensis* and *G. pusilla*) for biological control of purple loosestrife (*Lythrum salicaria*). Weed Sci. 42(1):134-40.

Frazier, B.E., and B.C. Moore. 1993. Some tests of film types for remote sensing of purple loosestrife *Lythrum salicaria* at low densities. Wetlands 13(3):145-52.

Gabor, T.S., and H.R. Murkin. 1990. Effects of clipping purple loosestrife seedlings during a simulated wetland drawdown. J. Aquat. Plant Manage. 28(2):98-100.

Gleason, H.A. 1952. Illustrated Flora of the Northern United States and Adjacent Canada. Hafner Press, New York.

Haworth-Brockman, M.J., H.R. Murkin, R.T. Clay, and E. Armson. 1991. Effects of underwater clipping of purple loosestrife in a southern Ontario wetland. J. Aquat. Plant Manage. 29:117-18.

Haworth-Brockman, M.J., H.R. Murkin, and R.T. Clay. 1993. Effects of shallow flooding on newly established purple loosestrife seedlings. Wetlands 13(3):224-27.

Hulten, E. 1971. The circumpolar plants. 2. Almqvist & Wiksell, Stockholm.

Lindgren, C.J., and R.T. Clay. 1993. Fertility of 'Morden Pink' *Lythrum virgatum* L. transplanted into wild stands of *L. salicaria* L. in Manitoba. Hortscience 28(9):954.

Mal, T.K. J. Lovett-Doust, L. Lovett-Doust, and G.A. Mulligan. 1992. The biology of Canadian weed. 100. *Lythrum salicaria*. Can. J. Plant Sci. 72(4):1305-30.

Malecki, R.A., and T.J. Rawinski. 1985. New methods for controlling purple loosestrife. NY Fish & Game J. 32:9-19.

Malecki, R.A., B. Blossey, S.D. Hight, D. Schroeder, L.T. Kok, and J.R. Coulson. 1993. Biological control of purple loosestrife. Bioscience 43(10):680-86.

Ottenbreit, K.A., and R.J. Staniforth. 1994. Crossability of naturalized and cultivated *Lythrum* taxa. Can. J. Bot. 72(3):337-41.

Purple Lythrum Task Force, Montana Weed Control Association. 1994. A Purple Lythrum Management Plan for Montana, Helena.

Rawinski, T. 1982. The ecology and management of purple loosestrife (*Lythrum salicaria* L.) in central New York. M.S. Thesis. Cornell University.

Rawinski, T.J., and R.A. Malecki. 1984. Ecological relationships among purple loosestrife, cattail, and wildlife at the Montezuma National Wildlife Refuge. NY Fish and Game J. 31(1):81-87.

Rees, N.E., P.C. Quimby, Jr., G.L. Piper, E.M. Coombs, C.E. Turner, N.R. Spencer, and L.V. Knutson (eds). 1996. Biological Control of Weeds in the West. West. Soc. Weed Sci., Bozeman, MT.

Stuckey, R.L. 1980. Distributional history of *Lythrum salicaria* (purple loosestrife) in North America. Bartonia 47:3-20.

Thompson, D.Q., R.L. Stuckey, and E.B. Thompson. 1987. Spread, Impact, and Control of Purple Loosestrife (*Lythrum salicaria*) in North American Wetlands. USDA Fish & Wildlife Service, Washington, D.C.

Torrey and Gray. 1840. A Flora of North America. Wiley and Putnam, New York.

Welling, C.H. and R.L. Becker. 1993. Reduction of purple loosestrife establishment in Minnesota wetlands. Wildlife Soc. Bull. 21(1):56-64.

Wilcox. D.A. 1989. Migration and control of purple loosestrife *Lythrum-salicaria* L. along highway corridors. Environ. Manage. 13(3):365-70.

Wright, J., and P.K. Fay. 1992. Personal Communication. Montana State University, Bozeman.

Rush Skeletonweed

Roger L. Sheley, Joseph M. Hudak, and Robert T. Grubb

Rush skeletonweed (*Chondrilla juncea*) is a non-indigenous perennial weed that infests millions of acres in the Pacific Northwest and California. Rush skeletonweed has the potential to dominate productive crop and grazing land. This weed thrives on well-drained, sandy-textured or rocky soils, along roadsides, in rangelands, pastures, and grain fields.

Identification

Rush skeletonweed is a herbaceous, relatively long-lived perennial member of the sunflower (Asteraceae) family. Its lifecycle begins in the fall with seed germination and seedling establishment as well as regrowth from perennial taproots. Plants usually overwinter as rosettes which closely resemble common dandelion (*Taraxacum officinale*). The hairless basal leaves are 2 to 5 inches (5 to 12.5 cm) long and $1/2$ to 2 inches (1.25 to 5 cm) wide. Rush skeletonweed grows anytime temperatures are above freezing, but usually begins growing rapidly in March or April.

During late spring, a spindly stem grows from 1 to 4 feet (30 to 120 cm) tall from the center of the rosette. At this time, the basal leaves have deep, irregular teeth that generally point backward toward the stem base. The stem has a few narrow, inconspicuous leaves that give the plant a skeleton-like appearance. An important characteristic of rush skeletonweed are the stiff, downward-pointing hairs on the lower 4 to 6 inches (10 to 15 cm) of the stem. The remainder of the stem is relatively smooth, or has a few rigid hairs. All plant parts, including the leaf, stem, and roots, exude a milky latex when cut or broken.

Flowering and seed development begin in early summer and continue until fall. Bright yellow flowers develop along the stem and branch tips, either singly or in clusters of two to five flower heads. Although flower heads are less than 1 inch (2.5 cm) in diameter and appear as a single flower, they consist of nine to 12 flowers.

Seeds mature 9 to 15 days after flowers open. An individual plant is capable of producing more than 20,000 seeds, but first-year plants usually produce from 250 to 350 seeds. The light brown or black-ribbed, pappus-bearing seeds grow to about $1/8$ inch (3 mm) long. These seeds are dispersed by wind to open sites, while parent plants die back to the soil surface. This life cycle is repeated with the arrival of fall precipitation.

Origin, History, and Distribution

Rush skeletonweed is native to Asia and the Mediterranean region, including North Africa. It has successfully invaded Australia, Argentina, Italy, Lebanon, New Zealand, Portugal, Spain, the United States, and the former Yugoslavia (Parsons and Cuthbertson 1992). Rush skeletonweed was first reported in the United States near Spokane, Washington, in 1938. It was found in Idaho and Oregon during the 1960s, and currently infests more than 6.2 million acres of rangeland in the Pacific Northwest and California.

Potential Invasion

Rush skeletonweed has the ability to invade both crop and rangelands throughout the western United States. Cool winters and warm summers with winter and spring rainfall, but without severe drought, are optimum conditions for the growth and reproduction of rush skeletonweed. Summer temperatures reaching at least 59° F (15° C) appear to be necessary for flower and seed production, but seed production can be limited by drought.

Rush skeletonweed can flourish in very dry to very wet environments. It has established in habitats from 9 to 59 inches (22 to 150 cm) of annual precipitation (Moore 1964). Where it is established, this weed dominates disturbed areas such as roadways, waste areas, and areas weakened by drought or improper grazing. Big sagebrush/needle-and-thread grass (*Artemisia tridentata/Stipa comata*), bluebunch wheatgrass/Sandberg's bluegrass (*Agropyron spicatum/Poa secunda*), and bitterbrush/

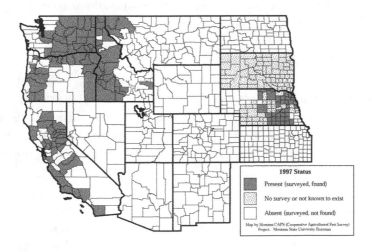

Distribution of rush skeletonweed by county in the western United States. Data were collected by surveying various weed authorities in each state.

bluebunch wheatgrass (*Purshia tridentata/A. spicatum*) are some of the habitat types susceptible to invasion by rush skeletonweed. Healthy native vegetation is seldom invaded by rush skeletonweed (McVean 1969).

Impacts
• *Detrimental*
Rush skeletonweed reduced wheat yield 80% in southeastern Australia (Groves and Cullen 1981). This weed competes against the crop for soil moisture and nutrients (primarily nitrogen), and the wiry stem interferes with harvesting. On rangeland, rush skeletonweed can form dense monocultures. It displaces indigenous plants, dramatically reduces rangeland forage production, and threatens the cattle industry. This species spreads from rangeland to adjacent cropland.

• *Beneficial*
In Australia, rush skeletonweed is a drought-tolerant pasture plant. It is palatable and nutritious for sheep in the rosette and early-flowering stage and has become a grazed component of low-quality pastures in many parts of southeastern Australia (Cuthbertson 1967). When rain is adequate, this species can be a major source of pollen for honeybees (Clemson 1985).

Biology and Ecology
• *Variability*
While more than 300 morphologically distinct forms of rush skeletonweed have been recognized in Australia, only three are widespread in the United States. These forms, designated A, B, and C, have narrow, intermediate, and broad rosette leaves, respectively. Rush skeletonweed plant forms also differ in inflorescence morphology, fruit characters, potential for regrowth from roots, and susceptibility to specific biological and chemical controls (Cullen 1974).

• *Germination and Emergence*
Rush skeletonweed seeds display virtually no dormancy. Seeds germinate within 24 hours under optimal conditions (59° to 86° F [15° to 30° C]). Buried seeds germinate within a year or two, even if less than $5/16$ inch (7.5 mm) of rain falls at one time. However, seedlings require continuous rainfall for three to six weeks for successful establishment. During drought, most seedlings die without emerging.

• *Roots*
Rush skeletonweed roots often reach depths of 8 feet (2.4 m). Roots exhibit little lateral growth, except in very sandy or gravelly soils where lateral roots are formed. When rush skeletonweed roots are severed, they produce shoots that can reach the soil surface from depths of 4 feet (Moore 1964). Taproot cuttings as small as $1/2$ inch

(1.25 cm) wide and 1 inch (2.5 cm) long can produce new plants under moist conditions. In general, the ability of shoots to emerge from roots increases with the size of root fragments and decreases with depth of burial.

Management

In smaller infestations, managing rush skeletonweed should focus on prevention and eradication. Existing infestations should be eradicated with diligence. Once the weed becomes widely established, an integrated strategy of cultural, chemical, and biological controls should be implemented to reduce the frequency of the weed to manageable levels.

• Prevention

To prevent rush skeletonweed invasion, seed production and dispersal must be stopped. Seeds are dispersed mainly by wind, water, trains, vehicles, and machinery. It is important to refrain from driving vehicles and machinery through rush skeletonweed-infested areas during the seeding period, and to wash the undercarriage of vehicles and machinery before leaving infested areas. Livestock should not graze weed-infested areas during seed formation. Before being moved to weed-free range, livestock grazing infested ranges should be transported to a holding area for 10 to 14 days after grazing.

Proper livestock grazing is essential to maintain competitive grass stands, which will help limit rush skeletonweed encroachment. A grazing management plan should be developed for any management unit involved in a rush skeletonweed prevention program. Management should include altering the season of use and stocking rates to achieve proper grass utilization. Grazing systems should include altering the season of use, rotating livestock to allow plants to recover before being regrazed, and promoting plant litter accumulation.

An integral part of any weed prevention program is to contain neighboring weed infestations. It is critical to contain rush skeletonweed along highways, railways, and waterways to prevent seed transportation. This may require annual applications of picloram.

Detecting new infestations and implementing eradication programs is the second step to preventing the invasion of rush skeletonweed. Systematic surveys along weed dispersal corridors are necessary to detect weed infestations early. Once an infestation is found, an eradication plan should be designed and implemented. The plan should include an outline of the infestation boundaries, control treatments, control schedule, revegetation plans, follow-up monitoring, and costs.

• Mechanical Control

Diligent hand-pulling or grubbing can provide effective control of very small infestations. Successful hand-pulling requires removal of plant growth two or three times per year for 6 to 10 years because new plants will emerge from severed roots

311

and buried seeds. Removing rush skeletonweed plants is easier when the soil is wet. Pulled plants should be destroyed by burning in a very hot fire to ensure seed and root destruction (Cuthbertson 1967).

Mowing and cultivation are ineffective methods for controlling rush skeletonweed. Mowing does not affect carbohydrate reserves, and only limits seed production in very dry years. Cultivation spreads root fragments and may actually increase the infestation (Myers and Fitzsimon1965).

• Cultural Control

Planting competitive legumes, such as alfalfa (*Medicago sativa*), has increased soil fertility and effectively reduced populations of rush skeletonweed in crop-pasture rotations (Wells 1969). Dense stands of legumes compete for soil moisture and shade rush skeletonweed plants. However, the level of pasture management needed to effectively control the weed is difficult to achieve. Integrating competitive plantings with biological controls has proven effective in Australia (Wells 1969).

Proper grazing by sheep can reduce or prevent production of rush skeletonweed rosettes and seed. Continuous (rather than rotational) grazing produces the lowest densities of the weed. Moderate grazing is as effective as heavy grazing in controlling rush skeletonweed because heavy grazing decreases the competitive ability of desired plant species. Integrating the use of competitive plantings, sheep grazing, and biological control agents appears to have potential for managing rush skeletonweed infestations (Groves and Cullen 1981).

• Chemical Control

Rush skeletonweed is difficult to control using herbicides. Successful chemical control depends on specific conditions of the site and usually requires an aggressive re-application program. Historically, picloram has been applied at 2 lb active ingredient per acre (2.24 kg ai/ha) to rosettes to control rush skeletonweed. An application of 2,4-D amine at a rate of 1 lb ai/acre (1.12 kg ai/ha) provides some control. In Idaho, picloram at 1 lb ai/acre (1.12 kg ai/ha) plus 2,4-D at 1 lb ai/ac (1.12 kg ai/ha) gave the best control (Cheney et al. 1980). In Australia, recent studies showed that a single application of clopyralid at 0.2 lb ai/acre (0.22 kg ai/ha) reduced rush skeletonweed shoots approximately 60% three years after application (Heap 1993). Mixing clopyralid at 0.2 lb ai/acre (0.22 kg ai/ha) with dicamba at 1 lb ai/acre (1.12 kg ai/ha) gave the best long-term control, reducing the number of shoots 75% three years after application. Annual applications were necessary to provide 95% control of rush skeletonweed. Herbicides are most effective when applied to plants infected with biological control agents.

High rates of nitrogen fertilizer minimized the effect of rush skeletonweed upon both wheat and pasture yields under moist conditions (Myers and Fitzsimon 1965). Nitrogen increased the size of rush skeletonweed plants, but their density decreased. Apparently, nitrogen reduces weed density by increasing competition.

• Biological Control

Three biological control agents have been released for control of rush skeletonweed in North America: a rust, a mite, and a midge (Cullen 1974). The rust, *Puccinia chondrillina*, infects Form A of skeletonweed, causing pustules that erupt through the leaf and stem surface. This reduces the plant's ability to photosynthesize, and desiccates leaves. Severe rust infections can control Form A of rush skeletonweed, while light infections reduce seed production and viability.

The rust spores are carried by wind and rain. The disease moved about five miles (8 km) within four generations, and 200 miles (322 km) after 12 generations (Cullen 1974). The spores can be collected and released on new weed infestations. Spores require six hours of both dew and darkness to germinate and establish a rust infection.

Several strains of rust specific to Form B have been collected. However, they have not proven effective under field conditions.

The gall mite *Aceria chondrillae* induces the vegetative and floral buds to form leafy galls causing stunting of the plant and greatly reducing seed production (Cullen 1974). This small parasite is the most damaging of the three biological control agents, but is only effective on Form A plants. The gall overwinters in the central bud of the rosettes without inducing gall formation. As the stem elongates, the mites colonize newly formed floral buds. As females reproduce, the galls swell. As the gall drys, the mites emerge and crawl to other buds or rush skeletonweed plants. The plant can be covered with as many as 4,000 galls when four or five generations of the insect occur per year (Groves and Cullen 1981).

The only biological control agent that attacks all three forms of rush skeletonweed is the gall midge *Cystiphora schmidti* (Groves and Cullen 1981). The midge deforms plants and reduces seed production by feeding on the rosettes, stem leaves, and stems of rush skeletonweed. The gall midge overwinters in the rosettes, emerges in April, and is active through October. Females lay eggs in plant tissue, which causes some obstruction of nutrient movement within the plant. Despite a relatively short generation time, the gall midge impact is less than either the rust or mites, and the midges' sensitivity to climatic variation is high. Therefore, the gall midge may not overwinter well in cold climates (Groves and Cullen 1981).

• Integrated Weed Management

No single treatment provides long-term control of rush skeletonweed, so an integrated strategy must be adopted. The first line of defense is to prevent introductions of the weed. Systematic surveys, early detection, and the implementation of an eradication program on small infestations is the second line of defense. Once the weed becomes established, integrating various combinations of competitive plantings, crop-pasture rotations, sheep grazing, biological control agents, herbicides, and possibly fertilizers can reduce rush skeletonweed to manageable levels. The key component of any successful weed management program is sustained effort, constant evaluation, and the adoption of improved strategies.

Literature Cited

Cheney, T.M., W.S. Belles, and G.A. Lee. 1980. Herbicidal control of rush skeletonweed (*Chondrilla juncea* L.) Proc. Weed Sci. Soc. Am. 33:52-56.

Clemson, A. 1985. Honey and Pollen flora. Inkata Press, Melbourne, Australia.

Cullen, J.M. 1974. Seasonal and regional variation in the success of organisms imported to combat skeleton weed (*Chondrilla juncea* L.) in Australia. Miscellaneous Publication, Commonwealth Biology Control No. 8, 111-17.

Cuthbertson, E.G. 1967. *Chondrilla juncea* in Australia. Distribution and control. New South Wales Dept. Agric. Bull. No. 68.

Groves, R.H., and J.M. Cullen. 1981. *Chondrilla juncea*: the ecological control of a weed. *In:* The Ecology of Pests. R.L. Kitching and R.E. Jones (eds.), Commonwealth Scientific and Industrial Research Organization, Melbourne, Australia, 7-17.

Heap, J.W. 1993. Control of rush skeletonweed (*Chondrilla juncea*) with herbicides. Weed Technol. 7:954-59.

McVean, D.N. 1969. Ecology of *Chondrilla juncea* L. in south-eastern Australia. J. Ecol. 54:345-65.

Moore, R.M. 1964. *Chondrilla juncea* L. (skeleton weed) in Australia. Proc. 7th British Weed Control Conf., pp. 563-68.

Myers, L.F., and D. Fitzsimon. 1965. The effect of fallow herbicide sprays and cultivation on the yield of wheat in the presence of skeleton weed, *Chondrilla juncea* L. Field Station Records, Div. Plant Ind., CSIRO 4:71-78.

Parsons, W.T., and E.G. Cuthbertson. 1992. Noxious Weeds of Australia. Inkata Press, Melbourne and Sydney, Australia.

Wells, G. J. 1969. Skeleton weed (*Chondrilla juncea*) in the Victorian Mallee. I. Competition with legumes. Australian J. Exper. Agric. and Anim. Husb. 9:521-27.

Russian Knapweed

Tom D. Whitson

Russian knapweed (*Centaurea repens*) is a perennial noxious weed that establishes as monocultures. It is allelopathic, very competitive, and continuously fills in as other perennial plants are overgrazed or eliminated by disturbances (Bottoms and Whitson 1997). Russian knapweed has protein values equal to alfalfa hay, but cannot be eaten by livestock because of its bitter flavor. It can be controlled by herbicides such as picloram or clopyralid for three to five years, but will reinvade the site if cool-season grasses cannot be established on land previously occupied by monocultures of Russian knapweed. Russian knapweed that is controlled with a herbicide should be followed by the complete removal of all Russian knapweed plant residues before seeding other vegetation (Benz et al. 1996). A single control approach such as the application of herbicides will not provide sustainable control of Russian knapweed; therefore, a system of management should be used to manage Russian knapweed.

Identification

Russian knapweed is perennial, forming dense colonies by adventitious shoots from widely spreading black, bark-covered roots. Stems are covered with fine, white hairs that give the stem a blue-green color. Stems are erect, openly branched, with plants standing up to 3 feet (0.9 m) tall. Lower leaves are often lobed and toothed, and up to 4 inches (10 cm) long; upper leaves are both entire and serrated on upper leaf margins, but are attached to the stem without a stalk at their base. Flowerheads are $^1/_4$ to $^1/_2$ inch (0.6 to 1.3 cm) in diameter and are always found at the tip of the plant's leafy branches. Flowers, pink to lavender, are found in June to September. Involucral bracts surround the seeds as they are developing, and have a pearly, rounded appearance with papery margins (Whitson et al. 1996). Seeds develop in late summer. Oblong, grey-brown seeds are 2.1 to 2.4 mm in length by 0.6 to 0.7 mm width. They are covered with many fine, white hairs with a ring of bristles on the apex of the seed. There are several ridges lengthwise on each seed face (Davis 1993). Seeds of Russian knapweed germinate for up to five years after being shed, with up to 80% of the seeds germinating the first year (Anderson 1968).

Origin, History, and Distribution

Russian knapweed is a native of Eurasia, probably introduced to North America about 1898. It is widely established in the western United States, and forms dense colonies in cultivated fields, orchards, and pastures, along streambanks and on roadsides. It frequently occurs in locations that have shallow water tables or have extra water from irrigation. Plants spread by seeds and by roots that can penetrate soils to more than 8 feet (2.4 m).

In 1998, Russian knapweed was considered a noxious weed in 412 counties within 21 western U.S. states. Range and weed scientists consider it a serious habitat invader because of its aggressive nature and allelopathic properties. In Wyoming, infestations increased from 28,300 acres (11,300 ha) in 1959 to 117,300 acres (46,500 ha) in 1987. Infestations were first reported in Colorado in 1928. Reports indicate that more than 49,500 acres (20,000 ha) of Colorado rangeland were occupied by Russian knapweed in 1998.

Potential Invasion

Russian knapweed invades open, disturbed land, suppresses growth of surrounding plants, and once established, forms a single-species stand (Watson 1980).

In 1964, Selleck observed that infestations spread in dry locations while decreasing in moist areas because of competition with perennial grasses.

Impacts

The Bureau of Land Management estimated the average annual rate of spread to be 8% in the northwestern United States, with an annual loss of 55% in livestock carrying capacity (Simmons 1985). Wyoming infestations have increased annually by an 11% average rate, occupying about 116,000 acres (46,500 ha) by 1987.

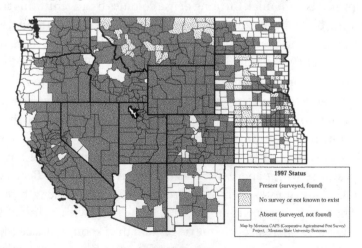

Distribution of Russian knapweed by county in the western United States. Data were collected by surveying various weed authorities in each state.

Russian knapweed infestations in 1998 were estimated at 1,380,000 acres (557,000 ha) in the western United States, which includes: 30,000 acres (12,000 ha) in Colorado, 900,000 acres (363,300 ha) in Idaho, 48,500 acres (19,600 ha) in Montana, 250 acres (101 ha) in North Dakota, 15,000 acres (6,000 ha) in Oregon, 3,000 acres (1,200 ha) in South Dakota, 152,000 acres (61,400 ha) in Utah, 8,300 acres (3,300 ha) in Washington, and 200,000 acres (80,800 ha) in Wyoming (Roché and Roché 1991). In addition to reducing forage for livestock, this perennial weed greatly reduces biodiversity for wildlife habitat (Kurtz et al. 1995). Plants ingested as fresh or dried forage are toxic to horses, causing the neurological disorder nigropallidal encephalomalacia (Young et al. 1970a, b).

Cropland infested with Russian knapweed in past years has caused lands to be abandoned (Renney and Dent 1958, Berezovskii and Raskin 1971, Maddox et al. 1985).

Biology and Ecology

Russian knapweed is an aggressive perennial weed that reproduces from seed and adventitious buds on a creeping root system (Fletcher and Renney 1963, Moore and Frankton 1974). Infestations increase primarily by adventitious roots; Russian knapweed does not reproduce extensively from seed (Watson 1980). Roots grow 6 feet 6 inches to 8 feet 3 inches (2 to 2.5 m) deep the first season and 16 feet 6 inches to 23 feet (5 to 7 m) deep in second season (Frazier 1944, Ivanova 1966, Agadzhanyan and Agadzhanyan 1967). Russian knapweed develops radially, with single plants covering as much as 14.4 yd^2 (12 m$^{2)}$ in two seasons (Frazier 1944, Selleck 1964, Ivanova 1966). A single plant may produce 1,200 seeds, which remain viable two to three years (Ivanova 1966). Russian knapweed competes with desirable vegetation for soil moisture and nutrients (Berezovskii and Raskin 1971, Popov et al. 1973).

Management

Russian knapweed might be managed temporarily with herbicides or, in the future, with insects, but long-term reductions must include planting competitive plant species to occupy bare ground once infested by the weed. Although Russian knapweed is allelopathic (Renney and Dent 1958, Anderson 1960, Fletcher and Renney 1963, Berezovskii and Raskin 1971, Evstratova et al. 1973, Stevens and Merrill 1985), control through plant competition should be exploited. Four years of winter rye (*Secale cereale*) or wheat (*Triticum aestivum*) monoculture reduced Russian knapweed by 99% and 78%, respectively, when crops were harvested for silage or grain (Sulima 1968). Russian knapweed is sensitive to light competition. Root and shoot dry matter and flower production declined, and leaf area increased, as light intensity was decreased (Dall'Armellina and Zimdahl 1988). Preliminary studies done at the University of Wyoming suggest that plant competition could be an important part of a Russian knapweed management system.

317

However, there is limited data on interference between Russian knapweed and rangeland grasses. Research conducted at Colorado State University indicates that western wheatgrass (*Agropyron smithii*) germination may be suppressed but not eliminated when exposed to Russian knapweed aqueous extracts, whereas smooth brome (*Bromus inermis*) germination was not reduced. Interference experiments between Russian knapweed and western wheatgrass or smooth brome indicate that Russian knapweed and smooth brome competed with one another for limited resources, but Russian knapweed and western wheatgrass did not compete.

At the 1989 Knapweed and Leafy Spurge Symposium, scientists agreed that integrated weed management systems need to be developed to recover land infested by knapweed and leafy spurge. However, research on the development of such systems has been limited.

Many improved grass species can be seeded in late fall or winter when seedbeds are properly prepared. An initial herbicide treatment is important to suppress problem perennial weeds before seeding. Pasture and hayland seedings without tillage have been successful, but there has been little work in which grasses were seeded into perennial weed-infested rangeland using current technology (Koch et al. 1984, Mueller-Warrant and Koch 1980). Whitson et al. (1989) reported that some grass species established more successfully than others without tillage in a leafy spurge-infested range previously treated with glyphosate. With Russian knapweed, however, tillage of surface residue will be necessary to hasten decomposition of allelochemicals which accumulate from foliage (Fletcher and Renney 1963).

Important grass characteristics to be considered for long-term control of problem weeds such as Russian knapweed include 1) adaptation to the soil and climate; 2) ease of establishment; 3) competitiveness with weeds; 4) palatability and nutritive value, particularly for late-season use; 5) dry matter productivity; and 6) stand longevity.

The use of herbicides to control Russian knapweed before establishing perennial grasses is an important part of a management system. To determine proper time of herbicide application, the best choice of herbicides, and their lowest possible use rates to provide adequate control of Russian knapweed, three experiments were established by Whitson and Baker in 1989 in northcentral Wyoming. Control with herbicides in the experiment was greater if applications were made when Russian knapweed was either at the bloom or seed stage, rather than the rosette or early growth stage. Effective controls for two years after treatments of greater than 95% were obtained with applications of picloram at 0.38 lb ai/ac and above, clopyralid at 0.25 lb ai/ac and above, and the combination of clopyralid plus 2,4-D plus picloram at 0.18 + 1.0 + 0.25 lb ai/ac (Whitson et al. 1991).

Two studies were initiated using five perennial cool-season grasses after areas were treated with three different herbicides, as well as the non-herbicide control methods of burning and mowing. The studies, located on Lander Complex sandy loam soils near Riverton and Ft. Washakie, WY, were treated with various herbicides

Table 1. Comparison of percent live canopy cover of Russian knapweed and seeded to five grasses in five weed control treatments (avg. two locations).

	Metsulfuron 0.5 oz		Clopyralid 0.29 lb+ 2,4-D 1.5 lb		Picloram 0.25 lb		Mow		Burn	
	RK*	Gr*	RK	Gr	RK	Gr	RK	Gr	RK	Gr
	%		%		%		%		%	
Streambank wheatgrass	44.4	9.7	13.1	24.2	13.3	28.3	35.8	6.0	38.7	7.8
Thickspike wheatgrass	50.0	12.8	16.5	24.7	15.5	29.2	41.7	4.9	42.9	3.9
Crested wheatgrass	44.3	8.4	16.2	12.9	18.0	13.8	39.2	3.8	41.4	2.3
Western wheatgrass	48.4	6.5	14.8	16.8	20.1	15.8	38.9	3.9	44.5	4.2
Russian wildrye	46.0	8.2	14.4	19.0	20.4	16.8	40.5	3.4	39.0	1.4

* RK = Russian knapweed; Gr = grass

Table 2. Production and hay forage values in plots treated with clopyralid plus 2,4-D and planted to various wheatgrass species, as well as Russian wildrye.

	Streambank	Thickspike	Crested	Western	Russian
Clipped 6/15/94					
Dry Matter (lb/A)	855	922	994	600	778
TDN (lb/A)	525	428	492	278	398
Protein (lb/A)	117	95	127	65	79
Value	$ 34	$ 37	$ 40	$ 24	$ 31
Clipped 8/4/94					
Dry Matter (lb/A)	1,298	956	1,829	796	1,024
TDN (lb/A)	515	380	807	356	449
Protein (lb/A)	116	78	126	73	98
Value	$ 52	$ 38	$ 73	$ 32	$ 41

in early October 1991. Plots were tilled with a rototiller 10 days later. Metsulfuron and clopyralid were applied in August 1992, and all herbicides were reapplied in August 1994. Russian knapweed had started into winter dormancy during the 1991 application and was in late bloom in 1992 and early bloom in 1994. All plots were seeded with Sodar streambank wheatgrass (*Elymus lanceolatus*), Critana thickspike wheatgrass (*Agropyron dasystachyum*), Hycrest crested wheatgrass (*A. cristatum*), Rosana western wheatgrass (*Paseopyrum smithii*) and Bozoisky Russian wildrye (*Psathyrostachys juncea*) at 10 lbs PLS/acre, except Russian wildrye which was seeded at 6 lbs/acre in April 1992.

Russian knapweed live canopy was reduced from over 50% in the untreated check to 10% in the areas treated with picloram and clopyralid plus 2,4-D. Stands of the five perennial grasses averaged 27% live canopy cover in the clopyralid plus 2,4-D treatments and 33% in the areas treated with picloram. Neither burning nor mowing reduced the live canopy cover of Russian knapweed. The two grasses having the greatest overall establishment were Critana thickspike wheatgrass, with an average of 19.7% live canopy cover, and Sodar streambank wheatgrass with 18.2% live canopy cover. The lowest amount of Russian knapweed (13.1%) and the highest percent live canopy of grasses (24.2%) were found in areas treated with the three clopyralid-plus-2,4-D treatments and seeded to Sodar streambank wheatgrass (Table 1).

Grasses were clipped twice during the growing season and analyzed for Total Digestible Nutrients (TDN) and protein. Hay values were then derived from the TDN, protein, and production for each species (Table 2). Crested wheatgrass had the highest yield and value ($73/ac) with 1,829 lbs of air-dried forage per acre. Other values ranged from $32 to $52 per acre (Table 2).

The first biological control organism introduced for Russian knapweed control was the gall-forming nematode *Subanguina picridis,* released from the former USSR. Other insects and mites being studied in 1998 for possible release include the gall-forming mite *Aceria acroptiloni,* a wasp (*Aulacida acroptilonica*), a stem gall-forming insect, and the rust fungus *Puccinia acroptili* which infects the foliage. Native fungi that have been found on Russian knapweed include: *Alternaria* spp., *Puccinia acroptili,* and *Sclerotinia sclerotiorum,* all of which attack leaves of Russian knapweed.

The nematode *Subanguina picridis,* released in the United States, is being mass-reared in Uzbekistan. Releases made in Canada have had limited success, but its success might be increased when plant competition can also be used (Rees 1995).

Literature Cited

Agadzhanyan, G.Kh. and A.A. Agadzhanyan. 1967. The distribution of and damage caused by Russian knapweed (*Acroptilon picris* C.A.M.) in Armenia and its control [in Russian, English summary]. Biol. Zh. Armenii 20:15-21.

Anderson, E.G. 1960. A national organization for weed control and some significant results. Proc. 14th Northeastern Weed Control Conf. 7-15.

Anderson, Robert N. 1968. Germination and Establishment of Weeds for Experimental Purposes. Weed Sci. Soc. of Amer. Handbook, 40.

Benz, L.J., K.G. Beck, T.D. Whitson, and D.W. Koch. 1996. The economic feasibility of developing a management system to reclaim Russian knapweed infested rangeland for multiple use. Proc. West. Soc. Weed Sci., 60-61.

Berezovskii, M. Ya. and M.S. Raskin. 1971. Some biological characteristics of Russian knapweed [in Russian, English summary]. Dokl. TSKhA 168:179-83.

Bottoms, R.M. and T.D. Whitson. 1997. A systems approach for the control of Russian knapweed. Brighton Crop Prot. Conf.-Weeds. Vol. 2, 699-704.

Dall'Armellina, A.A. and R.L. Zimdahl. 1988. Effect of light on growth and development of field bindweed and Russian knapweed. Weed Sci. 36(6):779-83.

Davis, Linda W. 1993. Weed Seeds of the Great Plains. University Press of Kansas, Lawrence.

Evstratova, R.I., V.I. Shiechenko, and K.S. Rybalko. 1973. The structure of acroptilin-a sesquiterpene lactone from *Acroptilon repens* [in Russian, translated]. Kh. Prirod. Soedin. 2:161-67.

Fletcher, R.A. and A.J. Renney. 1963. A growth inhibitor found in *Centaurea* spp. Can. J. Plant Sci. 43:475-81.

Frazier, J.C. 1944. Nature and rate of development of root system of *Centaurea picris*. Bot. Gaz. 105:345-51.

Ivanova, T.S. 1966. Biological control of mountain bluet (*Acroptilon picris* C.A.M.) [in Russian]. Azv. Acad. Nauk. Tadzhik. SSR. (Otdel Biol. Nauk.) 2:51-63. [translation—Translation Bureau, Can. Dep. Secretary of State, No. 3793].

Koch, D.W., G.W. Mueller-Warrant, and J.R. Mitchell. 1984. Sod seeding of forages. I. No-till as an alternative to conventional establishment of legumes. N.H. Agric. Exp. Sta. Bull. 525.

Kurz, G.L., R.A. Olson, and T.D. Whitson. 1995. Ecological implications of Russian knapweed (*Centaurea repens* L.) infestation: Small mammal and habitat associations. Proc. West. Soc. Weed Sci., 56.

Maddox, D.M., A. Mayfield, and N.H. Foritz. 1985. Distribution of yellow starthistle (*Centaurea solstitialis*) and Russian knapweed (*Centaurea repens*). Weed Sci. 33:315-27.

Moore, R.J. and C. Frankton. 1974. The thistles of Canada. Can. Dept. Agric. Monogr. No. 10, Information Canada, Ottawa, Ont.

Mueller-Warrant, G.W. and D.W. Koch. 1980. Establishment of alfalfa by conventional and minimum tillage seeding techniques in a quackgrass-dominant sward. Agron. 72:883-89.

Popov, V.G., M.S. Raskin, and G.S. Gruzdev. 1973. The noxious effect of Russian knapweed [*Acroptilon repens* (L.) D.C.] in field plant associations [in Russian, English summary]. Iz. Timiryaz. Sel'sk. Adad. 2:221-23.

Rees, N.E., et al. (eds.). 1995. Biological Control of Weeds in the West (Knapweed section). West. Soc. Weed Sci., USDA-ARS, Montana St. Univ., Bozeman, MT.

Renney. A.J. and W.J. Dent. 1958. Growth inhibition caused by Russian knapweed (*Centaurea repens* L.). Res. Rep. Natl. Weed Comm. West. Sec., 122-23.

Roché, B.F., and C.T. Roché. 1991. Identification, introduction, distribution, ecology, and economics of *Centaurea* species. *In*: Noxious Range Weeds, Westview Press, Boulder, Co, 274-91.

Selleck, G.W. 1964. A competition study of *Cardaria* spp. and *Centaurea repens*. Proc. 7th British Weed Control Conf., 569-76.

Simmons, R.G. 1985. Northwest Area Noxious Weed Control Program—Environmental Impact Statement, 152-56. U.S. Dept. of Interior, Bureau of Land Management.

Stevens, K.L. and G.B. Merrill. 1985. Sesquiterpene lactones and allelochemicals from *Centaurea* species. *In:* A.C. Thompson, et al. The Chemistry of Allelopathy. Biochemical Interactions Among Plants. Amer. Chem. Soc., Washington, D.C, 83-98.

Sulima, A.G. 1968. Agronomical methods of controlling Russian knapweed [in Russian, English summary]. Zashch. Rast. 13:49-50.

Watson, A.K. 1980. The biology of Canadian weeds. 43. *Acroptilon* (*Centaurea*) *repens* (L.) DC. Can. J. Plant Sci. 60:993-1004.

Whitson, T.D., J.L. Baker, R.D. Cunningham, and T.E. Heald. 1991. Control of Russian knapweed with various herbicides applied at three growth stages. West. Soc. of Weed Sci. Res. Prog. Rep. 88-89.

Whitson, T.D., D.W. Koch, A.E. Gade, and M.E. Ferrell. 1989. The control of leafy spurge (*Euphorbia esula* L.) by the integration of herbicides and perennial grasses. Univ. of Wyo. Rangeland Research and Extension Demonstration. Coop. Ext. Service, Agric. Exp. Sta., College of Agric, 42-47.

Whitson, T.D., (ed.). 1996. Weeds of the West. 5th Edition. West. Soc. Weed Sci. 92-93.

Young, S., W.W. Brown, and B. Klinger. 1970a. Nigropallidal encephalomalacia in horses fed Russian knapweed (*Centaurea repens* L.). Amer. J. Vet. Res. 31:1393-404.

Young, S., W.W. Brown, and B. Klinger. 1970b. Nigropallidal encephalomalacia in horses fed Russian knapweed (*Centaurea repens* L.). Amer. J. Vet. Res. 1157:1602-05.

Snakeweeds

Tracy M. Sterling, David C. Thompson, and Kirk C. McDaniel

The perennial snakeweeds—broom snakeweed (*Gutierrezia sarothrae*) and threadleaf snakeweed (*Gutierrezia microcephala*)—are members of the sunflower (Asteraceae) family. Snakeweeds are indigenous to North America and are wide-spread throughout the western United States. These weeds reduce grassland forage production and are toxic to livestock. They are shallow-rooted, low-growing, often short-lived, perennial shrubs growing on a wide variety of soils and across a broad spectrum of climates and vegetation types. Broom snakeweed is distributed throughout the western United States from northern Mexico to southern Canada, while threadleaf snakeweed is confined mainly to the southwestern United States and northern Mexico (Lane 1980).

Identification

The genus *Gutierrezia* was first described in 1806 by Lagasca (Solbrig 1960). The species name commemorates the Spanish noble family Gutierrez. The genus *Gutierrezia* contains 11 species in South America and 16 species in North America (Solbrig 1960, Lane 1982). Broom and threadleaf snakeweed are similar to one another in plant canopy and root characteristics (McDaniel et al. 1984), although the plant canopy of broom snakeweed is more compact than that of threadleaf snakeweed. The shape of the snakeweed plant canopy has been described mathematically as a half sphere by Ludwig et al. (1975). Both species have shallow, fibrous root systems around a main taproot. They have sparse foliage (Depuit and Caldwell 1975). Lower branches are leafless, and support numerous upright woody stems that terminate as short, panicled stalks with two to five very small flower heads borne in clusters. Nodes on lower stems will root if touching the soil surface. Leaves are alternate and linear to lanceolate in shape with entire margins.

Broom and threadleaf snakeweeds are most distinguishable at flowering. These species produce canopies of bright yellow flowers in the late summer and early fall. Those growing at higher latitudes flower earlier than those at lower latitudes. The flower heads of broom snakeweed are larger than those of threadleaf snakeweed, which are more cylindrical (McDaniel et al. 1984). The characteristic that distinguishes these two species from one another at flowering is that broom snakeweed has two or more ray florets per flower head, while threadleaf snakeweed usually has only one ray floret per flower head.

Origin, History, and Distribution

Studies of fossilized pack rat middens indicate *Gutierrezia* species have been present on southwestern U.S. rangelands for at least 15,000 years (Van Devender and Everitt 1977, Van Devender et al. 1978). Broom snakeweed is the most widely distributed North American species of the genus *Gutierrezia*, growing from cold-temperate climates in southern British Columbia and Saskatchewan, Canada, to subtropical areas in Nuevo León and Sinaloa, Mexico, between the eastern foothills of the Rocky Mountains and the Great Plains (Solbrig 1960, Lane 1985). Broom snakeweed habitats range from 2,300 to 9,500 feet (700 to 2,900 m) in elevation, and are well-drained, high montane slopes, mesic or semi-arid grasslands, and deserts (Lane 1980, 1985). Threadleaf snakeweed grows mainly in the Mojave, Sonoran, and Chihuahuan deserts of the southwestern United States and northern Mexico at similar elevations (3,280 to 8,200 ft [1,000 to 2,500 m]), but in typically sandier soils than broom snakeweed. It is thought that broom snakeweed is the species most directly related to the ancestral population originating in Mexico (Solbrig 1960).

Snakeweed plants have a broad ecological amplitude, and can be found from low-lying deserts up to coniferous forests. The projected mean annual air temperature within the distribution range of broom snakeweed varies from 39° to 70° F (4° to 21° C), and growing season length from 100 to 300 days (Solbrig 1964). Average annual rainfall ranges from 8 to 20 inches (20 to 50 cm). Broom snakeweed and threadleaf snakeweed have slightly different germination requirements, which partially explains the relative distribution of the species. Broom snakeweed has a lower temperature requirement than threadleaf snakeweed for germination and has a more northern

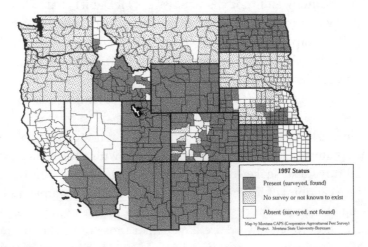

Distribution of snakeweeds by county in the western United States. Data were collected by surveying various weed authorities in each state.

distribution (Mayeux and Leotta 1981). At northern latitudes, broom snakeweed is widespread, but usually it is not abundant on the Great Plains where summers are mild and winters are cold. Exceptionally dense populations of broom snakeweed grow on southwestern grasslands, where summers are hot but winters mild (Pieper and McDaniel 1989).

Potential Invasion

Perennial snakeweeds are native to and widely distributed in western North America. Estimates vary as to the importance of snakeweed under undisturbed conditions, but Campbell and Bomberger (1934) suggested snakeweeds should contribute up to 15% of the plant composition on Chihuahuan desert grasslands. Sosebee (1983) reported broom snakeweed should contribute 5% to 10% of the plant composition on pristine plains grasslands. Under undisturbed conditions the amount present will change through time, according to climatic conditions. The relative number of plants would be expected to increase depending on the level of disturbance in a given area. Areas with a high amount of exposed soil are more likely to support a densely populated stand of snakeweed than are areas with greater grass cover (McDaniel et al. 1982).

Drought, grazing, and fire are principal factors creating an increased amount of exposed soil, and reducing other plant species that may compete with snakeweed. On the Fort Stanton Experimental Ranch in central New Mexico, broom snakeweed numbers increased following droughts in 1970-71, 1974, 1976, and 1980, but the degree of infestation was less on moderately-grazed pasture with high blue grama (*Bouteloua gracilis*) cover than on heavily-grazed pasture with lower grass cover (McDaniel et al. 1982). Fire can create a habitat highly suitable for broom snakeweed invasion, depending on the timing of the burn and the growth of associated plants. For example, prescribed fires conducted in the summer near Corona, NM, eliminated most mature snakeweed plants. However, blue grama was highly damaged also, and the following year snakeweed seedlings established in numbers well above the original population (Hart 1992). Spring burning in the same area resulted in less blue grama disturbance and fewer snakeweed seedlings the next year.

Impacts

The snakeweeds have little or no grazing value to livestock, and are not consumed in large quantities except in areas of extreme range deterioration or during the early spring when other green vegetation is scarce (Huddleston and Pieper 1989). Snakeweed is occasionally found in the diets of sheep, goats, blacktailed jackrabbits, pronghorn sheep, mule deer, and game birds (Pieper 1989).

Snakeweeds contain saponins, alkaloids, terpenes, flavonols and other substances that could be toxic to livestock (Smith et al. 1991). Saponins have been identified as the primary toxic substance in snakeweed. Losses caused by abortion and birth-

325

related abnormalities in cattle, sheep, and goats are sporadic (Norris and Valentine 1957, Martinez et al. 1993, Williams et al. 1993) and may be the result of the genetic variability of snakeweed (Hou and Sterling 1995) or the fluctuating climatic and range conditions under which it grows. Snakeweed grows abundantly on loam soils, but is more toxic when growing on sandy soils (Dollahite and Anthony 1957). Chavez-Gonzales (1988) found the saponin content to vary from highest to lowest in the leaves, small branches, large branches and roots, respectively. Mathews (1936) found no decrease in toxicity as the plant matures.

Economic losses from snakeweed vary depending on beef prices and production costs, but on the 23 million acres (9.3 million ha) of rangelands in the western United States infested with snakeweed, losses are substantial (McDaniel and Torell 1987). Grass production can be decreased by an estimated 90% in dense snakeweed stands (McDaniel et al. 1982).

Biology and Ecology
• *Seed Production*
Broom snakeweed is often described as a short-lived, outcrossing, perennial plant that is dependent upon seed production for survival (Solbrig 1960). Seed numbers per plant are highly variable and dependent on plant age, interspecific competition, and seasonal growing conditions. Mature plants growing in dense stands produce fewer seeds than younger plants growing in open areas with less interference. Plants restricted in size (less than 6 inches [15 cm] tall) by shallow surface soil and limited soil moisture may produce fewer than 1,000 seeds per plant. Robust plants (more than 12 inches [30 cm] tall) growing in deeper soils may produce 10,000 to 30,000 seeds. During drought, plants restrict and occasionally eliminate flowering to divert available resources to individual plant survival.

• *Seed Dispersal*
Flowering usually occurs earlier and over a shorter period in northern latitudes (Wyoming) than in southern regions (New Mexico). Flower production ceases when minimum air temperatures fall below 41° F (5° C). Seeds begin to be released when freezing temperatures loosen the flower head. Dispersal is gradual and continues as long as flower material remains on the plant. In the Southwest, dispersal may continue into the next summer, but the majority of seeds (82%) are dispersed between November and January (Wood et al. 1997). Abundant seedfall coincides with exceptionally high winds (above 26 mi/h [12 m/sec]), snowfall, and intense rain that dislodges seeds.

Broom snakeweed seeds have no specialized structures, such as wings, to facilitate long-range dispersal; thus, the progeny from parent plants is mostly distributed in an elliptical pattern outward less than 3 feet (1 m) from the shrub canopy (Wood et al. 1997). The dispersal direction is with the prevailing winds, which generally are toward the northeast throughout much of western North America (Young and Evans 1989).

• *Viability and Germination*

The viability of broom snakeweed seeds is reportedly longer than four years under laboratory storage conditions (Mayeux 1983); however, seed reserves in the soil are virtually depleted before the next seed drop, according to bioassays (Berry 1992). A small fraction of seeds are undoubtedly longer-lived, but the vast majority of seeds succumb to insect predation, dehydration, pathogens, and other causes within months of dispersal (Wood et al. 1997).

Propagation from seeds in the fall is rare, and is usually delayed until late winter and spring (February to May) when minimum seedbed temperatures and soil moisture become sufficient to permit germination. Laboratory studies suggest optimal seed germination (about 90%) occurs at 50° to 68° F (10° to 20° C), and little germination occurs below 39° F (4° C) or above 90° F (32° C) (Kruse 1979, Mayeux and Leotta 1981). Seeds lying on the soil surface are more likely to germinate (88%) than seeds immediately below the soil surface (33%) or seeds buried $^3/_8$ to $^{13}/_{16}$ inch (1 to 2 cm) deep (3% and 0%, respectively) (Mayeux 1983). Snakeweed seeds will not successfully germinate unless the soil surface is saturated for three to four days, to allow seeds to imbibe and secure themselves in the soil (Wood et al. 1997).

Broom snakeweed seedlings that survive the first year often germinate in late winter in the Chihuahuan desert and in late spring in the northern plains. Years may pass without germination or seedlings becoming established (Pieper and McDaniel 1989). Both the abundance and timing of rainfall are crucial to seedling survival. In New Mexico, 75% of plants that emerge in April survive into the fall, provided at least $^1/_8$ inch (2.5 mm) of rain is received within 12-day intervals through the usually hot, dry months of May and June (McDaniel 1989). When an equal amount of rainfall is received at intervals less frequent than every 12 days, seedling survival is usually less than 10%.

During its first growing season, a seedling produces a taproot that extends about 11 inches (27 cm) deep (Osman and Pieper 1988). As the plant matures, a more extensive, fibrous root system develops that roughly mirrors the size and shape of the canopy above.

• *Phenology*

Snakeweeds die back to a low, woody central stem every year. In their northern range, snakeweed plants drop most of their leaves in winter; however, in their southern range, where winters are more mild, primordial leaves on lower woody stems remain green year-round (Breide 1990). After winter dormancy, axillary shoots formed the previous fall produce large, fleshy leaves as soils warm (mid- to late May) (Depuit and Caldwell 1975). In early summer (mid-June), smaller, denser leaves begin to develop at stem nodes and floral buds emerge. In late June and July,

327

the large, fleshy leaves die almost completely, replaced by the smaller, denser leaves, which now serve as the primary foliage. Flowers develop in August and September; fruits develop in early to mid-October until seeds are released.

Snakeweed populations are considered cyclical (Torell et al. 1992). When the high mortality of seedlings is included, the average life span of snakeweed plants is about 2.4 years (Dittberner 1971). The life span of those plants reaching one year of age is as long as 15 years, with an average life span of four years.

Management

Many methods have been used to control snakeweed in the past, and several hold promise for the future. Chemical, mechanical, fire, and biological methods have been examined and implemented with differing degrees of success (Huddleston and Pieper 1989). Although herbicides are more widely used than other methods for broom snakeweed control (McDaniel et al. 1986), an integrated management strategy is necessary for long-term control of snakeweed.

• *Chemical Control*

Herbicides have been studied for control of snakeweed since the 1940s, but only during the early 1980s was the technology developed to achieve predictable snakeweed mortality (Williams 1990). Herbicide activity is influenced by environmental conditions and the physiological, morphological, and phenological stage of the plant (Sosebee 1983, McDaniel 1989). The primary herbicide used to control snakeweed has been picloram. McDaniel and Duncan (1987) found that fall applications of picloram at a rate of 0.25 lb acid equivalent per acre (0.28 kg ae/ha) controlled greater than 95% of the broom snakeweed. Picloram at other times of the year is usually not as effective at comparable application rates. Although picloram is most widely used (Townsend 1995), other herbicides such as metsulfuron (0.03 lb ai/ac [0.034 kg ai/ha]) are also effective in controlling snakeweed. However, after the initial herbicide treatment, five years of benefit must be realized before a rancher can justify the cost of $9 per acre ($22 per hectare) for control (Torell et al. 1988, Carpenter et al. 1991).

• *Mechanical Control*

Mechanical methods for snakeweed control such as grubbing or disking often result in high levels of soil disturbance which creates an ideal environment for reinvasion of snakeweed and other weeds. Seeding improved varieties of grasses may be needed in mechanically disturbed areas. In areas where obstacles are not a factor, shredding or mowing may be used. However, when snakeweed top growth is removed, new stems often arise from the nodes of remaining branches making repeated mowing necessary. Mowing should be conducted when vegetative growth is complete, and soils are hot and dry. Shredding before flowering or seed set eliminates seed production and reduces the number of seedlings that may subsequently invade a site (McDaniel and Torell 1987).

• Burning

Broom snakeweed is susceptible to fire (Humphery and Everson 1951, Carroll 1994). Factors affecting the efficiency of fire include wind speed, soil moisture, air temperature, relative humidity, fuel moisture, and load (McDaniel et al. 1989, Hart 1992). A fire can reduce the standing crop of snakeweed, but seeds in the soil may not be killed. Seedlings may establish soon after burning, especially when grass competition is low (Vallentine 1980, Carroll 1994). The effectiveness of prescribed fire is highly dependent on the amount and characteristics of fine fuel present at the time of the burn. For example, on blue grama grasslands infested with snakeweed, McDaniel et al. (1997) recommended that the fire-fuel load should exceed 450 lb/ac (500 kg/ha) and broom snakeweed biomass should be below 270 lb/ac (300 kg/ha); burning should be conducted when air temperatures range from 68° to 86° F (22° to 28° C), and relative humidity is between 10% and 20%, wind speed is between 4 and 18 mi/hour (2 and 8 m/sec), soil moisture is between 3% and 10%, and grass moisture is less than 15%.

• Biological Control

Biological control of snakeweed continues to be studied. A non-indigenous weevil, *Heilipodus ventralis,* from Argentina, was released by C.J. DeLoach of the USDA Agricultural Research Service's Biological Control of Weeds Laboratory in Temple, TX, on snakeweeds in the southwestern United States. This release is the first time that a non-indigenous biological control agent has been released on a native weed in the United States. Unfortunately, as of 1998, this weevil had failed to establish at any release site (DeLoach and Cuda 1998, Cordo et al. 1998). A root-feeding moth, *Carmenta haematica,* is currently in quarantine in Texas (Cordo et al. 1995) and a seed-feeding fly is being studied in Argentina. Neither has been approved for release in the United States.

More than 400 native insects and eight native pathogens have been collected from snakeweed (Foster et al. 1981, Farr et al. 1989, Liddell 1989, Richman and Thompson 1998). The importance of these native insects and pathogens associated with snakeweed is being quantified (Richman et al. 1992; Liddell et al. 1993; Thompson et al. 1995, 1996). The grasshopper *Hesperotettix viridis* feeds almost exclusively on snakeweeds, preferring seedlings and young plants. More than four grasshoppers per snakeweed plant results in a significant decrease in aboveground photosynthetic biomass (Thompson et al. 1995). In addition, the standing crop of grasses increased 23% by the end of the treatment year, and 44% one year after treatment, compared to plots with no grasshoppers (Thompson et al. 1996). Root-borers, such as the root-boring beetle *Crossidius pulchellus* LeConte, primarily attack larger (greater than 6 inches [15 cm] tall) and older plants by hollowing out the taproot and killing most plants that are attacked (Richman and Huddleston 1981). Grasshoppers or beetles acting alone are often unable to suppress snakeweed populations completely, possibly because of environmental conditions, current range management techniques

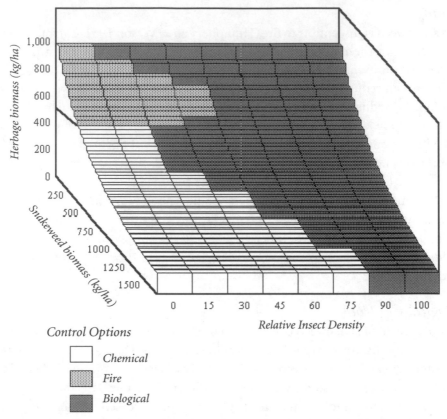

Figure 1. Conceptual perspective of an integrated weed management program for snakeweed relating snakeweed biomass, herbage biomass, and relative insect density. Shaded areas represent the situations under which each method of control is most appropriate.

(e.g., rangeland insect control programs, stocking rates, etc.) or density-dependent natural control factors (e.g., predators, parasites, or pathogens).

• Integrated Weed Management

The most difficult aspect of snakeweed control is the unpredictability of reinfestation and natural die-off (Torell et al. 1989, Bain 1992, Townsend 1995). Integrated management strategies are necessary to sustain our rangeland ecosystems as truly renewable resources. The key is to estimate snakeweed production, grass production, and the densities and types of potential biological control agents present, before implementing a management program (Figure 1). This knowledge improves management decisions. For example, if root-boring insects are infesting more than 50% of the plants and/or grasshopper densities exceed one per plant, chemical control may be unnecessary because these plants may die shortly from natural causes. If snakeweed biomass is especially dense (exceeding 450 lbs/ac [500 kg/ha]), but no beneficial insects are present, then chemical control may be warranted. With this amount of snakeweed, grass production will probably be insufficient to allow prescribed burning as an option (McDaniel et al. 1997). However, if a light to moderate stand of snakeweed (less than 270 lb/ac [300 kg/ha]) is mixed with a uniform grass cover, then fire may be a very effective management tool. Chemical control may be viewed as a reclamation tool to reduce high densities of snakeweed, whereas burning and biological controls are tools for maintaining non-economic populations of snakeweed.

Finally, the most critical and often overlooked factor in establishing an integrated snakeweed management program is the grazing deferment required with any management technique (McDaniel et al. 1982). Sites with little grass production treated with herbicide should be deferred for one, and preferably two, growing seasons, to allow grasses to recover. A fall application of picloram kills snakeweed slowly, allowing many biological control agents to move or complete their life cycles before running out of food. When prescribed burning is used, the site must be deferred at least two growing seasons: one season to build up adequate fine-fuel loads, and a second season for grass recovery. Spring burns are the least damaging to snakeweed biological control agents because most agents are underground at that time or are in a fire-resistant, resting stage to survive the winter. Thus, integrated management of snakeweed appears to be a viable approach to managing native snakeweeds on western rangelands.

Literature Cited

Bain, J.W. 1992. Profit maximizing management alternatives for dealing with drought and snakeweed infestations in eastern New Mexico. M.S. Thesis. New Mexico State Univ., Las Cruces.

Berry, P.N. 1992. Influences of soil mositure on propagation and survival of broom snakeweed (*Gutierrezia sarothrae*). M.S. Thesis, New Mexico State Univ., Las Cruces, NM.

Breide, J.W. 1990. The physiological ecology of *Gutierrezia sarothrae* (Pursh) Rusby in the Chihuahuan Desert of southern New Mexico. Ph.D. Dissertation, New Mexico State Univ., Las Cruces, NM.

Campbell, R.S., and E.H. Bomberger. 1934. The occurrence of *Gutierrezia sarothrae* on *Bouteloua eriopoda* ranges in southern New Mexico. Ecology 25:312-26.

Carpenter, B.D., D.E. Ethridge, and R.E. Sosebee. 1991. Economics of broom snakeweed control on the southern plains. J. Range Manage. 44:232-37.

Carroll, D.B. 1994. Broom snakeweed [*Gutierrezia sarothrae* (Pursh) Britt. & Rusby] seedling response to spring and summer burning in central New Mexico. M.S. Thesis. New Mexico State Univ., Las Cruces, NM.

Chavez-Gonzales, V. 1988. Saponin content in broom snakeweed plant parts during the growing season. M.S. Thesis. New Mexico State Univ., Las Cruces, NM.

Cordo, H.A., C.J. DeLoach, and R. Ferrer. 1995. Host range of the Argentine root borer *Carmenta haematica* (Ureta) (Lepidoptera: Sesiidae), a potential biocontrol agent for snakeweeds (*Gutierrezia* spp.) in the United States. Biol. Cont. 5:1-10.

Cordo, H.A., C.J. DeLoach, and D.H. Habeck. 1998. Biology of *Heilipodus ventralis* (Coleoptera: Curculionidae), an Argentine weevil for biological control of snakeweeds (*Gutierrezia* spp.) in the United States. Biol. Cont. (*In press*).

DeLoach, C.J., and J.P. Cuda. 1998. Host specificity of the Argentine root-boring weevil, *Heilipodus ventralis*, a potential biocontrol agent for snakeweeds (*Gutierrezia*: Asteraceae) in western North American rangelands: Quarantine tests. Biol. Cont. (*In press*).

Depuit, E.J., and M.M. Caldwell. 1975. Gas exchange of three cool semi-desert species in relation to temperature and water stress. J. Ecol. 63:835-58.

Dittberner, P.L. 1971. A demographic study of some semidesert grassland plants. M.S. Thesis. New Mexico State Univ., Las Cruces, NM.

Dollahite, J.W., and W.V. Anthony. 1957. Poisoning of cattle with *Gutierrezia microcephala*, a perennial broomweed. J.A.V.M.A. 130:525-30.

Farr, D.F., G.F. Bills, G.P. Chamuris, and A.Y. Rossman. 1989. Fungi on plants and plant products in the United States. APS Press, MN.

Foster, D.E., D.N. Uekert, and C.J. DeLoach. 1981. Insects associated with broom snakeweed (*Xanthocephalum sarothrae*) and threadleaf snakeweed (*Xanthocephalum microcephala*) in west Texas and eastern New Mexico. J. Range Manage. 34:446-54.

Hart, C.R. 1992. Broom snakeweed (*Gutierrezia sarothrae*) and associated herbage response to seasonal burning in New Mexico. Ph.D. Dissertation, New Mexico State Univ., Las Cruces, NM.

Hou, Y., and T.M. Sterling. 1995. Isozyme variation in broom snakeweed. Weed Sci. 43:156-65.

Huddleston, E.W., and R.D. Pieper. 1989. Snakeweed: Problems and Perspectives. New Mexico Agric. Exp. Stn. Bull. 751.

Humphery, R.R., and A.C. Everson. 1951. Effects of fire on a mixed grass-shrub range in southern Arizona. J. Range Manage. 4:264-66.

Kruse, W.H. 1979. Temperature and moisture stress affect germination of *Gutierrezia sarothrae*. J. Range Manage. 23:143-45.

Lane, M.A. 1980. Systematics of *Amphiachyris, Creenella, Gutierrezia, Gymnosperma, Thurovia* and *Xanthocephalum* (Compositae: Astereae). Ph.D. Dissertation, Univ. Texas, Austin, TX

Lane, M.A. 1982. Generic limits of *Xanthocephalum, Gutierrezia, Amphiachyris, Gymnosperma, Greanella*, and *Thurovia* (Compositae: Astereae). Syst. Bot 7:405-16

Lane, M.A. 1985. Taxonomy of *Gutierrezia* (Compositae: Astereae) in North America. Syst. Bot. 10:7-28.

Liddell, C.M. 1989. Plant pathogens as biological control agents of broom snakeweed in New Mexico. *In:* Snakeweed: Problems and Perspectives. E.W. Huddleston and R.D. Pieper (eds.). New Mexico Agric. Exp. Stn. Bull. 751, 195-202.

Liddell, C., C. Waddell, J. McEntee, and D. Hansen. 1993. Snakeweed rust pathogen kills broom snakeweed. *In:* Sterling, T.M. and D.C. Thompson (eds.). Snakeweed Research: Updates and Highlights. New Mexico Agric. Exp. Stn. Res. Report #674, 26-27.

Ludwig, J.A., J.F. Reynolds, and P.D. Whitson. 1975. Size-biomass relationships of several Chihuahuan desert shrubs. Amer. Midl. Nat. 94:451-61.

Martinez, J.H., T.T. Ross, K.A. Becker, J.L. Williams, D. Campos, and G.S. Smith. 1993. Snakeweed toxicosis in late gestation ewes and heifers. *In:* Sterling, T.M. and D.C. Thompson (eds.). Snakeweed Research: Updates and Highlights. New Mexico Agric. Exp. Stn. Res. Report #674, 48-49.

Mathews, F.P. 1936. The toxicity of broomweed (*Gutierrezia microcephala*) for sheep, cattle and goats. J.A.V.M.A. 88.56-61.

Mayeux, H.S., Jr. 1983. Effects of soil texture and seed placement on the emergence of four subshrubs. Weed Sci. 31:380-84.

Mayeux, H.S., Jr., and H. Leotta. 1981. Germination of broom snakeweed (*Gutierrezia sarothrae*) and threadleaf snakeweed (*G. microcephalum*) seed. Weed Sci. 29:530-34.

McDaniel, K.C. 1989. Use of herbicides in snakeweed management. *In:* Snakeweed: Problems and Perspectives. E.W. Huddleston and R.D. Peiper (eds.). New Mexico Agric. Exp. Stn. Bull. 751, 85-99.

McDaniel, K.C., and K.W. Duncan. 1987. Broom snakeweed (*Gutierrezia sarothrae*) control with picloram and metsulfuron. Weed Sci. 35:837-41.

McDaniel, K.C., and L.A. Torell. 1987. Ecology and management of broom snakeweed. *In:* Integrated pest management on rangeland, a shortgrass prairie perspective. J.L. Capinera (ed.). Westview Press. Boulder, CO, 101-15.

McDaniel, K.C., R.D. Pieper, and G.B. Donart. 1982. Grass response following thinning of broom snakeweed. J. Range Manage. 35:142-45.

McDaniel, K.C., R.D. Pieper, L.E. Loomis, and A.A. Osman. 1984. Taxonomy and ecology of perennial snakeweeds in New Mexico. New Mexico State Univ. Agric. Exp. Stn. Bull. 711, Las Cruces, NM.

McDaniel, K.C., L.A. Torell, J.M. Fowler, and K.W. Duncan. 1986. Brush control on New Mexico rangeland. New Mexico State Univ. Coop. Ext. Serv. Bull. 400b-18.

McDaniel, K.C., C. Hart, and K.W. Duncan. 1989. Use of fire in snakeweed management. *In:* Snakeweed: Problems and Perspectives. E.W. Huddleston and R.D. Pieper (eds.). New Mexico Agric. Exp. Stn. Bull. 751, 101-11.

McDaniel, K.C., L.A. Torell, and J.W. Bain. 1993. Overstory-understory relationships for broom snakeweed-blue grama grasslands. J. Range Manage. 46:506-11.

McDaniel, K.C., C. Hart, and D.B. Carroll. 1997. Broom snakeweed control with fire on New Mexico blue grama rangeland. J. Range Manage. 50:652-59.

Norris, J.J., and K.A. Valentine. 1957. Principal livestock-poisoning plants. New Mexico A&M College, Agric. Extension Serv. Cir. 274.

Osman, A., and R.D. Pieper. 1988. Growth of *Gutierrezia sarothrae* seedlings in the field. J. Range Manage. 41:92-93.

Pieper R.D. 1989. Broom snakewood content of herbivore diets. *In:* Snakeweed: Problems and Perspectives. E.W. Huddleston and R.D. Pieper (eds.). New Mexico Agric. Exp. Stn. Bull. 751, 203-10.

Pieper, R.D., and K.C. McDaniel. 1989. Ecology and management of broom snakeweed. *In:* Snakeweed:Problems and Perspectives. E.W. Huddleston and R.D. Pieper (eds.). New Mexico Agric. Exp. Stn. Bull. 751, 1-12.

Richman, D.B., and E.W. Huddleston. 1981. Root feeding by the beetle *Crossidius pulchellus* LeConte and other insects on broom snakeweed (*Gutierrezia* spp.) in eastern and central New Mexico. Environ. Entomol. 10:53-57.

Richman, D.B., and D.C. Thompson. 1998. Insect associations with woody snakeweeds in New Mexico, Texas, and Arizona. (*In press*) *In:* Papers from the Fourth Symposium on Resources of the Chihuahuan Desert Region. Chihuahuan Desert Research Institute, Alpine, TX.

Richman, D.B., D.C. Thompson, and J. O'Mara. 1992. Effects of leaftiers on broom snakeweed in central New Mexico. Southwestern Entomol. 17:187-89.

Smith, G.S., T.T. Ross, G.I. Flores-Rodriguez, B.C. Oetting, and T.S. Edrington. 1991. Toxicology of snakeweeds, *Gutierrezia microcephala* and *G. sarothrae*. *In:* James, L.F., J.O. Evans, M.H. Ralphs, and R.D. Childs (eds.). Noxious Range Weeds, Westview Press, Boulder, CO, 236-46.

Solbrig, O.T. 1960. Cytotaxonomic and evolutionary studies in the North American species of *Gutierrezia* (Compositae). Contrib. Gray Herb. 188:1-61.

Solbrig, O.T. 1964. Intraspecific variation in the *Gutierrezia sarothrae* complex (Compositate-Asteraceae). Contrib. Gray Herb. 193:67-115.

Sosebee, R.E. 1983. Physiological, phenological and environmental considerations in brush and weed control. *In:* Proc. Brush Management Symp., Soc. Range Manage. Denver, CO, 27-43.

Thompson, D.C., K.C. McDaniel, L.A. Torell, and D.B. Richman. 1995. Damage potential of *Hesperotettix viridis* (Orthoptera: Acrididae) on a native rangeland weed, *Gutierrezia sarothrae*. Environ. Entomol. 24:1315-21.

Thompson, D.C., K.C. McDaniel, and L.A. Torell. 1996. Feeding by a native grasshopper reduces densities and biomass of broom snakeweed. J. Range Manage. 49:407-12.

Torell, L.A., H.W. Gordon, K.C. McDaniel, and A. McGinty. 1988. Economic impact of perennial snakeweed infestations. *In:* The ecology and economic impact of poisonous plants on livestock production. James, L.F., M.H. Ralphs and D.B. Nielson (eds.). Westview Press. Boulder, CO, 57-69.

Torell, L.A., K. Williams, and K.C. McDaniel. 1989. Economics of broom snakeweed control. *In:* Snakeweed: Problems and Perspectives. E.W. Huddleston and R.D. Pieper (eds.). New Mexico Agric. Exp. Stn. Bull. 751, 113-38.

Torell, L.A., K.C. McDaniel, and K. Williams. 1992. Estimating the life of short-lived, cyclic weeds with Markov processes. Weed Technol. 6:62-67.

Townsend, C.L. 1995. The success and treatment life of commercial chemical snakeweed control projects in New Mexico. M.S. Thesis. New Mexico State Univ. Las Cruces, NM.

Vallentine, J.F. 1980. Range development and improvements. 2nd. ed. Brigham Young Univ. Press, Provo, UT.

Van Devender, T.R., and B.L. Everitt. 1977. The late pleistocene and recent vegetation of the Bishop's cap, south central New Mexico. Southwest Nat. 22:337-52.

Van Devender, T.R., C.F. Freeman, and R.D. Worthington. 1978. Full-glacial and recent vegetation of Livingston Hills, Presidio County, Texas. Southwest. Nat. 23:289-302.

Williams, K.E. 1990. Treatment life uncertainty and the economic decision to control broom snakeweed. M.S. Thesis. New Mexico State Univ. Las Cruces, NM.

Williams J.L., D Campos, T.T. Ross, G.S. Smith, J.M. Martinez, and K.A. Becker. 1993. Heifer reproduction is not impaired by snakeweed consumption. *In:* Sterling, T.M. and D.C. Thompson (eds.). Snakeweed Research: Updates and Highlights. New Mexico Agric. Exp. Stn. Res. Report #674, 46-47.

Wood, B.L. 1994. Dispersal, viability, and germination of *Gutierrezia sarothrae* seed. M.S. Thesis, New Mexico State Univ., Las Cruces, NM.

Wood, B.L, K.C. McDaniel, and D. Clason. 1997. Broom snakeweed (*Gutierrezia sarothrae*) dispersal, viability, and germination. Weed Sci. 45:77-84.

Young, J.A., and R.A. Evans. 1989. Dispersal and germination of big sagegrass (*Artemisia tridentata*) seeds. Weed Sci. 37:201-206.

Sowthistles

Richard K. Zollinger and Robert Parker

Annual and perennial sowthistles (*Sonchus* spp.) are members of the Cichoriaceae tribe in the sunflower (Asteraceae) family. Sowthistle occurs in temperate regions, but becomes troublesome and abundant in disturbed habitats of northern America and the prairie provinces of Canada. This noxious weed has a broad tolerance to variable environments, and adapts well to wet sites where there is little soil disturbance. This increases the potential for this species to survive different habitats. Pappus on seeds allow distant dissemination by wind. Seeds rapidly germinate and individual seedlings can become perennial a few weeks after emergence. Persistence of perennial sowthistle is due to continued shoot emergence from buds on lateral underground roots. Chemical control methods appear less effective on perennial sowthistle than other perennial noxious weeds.

Identification

Sowthistles are known by many different names, such as field sowthistle, field milk-thistle, corn sowthistle, milk thistle, swine thistle, tree sowthistle, dingle, or gutweed (Boulos 1973). Sowthistles begin early growth as a rosette of leaves that resemble a dandelion rosette. Sowthistles bolt in mid-summer, and produce erect plants with stems 3 to 7 feet (0.9 to 2 m) tall. Stems are dull green, furrowed, with a milky sap, unbranched and smooth in the lower part, branched in the upper part (Korsmo 1954). Leaves are 4 to 8 inches (10 to 20 cm) long, shining green, alternate, lobed, and lanceolate in outline. The lower leaves have tapering bases and the upper leaves have broadly cordate bases (Stevens 1926, Korsmo 1954). Flower heads are 1^1/$_2$ to 2 inches (4 to 5 cm) wide and yellow-orange (Stevens 1924, Korsmo 1954). Seeds are dark reddish-brown and bear a silky white pappus of many fine, barbed hairs (Stevens 1926, Korsmo 1954).

Two annual and two perennial sowthistles are common. The perennial forms of sowthistle are perennial sowthistle (*Sonchus arvensis*) and marsh sowthistle (*Sonchus uliginosus*) (Zollinger 1989). Perennial types reproduce by seed and shoots from adventitious roots from both horizontal and vertical thickened underground roots (Korsmo 1954). Perennial sowthistles produce leaves 4 to 8 inches long, alternate, entire to irregularly lobed. Flower heads of marsh sowthistle are covered with gland-tipped hairs, whereas perennial sowthistles have none (Stevens 1924, Stevens 1926, Korsmo 1954, Hsieh et al. 1972). Annual sowthistle (*Sonchus oleraceus*) and spiny sowthistle (*Sonchus asper*) have a taproot characteristic of annual plants, and are similar in appearance except the leaves of spiny sowthistle are prickly-edged, and the base of leaves attached to the stem have large, rounded, ear-like lobes (Hutchinson et al. 1984). Leaves of annual sowthistles are more deeply lobed than spiny, marsh, or perennial sowthistles, and have deeper cuts in each lobe that end with a large, pointed end segment.

Origin, History, and Distribution

Sowthistles are of European origin (Stevens 1926). Sowthistles have been reported all over Europe, Asia, Africa, New Zealand, Australia, and several islands (Hooker 1892, Hegi 1929, Kitamura 1960, Pegtel 1972, Peschken 1982). They were introduced to North America by the transport of contaminated crops coming from Europe (Small 1922, Boulos 1961). Pursh (1814) first reported the glandular-pubescent plants in 1814 from Pennsylvania. Sowthistles began to attract attention toward the end of the century when they were recorded as "alarmingly abundant in the maritime provinces of Canada" (Fletcher 1894). The glabrous form was recorded in Maine in 1894 (Fernald and Weigand 1910). In 1922, Stevens (1926) found it had covered the New England states, all of New York, Michigan, Wisconsin, Minnesota, North Dakota, most of Pennsylvania, and the northern parts of Ohio, Indiana, Illinois, Iowa, and South Dakota. Recently, it has been reported in all provinces of Canada and throughout the entire northern half of the United States (USDA 1970). It has become particularly abundant and troublesome in the agricultural regions of Quebec, Ontario, and the prairie provinces of Canada (Frankton and Mulligan 1970).

Potential Invasion

Sowthistles are pioneering species invading natural habitats and any disturbed sites. Pappus-borne seeds allow long-distance travel, and rapid germination causes rapid establishment in diverse environments. Sowthistle thrives and becomes a serious weed on low, heavy, moist soils (Stevens 1924, Stevens 1926, Groh 1942, Pegtel 1976, Zollinger and Kells 1991). Fernald and Wiegand (1910) found perennial sowthistle in calcareous regions in eastern North America where conditions were slightly basic or alkaline, or on neutral soils. He did not observe this species in highly alkaline areas. Braidek et al. (1984) carried out a field study to survey and collect native and introduced species tolerating or adapting to saline habitats. Perennial sowthistle was encountered in 29% of the sites surveyed, and was listed as one of 10 species occurring in 25% or more of the sites. Zollinger and Kells (1991) found that soil pH had little effect on sowthistle growth and development. However, at low soil pH, plants had less total dry weight even though plant height was unaffected. Sowthistle producing tall but lighter stems can adapt to low soil pH by growing to the top of the canopy with less dry matter production. As a consequence, plants grown at low pH produced less root mass than plants at higher soil pH. Less root growth may significantly decrease overwintering survivability through less accumulation of storage reserves in roots, and may help to explain the limitation of sowthistle vegetative propagation in acid soils.

Sowthistle is most competitive under abundant precipitation and moderate climates. In Canada, sowthistle occurs in areas that receive average annual precipitation of 12 to 120 inches (30 to 300 cm) and have a growing season ranging from 160 to 210 days. Sowthistle is commonly found in poorly drained soils and soils with a high water-holding capacity. Perennial sowthistle demonstrates a continuous positive response to increasing soil moisture, with the greatest growth occurring at complete saturation (Zollinger and Kells 1991). Plants grown at complete soil saturation produced 32% more leaves, 6% greater plant height, and 133% more flower heads than plants grown at field capacity. The same response occurs in vegetative rosette growth. Growth is severely inhibited at soil moisture levels below field capacity. With high soil moisture, sowthistle has a significant competitive advantage. However, sowthistle can tolerate low moisture levels until significant rains occur.

The cosmopolitan range where sowthistle can be found indicates a broad tolerance of climatic variation. However, sowthistle thrives in temperate climates. Zollinger and Kells (1991) found sowthistle grown at 68/59° F (20/15° C) day/night temperature produced 61% more leaves, 38% greater rosette diameter, 42% greater height, and 20% more flower heads than plants grown at 86/77° F (30/25° C) day/night temperature. Growth of plants grown at 50/41° F (10/5° C) temperature regime was severely inhibited for several weeks after planting. These results suggest that spread

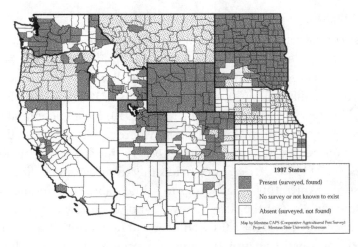

1997 Status

Present (surveyed, found)

No survey or not known to exist

Absent (surveyed, not found)

Map by Montana CAPS (Cooperative Agricultural Pest Survey) Project. Montana State University-Bozeman

Distribution of perennial sowthistles by county in the western United States. Data were collected by surveying various weed authorities in each state.

of sowthistle in North America would be limited to regions with moderate temperatures. Sowthistle would be at a competitive disadvantage in higher temperature climates, acting more as an annual with a limited growth period, reduced underground root growth, and increased flower head production. In moderate climates, sowthistle has a competitive advantage by increased root growth, which would ultimately increase the potential number of shoots that roots could produce.

Studies also showed that plants grown under full light intensity produced 2.3 times more leaves, and four times more flower heads than plants grown at low light levels. However, plants grown at the lowest light intensity produced a rosette 66% larger than plants grown under full light, but delayed the reproductive phase by four weeks. These results suggest that plants grown under less than full light produce fewer but larger leaves. At low light, plants continue stem and leaf growth at the expense of root and flower head production. Applying these results to cultural practices, planting desirable plants that exhibit early germination, strong seedling vigor and growth, and limited time until canopy closure would give the desired species a competitive advantage over sowthistle. If sowthistle growth is suppressed until canopy closure through chemical and/or mechanical control, the desired species may effectively reduce weed growth and spread.

Sowthistle is a minor element in the diet of some North American birds (Martin et al. 1951). Some achenes may germinate after ingestion and excretion by birds and animals, so these also serve as minor dispersing agents (Stevens 1926, Salisbury 1964). The pappus may tangle in the feathers of birds or the wool or hair of animals, aiding in long-range dispersal.

Impacts

Sowthistles are usually viewed as detrimental because of their pernicious, persistent, and noxious characteristics. Perennial sowthistles have been classified as a noxious weed in several states, including Idaho, Minnesota, Montana, Nevada, North Dakota, South Dakota, Washington, Wyoming, all prairie provinces of Canada, and Ontario (USDA 1970). Sowthistles are serious problems on rangeland, pastures, wasteland, roadsides, cultivated fields, canals, salt and brackish marshes, and disturbed habitats (Shumovich and Montgomery 1955, Pegtel 1976). Under ideal environments, this species can displace native plant communities. Surveys (Peschken et al. 1983) show perennial sowthistle commonly occurring in the northern Great Plains of North America. It was present at mean densities as high as 6 shoots/yd^2 (5.5 shoots/m^2) in up to 43% of the fields. It is an aggressive weed that may produce 1.2 tons of dry shoots/acre (3 tons of dry shoots/ha) (Abramov 1969).

Sowthistles have beneficial uses (Quisumbing 1951). The use of sowthistles as salad vegetables was a common practice in the Old World. Levy (1966) noted that young perennial sowthistle leaves and rosette hearts are rich in minerals, and are eaten as salad herbs by gypsies, people of the Swiss and French Alps, and by the Arabs. Consumption of perennial sowthistles was suggested for treatment of fever,

339

skin disorders, acid stomach, disorders of the intestinal tract, heart palpitations, heart tonic, and for anemia. The white milky juice from stem and leaves can be applied to sunburns. Intake of perennial sowthistle roots has been prescribed for jaundice (Chopra et al. 1956).

Sowthistles have been found to contain several chemical compounds that can be used for industrial and pharmaceutical purposes. Baruah et al. (1983) found sowthistle to contain several major lipids classified as glycosylglycerides. Mansour et al. (1983) found the coumarins scopoletin and esculetin as major constituents in several species in the *Sonchus* genera. Duke and Ayensu (1985) reported perennial sowthistle plants as having insecticidal properties and containing the compounds inositol, lactucerol, mannitol, taraxasterol, ceryl alcohol cholin, invert sugars, and tartaric acid. Aerial plant parts contain palmitic and stearic acids. White milky latex from stem and leaves contain rubber and alpha and beta lactucerols with the seeds containing approximately 31.5% oil.

Sowthistles have been examined as a potential source of natural rubber, energy, and waxes (Buchanan et al. 1978). Sowthistle has been evaluated as a crop for petrochemical production. Hooper et al. (1982) found sowthistle an excellent source of pentacyclic triterpenes. This class of compounds is seldom commercially available and has been shown to exhibit antitumor, antihypercholesteremic, antiinflammatory, anticonvulsant, antibacterial, analgesic, and antitussive properites. Buchanan et al. (1978) found perennial sowthistle to be an excellent source of fiber, protein, oil, and terpene hydrocarbons.

Biology and Ecology

• Life Cycle

Perennial and marsh sowthistle are perennial from both horizontal and vertical thickened roots. Aboveground portions of the plant die each winter, and new shoots are produced from buds on the roots each spring. Seeds of the annual taprooted species, annual and spiny sowthistle, germinate each spring, bolt, flower, set seed, and complete their life cycle each summer (Boulos 1960).

• Roots

Perennial sowthistles propagate rapidly by thickened horizontal roots, usually 2 to 4 inches (5 to 10 cm) below the surface. Numerous buds develop on the roots at irregular intervals, and pieces only 1/4 inch (0.64 cm) long are capable of producing a flowering plant in three months (Stevens 1924). Lateral roots can reach 18 feet (5.5 m) long or longer in one season, depending on environmental factors and conditions (Stevens 1926). Vertical roots can penetrate to a depth of 24 inches (0.6 m) or greater, and are able to produce new aerial growth from as much as 16 inches (0.4 m) below the soil surface.

A resting period for perennial sowthistle roots is required before new plant roots can expand (Stevens 1926, Hakansson 1969, Henson 1969). Hakansson (1972) found

340

reduced shoot emergence and growth from roots collected and planted from late August to mid-October. After storage at 36° F (2° C) from December to January, the material sprouted readily. The majority of shoots from roots collected in July and August did not show any inhibited growth.

Schimming and Messersmith (1988) found perennial sowthistle roots were easily able to withstand extended temperatures as low as 1.4° F (-17° C). Perennial sowthistle was more resistant to freezing temperatures than leafy spurge or Canada thistle roots, which are located deeper in the soil.

• Seeds

In 1924 Stevens (1926) found an average of 20 seedheads per stalk and nearly 2,000 seedheads/yd^2 in an ordinary patch of perennial sowthistle in North Dakota. An average of 30 seeds/head yielded 50,000 fruits/yd^2. Derscheid and Schultz (1960) found an average of 60 to 80 seeds on perennial sowthistle heads in 1956, whereas 20 to 40 developed in 1957. The reason for the differences in seed production was based on the cross-pollination requirement due to insect activity and adverse weather patterns during pollination and seed development. A high percentage of seeds will germinate from sowthistle heads that have been open for 8 to 10 days.

Sowthistle seeds that have been after-ripened will germinate better as temperature increases (Stevens 1926, Lauer 1953). Maguire and Overland (1959) found 100% germination from seeds one year after collection. Zilke and Derscheid (1959) found 55% seed germination from 2-year-old seeds. Stevens (1926) found that seeds retained their viability for no more than five years in dry storage. However, Hutchinson et al. (1984) reported that light (red or far-red in particular) may affect germination of sowthistle seeds.

• Leaves

Perennial sowthistle leaves vary widely in morphology (Bell et al. 1968) due to variable growing conditions and genetic diversity (Stevens 1926). Perennial sowthistle leaves also vary in texture. The leaves of some are thick and waxy, while others are thin and wax-free. Some young leaves are covered with many fine bristles, while others possess only a few. Leaf color can vary from light green to dark green, and some plants have red surrounding the leaf midrib.

• Adults

Seedlings appear from four days to three weeks after introduction to the soil surface, primarily depending on temperature. Seedlings move from the rosette stage to flowering by bolting and producing a stem. Flower buds form and flowering extends from July through October. Mature seeds are produced approximately 10 days after flowering. Bracts on flower heads decurve when the seeds are mature, and the pappus-born seeds are liberated and dispersed by the wind. Indeterminant growth and flowering normally end with the first frost.

• *Sowthistles in the Plant Community*

The competitive success of sowthistle is directly related to moisture, and to some extent to temperature. Plants may germinate and survive under dry conditions. However, abundant moisture allows established plants to expand rapidly, choking out natural and existing vegetation. Optimal growth occurs in moderate temperatures with high soil moisture conditions.

Sowthistle can dominate plant communities by allelopathic processes (Putnam and Tang 1986, Zollinger and Kells 1993). Substances that inhibit other plants' seed germination and plant growth are produced by underground roots and the accumulation of toxins from decaying residue from the previous year's growth. As sowthistle infestations expand, allelochems destroy existing vegetation, allowing sowthistle rapidly to capture soil space.

Management

Successful management of sowthistle infestations requires an integrated approach to prevent establishment, detect and eradicate new introductions, contain existing infestations, and manage plants and large-scale infestations.

• *Prevention*

Because of the ease with which seed is wind-disseminated, scouting and detection are keys to preventing plant establishment. Seedlings are easily controlled through mechanical and chemical methods. Planting weed-free crop seed and controlling weeds on field borders can prevent initial infestations.

• *Cultural Control*

An alfalfa or alfalfa-grass mixture, regularly cut for hay, can eliminate 90% of the sowthistles in three years (Derscheid et al. 1961). Sowthistles can compete vigorously with alfalfa. However, mowing weakens sowthistle, and allows strong, healthy stands of alfalfa to compete with sowthistle under good growing conditions (Waddington 1980). Bromegrass regularly cut for hay and sprayed twice a year (early June and mid-August) with 0.5 to 0.75 lb/ac (0.84 kg/ha) of 2,4-D or MCPA gave similar control (Derscheid 1962).

• *Mechanical Control*

Simulated tillage studies found mechanical disturbance broke roots and activated new shoot growth from root pieces. Repeated soil disturbance killed plants by causing a relative exhaustion of the vegetative system (Hakansson 1969, Hakansson 1972, Hakansson and Wallgren 1972). However, a long-term management strategy must be implemented, because intermittent tillage will simply spread roots, increasing patch establishment and size. Root segments 1 inch (2.5 cm) or smaller can produce new plants. It takes 10 to 15 days for new shoots to emerge after the roots have been cut. Another 10 to 15 days elapse before there are enough leaves to produce perennation of roots. Therefore, for the first three to four weeks after tillage, only a small amount of plant food is stored in the roots, and root reserves are being used for plant growth.

Cultivation every three to four weeks causes a continuous drain on root reserves. The food supply in the roots is eventually depleted and the plants die. Intensive cultivation (every three to four weeks) with a duckfoot field cultivator or one-way disk from spring until freeze-up will kill a high percentage of sowthistle (Derscheid 1962). Perennial sowthistle root reserves are decreased more by spring cultivation than at later growth stages. However, roots remaining on the soil surface during winter months have higher mortality than if buried. Fallowing for a year beginning in the fall and cultivating every three weeks in the spring reduces perennial sowthistle stands 75% to 90%. However, combining intensive cultivation with chemical application is generally more practical than cultivating for an entire season, which can contribute to erosion. Following tillage with a perennial or broadleaf crop can reduce infestations up to 80%.

Mowing before flowers have been open one week will prevent seed production (Stevens 1924, Groh 1942, Derscheid and Wallace 1959, Derscheid and Schulz 1960, Derscheid 1962, Sylwester 1962). However, mowing once per season will control only annual sowthistles. Perennial sowthistles require multiple mowings to control stem growth. Seed production can be prevented by applying selective herbicides before blooming.

• Chemical Control

Herbicides are the most effective method of sowthistle control available when mechanical options are not applicable to rangeland and other undisturbed sites. As of 1998, no biological alternatives exist. Chemical control can be accomplished using 2,4-D, clopyralid, dicamba, glyphosate, and picloram.

Picloram at 0.25 to 0.5 lb active ingredients per acre (ai/ac) can be used in pasture, rangeland and fallow, or up to 1 lb ai/ac can be used on patches in pastures. Dicamba at 0.5 lb ai/ac can be used in pasture and rangeland, or up to 2 to 4 lb ai/ac can be used on patches in pasture, non-cropland, and fallow. Clopyralid at 0.12 to 0.5 lb ai/ac can be used on pasture and grass rangeland, but clopyralid + 2,4-D at 0.19 to 0.25 lb ai/ac + 1 to 1.5 lb ai/ac can be applied on Conservation Reserve Program (CRP) land and pastures. 2,4-D at 1 to 2 lb ai/ac can be applied in pasture, rangeland, or near trees (amine formulations only). Glyphosate at 0.75 to 2.25 lb ae/ac is non-selective, and is used in most revegetation projects to prepare a seedbed for desirable grasses.

Applications should be made at the pre-bud or bud stage. Earlier applications are less effective (Wood 1949, Sylwester 1962, Derscheid 1962, Derscheid and Parker 1972). High herbicide rates are required to kill underground roots. Low or medium rates may control top-growth, but new infestations may appear soon after application. Grass plantings in which low rates are used result in unsatisfactory and sometimes erratic control (Derscheid et al. 1961, Vidme 1961, Sylwester 1962, Terry and Wilson

1964, Waddington 1980). Multiple applications during the same growing season may increase control. Early and late applications of 2,4-D and dicamba at high rates control perennial sowthistle one year after application. Control can be maintained throughout the second year with the reapplication of the same herbicides to existing plants and patches.

Herbicides, especially glyphosate, can result in variable sowthistle control (Horney 1982, Zollinger 1989). Sublethal concentrations remaining within the root system may release dormant buds and increase root and shoot growth (Fletcher 1894, Fyske 1975, Fyske 1976). New growth may occur throughout the growing season after each application, and especially the following spring. Vidme (1961) found that good wetting ability of the spray solution was essential for control, because some sowthistles have very waxy leaves. The use of low spray carrier volumes and the addition of ammonium sulfate is essential for increased control from glyphosate (Hakansson 1972, Zollinger 1989). The addition of 2,4-D or dicamba to glyphosate may reduce sowthistles more than glyphosate alone (Zollinger 1989).

The reason for difficulty in chemical control of sowthistle is lack of susceptibility. Canada thistle occupies the same habitat as sowthistle, and both can be sprayed at the same time. Sowthistle responds to herbicides generally less than Canada thistle (Frankton and Mulligan 1970, Bell et al. 1973, Boluri 1977, Devine and Vandenborn 1985). Research has shown that herbicide absorption and movement in sowthistle is slow and limited, and very little herbicide is translocated into roots and root buds (Devine and Vandenborn 1985, Zollinger 1992). Translocation is governed by factors other than net carbohydrate flow. However, rosettes treated in the fall transported more herbicide to roots than larger plants, or than in plants that were in the rosette stage earlier in the year (Fyske 1974, 1975, 1976). Also, natural tolerance (resistance) of sowthistle biotypes to many of the herbicides listed was observed as far back as 1973 (Bell et al. 1973).

• Grazing

Infestations in pastures can be suppressed effectively by grazing cattle or sheep. Sowthistle makes excellent feed, and can be pastured with cattle or sheep (Stevens 1924). Intensive grazing weakens the plants and facilitates other control practices. Livestock will eat new growth and sometimes roots, thereby serving as an effective means of surface cultivation.

• Revegetation

As sowthistle infestations are controlled, fast-growing, highly competitive desirable plants should be established to replace bare ground (Derscheid and Wallace 1959, Derscheid et al. 1961, Derscheid 1962, Sylwester 1962).

• Biological Control

Schroeder (1974) surveyed Europe for biological control agents and found six insect species that appeared restricted to sowthistles. Five species were classified as "seed destroyers," meaning they reduce seed production by feeding on developing tissue within the seedhead. Of those, *Tephritis dilacerata* was the most promising, and was released into Canada. This fly is of the family Tephritidae and is phytophagous in the larval stage. It develops in the flower heads of sowthistle, where it gains protection from desiccation and natural enemies. Flower heads become swollen and distorted, forming a gall. Larval feeding induces the bases of the involucre bracts to expand laterally, causing the mature flower to become button-shaped. The larvae consume tissue of the developing florets and receptacle, and prevent the flower heads from opening (Shorthouse 1980).

Cystiphora sonchi has been approved for release in Canada to control annual and perennial sowthistles by forming galls on vegetative parts of the plants. As many as 721 galls were noted to form on one plant of perennial sowthistle, and up to 245 on one leaf. There are a limited number of pathogens that appear to be specific to sowthistle in North America (Conners 1967). Biological agents specific to sowthistle include: *Marssonia sonchi* (Conners 1967), *Septoria sonchifolia* in Manitoba (USDA 1960, Conners 1967), *Septoria sonchi-arvensis* in Manitoba and Saskatchewan (Conners 1967), *Alternaria sonchi* in Massachusetts (USDA 1960), *Bremia lactucae* in Maryland (USDA 1960), *Coleosporium sonchi-arvensis* in Washington, and *Phylosticta sonchi* in North Dakota (USDA 1960). The only report of nematodes that affect sowthistle is *Meloidogyne incognita* in Indiana (Gaskin 1958). The only reference to virus interaction is *Clorogenus callistephi* in New York (USDA 1960).

• Integrated Pest Managment

Annual sowthistles can be managed with close monitoring, use of selective herbicides in pasture and rangeland, and prevention of seed production. Perennial sowthistles, however, are more difficult to manage. Seeds of perennial sowthistles spread easily by wind, and germinated plants establish rapidly in many environments. Chemical and mechanical control before seeding, in-crop, and after harvesting will minimize infestations for that season and into the next. The plant must be controlled, as well as all root fragments and viable seeds in the soil. Crop rotation, tillage, and herbicides can reduce establishment and propagation.

To manage perennial sowthistles, use intensive cultivation, competitive crops, selective herbicides, or a combination of cultivation, crops, and chemicals. Field research in South Dakota found combinations of cultivation, competitive cropping, and herbicides reduced the sowthistle stands 95% to 100% in two years (Derscheid et al. 1961, Derscheid 1962).

Literature Cited

Abramov, N.G. 1969. Control of *Cirsium arvense* and *Sonchus arvensis*. Len Konop. 14:34-35 [Abstr. in Weed Abstract. 20, No.87 (1971)].

Baruah, P., N.C. Baruah, R.P. Sharma, J.N. Baruah, P. Kulathaivel, and W. Herz. 1983. A monoacyl galactosyl-glycerol from *Sonchus arvensis*. Phytochem. 22:1741-44.

Bell, A.R., J.D. Nalaweja, A.B. Schooler, and S. Alam. 1968. Variations in shape of perennial sowthistle. North Dakota Farm Res. 25:6-7.

Bell, A.R., J.D. Nalaweja, and A.B. Schooler. 1973a. Response of perennial sowthistle selections to herbicides. Crop Sci. 13:191-94.

Bell, A.R., J.D. Nalcwaja, S. Alam, A.B. Schooler, and T.S. Hsieh. 1973b. Herbicide response and morphology of interspecific sowthistle crosses. Weed Sci. 21:189-93.

Boluri, H. 1977. Survey of weeds in wheat fields in the Esfahan area. Iran. J. Plant Path. 13:19.

Boulos, L. 1960. Cytotaxonomic studies in the genus *Sonchus*. 2. The genus *Sonchus*, a general systematic treatment. Bot. Notiser 113:400-20.

Boulos, L. 1961. Cytotaxonomic studies in the genus *Sonchus*. 3. On cytotaxonomy and distribution of *Sonchus arvensis* L. Bot. Notiser 114:57-64.

Boulos, L. 1973. Cytotaxonomic studies in the genus *Sonchus*. 4. Systematic revision of the genus *Sonchus*. Bot. Notiser 126:180-81.

Braidek, J.T., P. Fedec, and D. Jones. 1984. Field survey of halophytic plants of undisturbed sites on the Canadian Prairies. 1984. Can. J. Plant Sci. 64:745-51.

Buchanan, R.A., I.M. Cull, F.H. Otey, and C.R. Russell. 1978a. Hydrocarbon and Rubber-Producing Crops. Evaluation of U.S. plant species. Econ. Bot. 32:131-45.

Buchanan, R.A., F.H. Otey, C.R. Fussell, and I.M. Cull. 1978b. Whole plant oils, potential industrial raw materials. J. Am. Oil Chem. Soc. 55:657-62.

Chopra, R.N., S.L. Nayar, and I.C. Chopra. 1956. Glossary of Indian medicinal plants. Council of Scientific & Industrial Research. New Delhi, 230.

Conners, I.L. 1967. An annotated list of diseases in Canada, and fungi recorded on plants in Alaska, Canada, and Greenland. Res. Branch, Canada Dep. Agric. Publ. 1251, 381.

Derscheid, L.A. 1962. Control of thistles and leafy spurge. N. Central Weed Control Conf. Procedings, 3-5.

Derscheid, L.A., and R. Parker. 1972. Thistles. Canada thistle, perennial sowthistle. S. Dakota State Univ. Ext. Service Fact Sheet 450, 4.

Derscheid, L.A., and R.E. Schultz. 1960. Achene development of Canada thistle and perennial sowthistle. Weeds 8:55-62.

Derscheid, L.A., and K.E. Wallace. 1959. Control and elimination of thistles. S. Dakota State Univ. Ext. Serv. Fact Sheet. Circ. 147, 4.

Derscheid, L.A., R.L. Nash, and G.A. Wicks. 1961a. Thistle control with cultivation, cropping, and chemicals. Weeds 9:90-102.

Derscheid, L.A., W.H. Wallace, and L. Wrage. 1961b. TBA, PBA, amitrol-T, simazine, fenac, amiben, 2,4-D, and 2-methoxy-3-6-dichlorobenzoic acid for eliminating perennial sowthistle. N. Central Weed Control Conf. Res. Rep., 8.

Devine, M.D., and W.H. Vandenborn. 1985. Absorption, translocation, and foliar activity of clopyalid and chlorsulfuron in Canada thistle (*Cirsium arvensis*) and perennial sowthistle (*Sonchus arvensis*). Weed Sci. 33:524-30.

Duke, J.A., E.S. Ayensu. 1985. Medicinal Plants of the World. (Vol. 2) Reference Publications, Inc., 181.

Erlandsson, S. 1950. (Bei A.U.D. Love). In G. Tischler, Die Chromosomenzahlen der Gefasspflanzen Mitteleuropas, W. Junk's-Gravenhage.

Fernald, M.L., and K.M. Wiegand. 1910. A summer's botanizing in Eastern Maine and Western New Brunswick. Part II. Rhodora 12:101-46.

Fletcher, J. 1894. Canada Exp. Farms Rept. Ottawa.

Frankton, C., and G.A. Mulligan. 1970. Weeds of Canada. Canada Dept. Agric. Publ. 948, 217.

Fykse, H. 1974. Untersuchungen uber *Sonchus arvensis* L. I. Translokation von C-14-markierten Assimilaten. Weed Res. 14:305-12.

Fykse, H. 1975. Untersuchungen uber *Sonchus arvensis* L. II. Translokation von C-14-MCPA unter verschiedenen Bedingungen. Weed Res. 15:165-70.

Fykse, H. 1976. Untersuchungen uber *Sonchus arvensis* L. III. Metabolismus von MCPA. Weed Res. 16:309-16.

Gaskin, T.A. 1958. Weed hosts of *Meloidogyne incognita* in Indiana. Plant Disease Reporter 42:802.

Gill, N.T. 1938. The viability of weed seeds at various stages of maturity. Ann. Appl. Biol. 25:447-456.

Groh, H. 1942. Perennial sowthistle and its smooth variety in Canada. Sci. Agr. 23:127-30.

Hakansson, S. 1969. Experiments with *Sonchus arvensis* L. I. Development and growth, and the response to burial and defoliation in different developmental stages. Lantrhogsk. Ann. 35:989-1030.

Hakansson, S. 1972. Experiments with *Sonchus arvensis* L. II. Reproduction, plant development and response to mechanical disturbance. Swed. J. Agric. Res. 2:3-14.

Hakansson, S., and B. Wallgren. 1972. Experiments with *Sonchus arvensis* L. III. The development from reproductive roots cut into different lengths and planted at different depths, with and without competition from barley. Swed. J. Agric. Res. 2:15-26.

Hegi, G. 1929. Illustrierte Flora von Mittel-Europa, Vol. V1- 2. J.F. Lehmanns, Munchen, 549-1386.

Henson, I.E. 1969. Studies on the regeneration of perennial weeds in the glasshouse. I. Temperate species. Agric. Res. Council. Weed Res. Org. Oxford. Tech. Rep. No. 12, 23.

Hooker, J.D. 1892. *In:* Bentham and Hooker. Handbook of the British Flora. 6th ed. London, 1892.

Hooper, S.N., R.F. Chandler, E. Lewis, and W.D Jamieson. 1982. Simultaneous determination of *Sonchus arvensis* L. triterpenes by gas chromatography-mass spectometry. Lipids 17:60-63.

Horney, P. 1982. Perennial sowthistle control evaluations. 1982. Weed Control Results in Field Crops. Dept. of Crop and Soil Sci., Michigan State Univ., East Lansing, MI.

Hsieh, T.S., A.B. Schooler, A.R. Bell, and J.D. Nalewaja. 1972. Cytotaxonomy of three *Sonchus* species. Amer. J. Bot. 59:789-96.

Hutchinson, I., J. Colosi, and R.A. Lewin. 1984. The biology of Canadian weeds. 63. *Sonchus asper* (L.) Hill and *S. oleraceus* L. Can. J. Plant Sci. 64:731-44.

Kitamura, S. 1960. Flora of Afghanistan, Kyoto, 447-448. p. 44.

Korsmo, E. 1954. Anatomy of weeds. Grondahl & Sons Forlag. Kirstes Boktrykkeri, Oslo, Norway.

Lauer, E. 1953. Uber die Keimtemperatur von Ackerunkrautern und deren Einfluss auf die Zusammensetzung von Unkrautgesellschaften. Flora Allg. Bot. Zeit. 140:551-95.

Levy, J.B. 1966. Herbal handbook for everyone. Charles T. Branford Co. Newton, MA, 120-21.

347

Mansour, R.M.A., N.A.M. Saleh, and L. Boulos. 1983. A chemosystematic study of the phenolics of *Sonchus*. Phytochemistry 22:489-92.

Maguire, J.D., and A. Overland. 1959. Laboratory germination of seeds of weedy and native plants. Washington Agr. Exp. Sta. Circ. 349, 15.

Martin, A.C., H.A. Zim, and A.L. Nelson. 1951. American wildlife and plants. A guide to wildlife food habits. Dover, New York, 500.

Pegtel, D.M. 1972. Effects of temperature and moisture on the germination of two ecotypes of *Sonchus arvensis* L. Acta Bot. Neeri. 21:48-53.

Pegtel, D.M., 1976. On the ecology of two varieties of *Sonchus arvensis* L. Ph.D. Thesis, Rijksuniversiteit te Groningen, 148.

Peschken, D.P. 1982. Host specificity and biology of *Cystiphora sonchi* [DIP.: Cicidomyiidae], a candidate for the biological control of *Sonchus* species. Entomophaga 27:405-16.

Peschken, D.P., A.G. Thomas, and R.F. Wise. 1983. Loss in yield of rapeseed (*Brassica napus, B. campestris*) caused by perennial sowthistle (*Sonchus arvensis*) in Saskatchewan and Manitoba. Weed Sci. 31:740-44.

Pursh, F. 1814. Florae Americae Septentrionalis. London.

Putnam, A.R., and C.S. Tang. 1986. Allelopathy: State of the Science, John Wiley & Sons, Inc., New York, NY.

Quisumbing, E. 1951. Medicinal plants of the Philippines. Tech. Bulletin 16. Manila Bureau of Printing, 996-97.

Salisbury, E.J. 1964. Weeds and aliens. Collins, London.

Schimming, W.K., and C.G. Messersmith. 1988. Freezing resistance of overwintering buds of four perennial weeds. Weed Sci. 36:568-73.

Schroeder, D. 1974. The phytophagous insects attacking *Sonchus* spp. [Compositae] in Europe. Proc. 3rd. Int. Symp. Biol. Control weeds. Montpellier, France, 120-23.

Shorthouse, J.D. 1980. Modification of the flower heads of *Sonchus arvensis* (family Compositae) by the gall former *Tephritis dilacerata* (order Diptera, family Tephridae). Can. J. Bot. 58:1534-40.

Shumovich, W., and F.H. Montgomery. 1955. The perennial sowthistles in Northeastern North America. Can. J. Agr. Sci. 35:601-605.

Small, J.K. 1922. Another *Sonchus* for America. Torreya 21:100-101.

Stevens, O.A. 1926. The sowthistles. North Dakota Agric. Exp. Stn. Circ. 32, 16.

Stevens, O.A. 1924. Perennial sowthistle, growth and reproduction. North Dakota Agric. Exp. Stn. Bull. 181, 44.

Sylwester, E.P. (Summarizer) 1962. Canada thistle and perennial sowthistle. Res. Rep. 19th N. Central Weed Control Conf., 17-18.

Sylwester, E.P. (Summarizer) 1962. Canada thistle and perennial sowthistle. Proc. 19th N. Central Weed Control Conf., 105.

Terry, H.J., and C.W. Wilson. 1964. A field study of the factors affecting the herbicidal activity of ioxynil and bromoxynil and their tolerance by cereals. Weed Res. 4:196-215.

USDA. 1960. Index of plant diseases in the United States. Agric. Handb. No. 165, Agric. Res. Serv., 87.

USDA. 1970. Selected weeds of the United States. Agric. Handb. No. 366, Agric. Res. Serv., II, 463.

Vidme, T. 1961. Control of *Sonchus arvensis* L. with chemicals. Weed Res. 1:289-300.

Waddington, J. 1980. Chemical control of dandelion (*Taraxicum officinale*) and perennial sowthistle (*Sonchus arvensis*) in alfalfa (*Medicago sativa*) grown for seed. Weed Sci. 28:164-67.

Wood, H.E. (Summarizer). 1949. Use of herbicides in control of perennial herbaceous weeds. Perennial sowthistle (*Sonchus arvensis*). Res. Rep. 6th N. Central Weed Control Conf., 9-13.

Zilke, S., and L.A. Derscheid. 1959. Effects of environmental factors on thistle seed germination. North Central Weed Contr. Conf. (16th) West. Can. Weed Control Conf. (10th) Joint Proc., 20-21.

Zollinger, R.K. 1989. Perennial sowthistle (*Sonchus arvensis* L.)—Distribution, biology, and control. Ph.D. Dissertation. Dept. of Crop and Soil Sci., Michigan State Univ., East Lansing, MI.

Zollinger, R.K., and J.J. Kells. 1991. Effect of soil pH, soil water, light intensity, and temperature on perennial sowthistle (*Sonchus arvensis* L.). Weed Sci. 39:376-84.

Zollinger, R.K., and J.J. Kells. 1993. Perennial sowthistle (*Sonchus arvensis* L.) interference in soybean (*Glycine max*) and dry edible bean (*Phaseolus vulgaris*). Weed Technol. 7:52-57.

Zollinger, R.K., D. Penner, and J.J. Kells. 1992. Absorption, translocation, and foliar activity of clopyralid and tribenuron in perennial sowthistle (*Sonchus arvensis* L.). Weed Sci. 40:528-33.

Spotted Knapweed

Roger L. Sheley, James S. Jacobs, and Michael F. Carpinelli

Spotted knapweed (*Centaurea maculosa*), a perennial, is a taprooted Eurasian weed invading rangeland throughout the western United States and Canada. This weed had spread to 20 counties in the Pacific Northwest by 1960, and to 48 counties by 1980. Between 1980 and the present, the range of spotted knapweed rapidly increased to include 326 counties in the western United States, including every county in Washington, Idaho, Montana, and Wyoming. Spotted knapweed invasion is associated with reductions in biodiversity, wildlife and livestock forage, and increased soil erosion.

Identification

Spotted knapweed is a deeply taprooted, rosette-forming plant in the sunflower (Asteraceae) family. Basal rosette leaves are borne on short stalks and grow up to 8 inches (20 cm) long and 2 inches (5 cm) wide. Rosette leaves are deeply divided (once or twice) into lobes on both sides of the center vein. Lobes are oblong, with the broadest part above the middle. Flowering stems stand 8 inches to 4 feet (0.2 to 1.2 m) tall. Spotted knapweed stems branch on their upper half. Stem leaves are alternate, sessile, have few lobes, or are linear and entire, and are smaller toward the stem apex. The uppermost leaves are small and simple. Flowerheads are ovate to oblong, $^1/_4$ inch (6 mm) wide and $^1/_2$ inch (12 mm) long, and are solitary or born in clusters of two or three at the branch ends. Involucre bracts are leaf-like, egg-shaped, widest at the base, yellow-green to brown below. The center spine of the bracts is shorter than the lateral ones. Spotted knapweed also has a black margin at the bract tips, and obvious dark longitudinal veins. Spotted knapweed flowers are purple to pink, rarely white, with 25 to 35 flowers per head. Spotted knapweed flowers bloom from June to October. The flowerheads usually remain on the plant.

Seeds are $^1/_8$ inch (3 mm) long, oval, brown to black, with pale longitudinal lines. Seeds bear a pappus of simple bristles which are less than $^1/_8$ inch (1 to 2 mm) long and persistent.

Origin, History, and Distribution

The native range of spotted knapweed is central Europe and east to central Russia, Caucasia, and western Siberia (Rees et al. 1996). Spotted knapweed was introduced to North America in the late 1800s as a contaminant in alfalfa (*Medicago sativa*) (Muller et al. 1988). Early introductions were also through discarded soil used as ship ballast (Roché and Talbott 1986). The first record of spotted knapweed was in Victoria, British Columbia, in 1883 (Groh 1944). This weed spread further in domestic alfalfa seeds and hay before it was recognized as a serious problem (Roché and Talbott 1986).

The geographic spread of spotted knapweed in the western United States up to 1980 has been documented using herbarium records compiled by Forcella and Harvey (1980), and current distribution through interviews with weed authorities in the region. In the United States, spotted knapweed was limited to the San Juan Islands, Wash., until 1920. In 1997 it could be found in 14 western states, and in every county in Washington, Idaho, Montana, and Wyoming.

Potential Invasion

In Europe, spotted knapweed is most aggressive in the forest-grassland interface on deep, well-developed to dry soils, but can form dense stands in more moist areas on well-drained soils including gravel, and in drier sites where summer precipitation is supplemented by runoff. Spotted knapweed does not compete with vigorously growing grass in moist areas, nor with diffuse knapweed (*Centaurea diffusa*) in steppic grassland. It occurs on the moist end of the range of diffuse knapweed (Harrison and Cranston 1979). Lacey et al. (1995) reported that spotted knapweed has been observed at elevations ranging from 1,900 to more than 10,000 feet (578 to more than 3,040 m) and in precipitation zones ranging from 8 to 79 inches (20 to 200 cm) annually.

Impacts

Spotted knapweed reduces livestock and wildlife forage. Watson and Renney (1974) found that spotted knapweed infestations decreased bluebunch wheatgrass (*Pseudoroegneria spicata*) yield by 88%. Elk use, as estimated by pellet groups/acre, was reduced by 98% on spotted knapweed-dominated range compared to bunchgrass-dominated sites (Hakim 1979). Spoon et al. (1983) predicted a loss of 220 elk from winter range in Montana annually because of spotted knapweed infestations on winter range in Montana.

Spotted knapweed dominance on bunchgrass rangeland is also detrimental to water and soil resources. Lacey et al. (1989) determined that surface water runoff and stream sediment yield were 56% and 192% higher, respectively, for spotted knapweed-dominated sites compared to bunchgrass-dominated sites. Bare ground was greater and water infiltration rates were less on spotted knapweed sites than on

351

bunchgrass sites (Lacey et al. 1989). On the beneficial side, however, spotted knapweed provides nectar and pollen for domestic bees (*Apis mellifera*) (Watson and Renney 1974).

Biology and Ecology

• *Life Cycle*

Spotted knapweed is a perennial that lives up to nine years and is capable of producing seeds each year (Boggs and Story 1987). Seed production of spotted knapweed ranges from 5,000 to 40,000 seeds/m^2 per year (Shirman 1981). Site conditions and precipitation during the growing season have the greatest effect on the number of seeds produced per year. More seeds are produced during wet years.

Seeds germinate in the fall and early spring when moisture and temperature are suitable (Watson and Renney 1974). Fall and early spring germinating seedlings are capable of maturing into seed-producing adults in one year (Shirman 1981). Seedlings develop into rosettes; most root growth occurs at this stage (Watson and Renney 1974). If rosettes do not bolt, they die back to the root crown to overwinter. Root crowns form rosettes in the early spring and bolt in early May. Spotted knapweed produces one to six stems. Flowering buds are formed in early June and flowering occurs from July through September. Mature seeds are formed by mid-August. Most spotted knapweed seeds are shed upon maturity; very few overwinter in the seedheads.

• *Spread*

Spotted knapweed populations are largely extended through peripheral enlargement of existing stands (Watson and Renney 1974). Bracts of the flowerheads open when dehydrated, two to three weeks after maturity, and wind or passing animals can flick

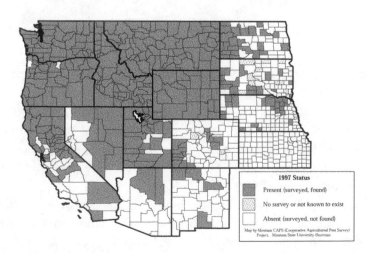

Distribution of spotted knapweed by county in the western United States. Data were collected by surveying various weed authorities in each state.

the loosely held seeds up to one yard from the parent plant. Long distance transport occurs when seeds become attached to passing animals, or by rodents and birds. Spotted knapweed flowerheads also become attached to the undercarriages of vehicles, are transported long distances in mud, and commonly become attached to or drop into shoes. Seeds of spotted knapweed spread through rivers and along watercourses and are transported in crop seed and hay.

• *Germination*

Spotted knapweed seeds germinate over a wide range of environmental conditions. Watson and Renney (1974) reported germination occurs at temperatures ranging from 45° to 93° F (7° to 34° C) with optimum germination at about 66° F (19° C). Seeds germinate in the light and dark, however, maximum germination occurs in an alternating light plus dark environment. They also found that 40% of spotted knapweed seeds germinated when removed from flowerheads at maturity, which improved to 80% 25 days after removal. Optimum seedling emergence occurs when seeds are at the soil surface, with little emergence when seeds are at depths below 2 inches (5 cm). Seedling mortality averages about 12%, but can be as high as 55% under dry conditions.

The longevity of spotted knapweed seeds has important management implications. When spotted knapweed was controlled over a seven year period, the soil seed bank decreased by 95%; however, approximately 98,800 viable seeds per acre (244,000/ha) remained (Davis et al. 1993). Wallander et al. (1995) reported that up to 22% of spotted knapweed seeds can remain viable after passing through the digestive tract of sheep and mule deer. In that study, viability of seeds passed through the digestive tracts decreased to zero three days after ingestion.

Spotted knapweed is able to extend lateral shoots below the soil surface that can form rosettes up to $^{3}/_{16}$ inches (3 cm) from the parent plant (Watson and Renney 1974). Multiple rosettes on a single spotted knapweed root crown are common.

• *Nutrient Value*

Spotted knapweed rosettes and young flower shoots have some nutritive value (Fletcher 1961). Nutrient analysis of spotted knapweed collected before flowering showed neutral detergent fiber at 24.2% to 53.0% (dry wt.), ether extract 3.1% to 9.0%, crude protein 6.2% to 18.2%, total non-structural carbohydrates 11.0% to 27.5%, ash 4.9% to 9.3%, *in vitro* dry matter digestibility 53.2% to 61.8%, and gross energy 4,088 to 4,539 cal/g (Kelsey and Mihalovich 1987). Kelsey and Mihalovich observed crude protein and nonstructural carbohydrates were most concentrated during the spring, becoming more fibrous and lower in proteins and carbohydrates as stems matured during the summer.

• *Allelopathy*

The allelopathic compound, cnicin, has been isolated from spotted knapweed leaves and shoots (Fletcher and Renney 1963). *In vitro* experiments show that cnicin reduces germination of crested wheatgrass (*Agropyron cristatum*), bluebunch wheatgrass, and rough fescue (*Festuca scabrella*) (Kelsey and Locken 1987). In spring and fall soil samples, Kelsey and Locken (1987) found cnicin in soil samples collected in August only, and in levels not toxic in *in vitro* experiments. Later they determined that aggressive resource competition was more important than allelopathy in determining knapweed success (Kelsey and Locken 1989).

• *Disturbance*

Spotted knapweed densities correlate to the degree of soil disturbance: the greater the disturbance, the higher the density (Watson and Renney 1974). Once established, spotted knapweed is able to form solid stands because the combination of seedlings, rosettes, and adults can occupy different soil rooting zones and occupy all available niches (Sheley and Larson 1996). Disturbance allows rapid establishment and spread; however, spotted knapweed is also capable of invading well-managed rangelands (Myers and Berube 1983, Tyser and Key 1988, Lacey et al. 1990).

Management

• *Management Objectives*

Land-use objectives must be developed before spotted knapweed management plans can be designed. Simply killing spotted knapweed, especially in large-scale infestations, is an inadequate objective. A broader objective might be to develop an ecologically healthy plant community that is weed-resistant and meets other land-use objectives such as livestock forage, wildlife habitat, or recreation. Once the desired plant community has been determined, an ecologically-based, integrated weed management strategy can be developed (Sheley et al. 1996b).

• *Inventory and Mapping*

The primary objective of weed inventory and mapping is to identify accurately and delineate lands infested with populations of unwanted plants (Cooksey and Sheley 1996). The goal is to record weed species present, areas infested, weed density, rangeland under threat of invasion, soil and range types, and other site factors pertinent to successful management of spotted knapweed-infested rangeland. Information from an inventory should be incorporated into a map that shows the location, type, and size of knapweed infestations. Accurate mapping is critical in developing an integrated knapweed management program.

• *Prevention*

Preventing spotted knapweed from spreading onto adjacent rangeland is the most cost-effective management strategy. Weed seed dispersal can be limited by not driving vehicles through weed-infested areas when seeds are present, by not grazing livestock in weed-infested areas during flowering and seeding (or by holding animals for seven days before moving to uninfested areas), and by using hay that is free of knapweed seeds (Sheley et al. 1996a). Knapweed spread can also be limited by detecting and eradicating weed introductions early, minimizing soil disturbance, and containing neighboring knapweed infestations (Sheley et al. 1996a). Bounty programs have been used to encourage the early detection of spotted knapweed in Montana.

• *Biological Control*

A variety of natural enemies have been released as biological control agents for spotted knapweed (Rees et al. 1996). Most biological control techniques use insect larvae to damage the root, shoot, leaf or flower, resulting in reduced seed production. Two seedhead-feeding flies (*Urophora affinis* and *U. quadrifasciata*) are well-established on spotted knapweed. These fly species lay their eggs inside the knapweed flower buds in June. Their larvae induce galls in the flowerheads, where they feed and reduce seed production by as much as 50% (Story et al. 1989). Larvae of the moth *Metzneria paucipunctella* feed on the flowers and seeds of spotted knapweed, and may reduce seed production by about 20% (Story et al. 1989).

Four root-mining insect species have been released on spotted knapweed. A root moth (*Agapeta zoegana*) and a root weevil (*Cyphocleonus achates*) have been observed to damage roots of spotted knapweed plants. Two other root moths, *Pelochrista medullana* and *Pterolonche inspersa*, have been released in Montana. Distribution of biological control agents is underway as of 1998.

In addition to insect biological control agents, fungal and bacterial pathogens infect spotted knapweed. *Sclerotinia sclerotiorum* is a common soil fungus native to North America which infects spotted knapweed and can cause wilt and death under some conditions, shifting the competition balance to associated grasses (Ford 1989, Jacobs et al. 1996). Infection of spotted knapweed by the bacteria *Pseudomonas syringae* pv. *syringae* has been reported (Kearing and Nowierski 1997). Pathogens are often associated with insect injury to the plant.

Biological control agents alone have not been successful in controlling spotted knapweed. They have, however, reduced seed production, and are believed to reduce spotted knapweed competitiveness. The most effective use of biological control may be in combination with other control methods, including herbicides, grazing, and revegetation with competitive plants.

• Burning

A single, low-intensity fire does not effectively control spotted knapweed because it is not hot enough to prevent resprouting from crowns or reestablishment from viable seeds in the soil (Renney and Hughes 1969). In northern Washington, a single, low-intensity fire increased the cover and density of this weed without improving the residual, desirable understory species (Sheley and Roché 1982). Similarly, spotted knapweed increased about six-fold within two years after a controlled fire on a forested site in Montana (S. Arno, unpublished data). Fires may create the type of disturbance that promotes the colonization of knapweeds.

Herbicide efficacy on spotted knapweed may increase when applied post-burn (Lacey et al. 1995). Picloram applied at 1 pint/acre (0.28 kg/ha) provided 100% control of this weed two years after post-burn application (Sheley and Roché 1982). Interestingly, residual understory grass cover and density on burned plots increased over unburned plots where picloram had been applied. The authors suggested that burning may have provided a flush of nutrients available to remaining desirable species.

• Cultivation

Cultivation to 7 inches (18 cm) eliminated spotted knapweed and encouraged subsequent vigorous grass growth (Popova 1960). However, Velagala (1996) found single cultivation to 8 inches (20 cm) increased spotted knapweed density over the control, but reduced spotted knapweed biomass one year after treatment. In that study, cultivation enhanced the establishment of intermediate wheatgrass (*Elytriga intermedia*) seedlings.

• Fertilization

Nitrogen fertilizer enhances spotted knapweed because spotted knapweed uses the nitrogen before neighboring desirable species (Popova 1960, Story et al. 1989). In a two-year study, Sheley and Jacobs (1997) found that fertilizer added to a picloram treatment did not increase the effectiveness or longevity of spotted knapweed control, or increase grass yield. However, fertilizer applied at 28+35 lb/acre (31.7+39.6 kg/ha) of nitrogen and phosphorous, respectively, increased grass yield on those sites with a substantial grass understory.

• Grazing

Spotted knapweed can tolerate defoliation; however, severe defoliation will reduce root, crown, and aboveground growth (Kennett et al. 1992). Low to moderate levels of grazing of spotted knapweed by cattle, sheep, and goats have been observed in Montana. Cattle appear to prefer grasses over spotted knapweed (Lacey et al. 1995). Although rosettes of the first year's growth are nutritious and edible, they are difficult for cattle to eat because they are closely appressed to the ground (Popova 1960). Mature knapweed plants are fibrous and coarse, which make them less desirable, and the rough flowering stems along with spines on the floral bracts can irritate the mouths and faces of grazing animals.

Limited grazing duration and repeated grazing by sheep of spotted knapweed when associated grasses were dormant reduced seedlings and rosettes without affecting older and larger plants, and reduced reproduction (Olson et al. 1997). Because of grass dormancy, there was minimal impact on associated grasses. Sheep consumed all one-year-old and a majority of two-year-old plants, suggesting that long-term grazing may provide control of spotted knapweed by limiting recruitment. Angora goats have been successfully used to control spotted knapweed on the Lee Metcalf National Wildlife Refuge in Montana (Gonzalez, personal communication, 1996).

• Hand-Pulling

Persistent and careful hand-pulling can control spotted knapweed (Lacey et al. 1995). Since regrowth can occur from crowns and viable seeds in the soil, entire plants must be removed before they produce seeds each year. Knapweed plants are best pulled when the soil is wet, because the crown is more completely removed. Flowering plants should be transported off site for disposal in a manner that ensures that seeds are not dispersed. Hand-pulling is only practical for new or very small infestations, or along streams and in sensitive areas.

• Herbicides

Picloram, clopyralid, dicamba, and 2,4-D effectively control spotted knapweed on rangeland. In an eight-year study, Davis (1990) found picloram at 0.25 lb/acre (0.28 kg/ha) provided nearly 100% spotted knapweed control for three to five years. The period of control tends to be shorter on coarse soils or as precipitation increases (Lacey et al. 1995). Application timing does not affect control (Duncan and Halstvelt, personal communication, 1997). During the period of control, residual grasses increased by 200% to 700% (Davis 1990). Sheley and Jacobs (1997) found that picloram increased grass yield by an average of 1,350 lb/acre (1,500 kg/ha) while providing nearly 100% spotted knapweed control for two years. Griffith and Lacey (1991) suggested that as site productivity, value of animal unit months, and rate of spotted knapweed spread to new areas increased, economic returns increased relative to picloram treatment costs. In contrast, picloram treatment became less cost-effective as knapweed use by livestock increased.

Clopyralid at 0.24 lb/acre (0.27 kg/ha) and clopyralid at 0.2 lb/acre (0.21 kg/ha) plus 2,4-D at 1 lb/acre (1.12 kg/ha) provided spotted knapweed control similar to picloram when applied during bolt or bud growth stages (Lacey et al. 1995). Percent control declined to below 83% when this herbicide combination was applied at rosette, flowering, and after flowering growth stages.

357

Dicamba at 1 lb/acre (1.12 kg/ha) and 2,4-D at 2 lbs/acre (2.24 kg/ha) provide inconsistent and short-term control of spotted knapweed (Lacey et al. 1995). For long-term control, these herbicides must be applied annually until the seed bank is depleted through attrition. Control with dicamba and 2,4-D was similar to that of clopyralid plus 2,4-D when applied at the bud stage, but lower than that of picloram three years after application.

• Mowing

Long-term effects of mowing on spotted knapweed population densities are unknown. A single mowing at the bud growth stage reduced the number of stems that produced seeds from about 34 (control) to below eight per square yard (Watson and Renney 1974). In that study, mowing at the flower stage reduced the number of flowering stems to one per square foot. Mowing at both the bud and flower stage had the same effect as mowing at either stage alone. Watson and Renney (1974) also found that mowing at the flowering stage or both bud and flowering stages reduced the percent germination of the seeds formed. In a greenhouse study, some spotted knapweed plants produced flowers even through they were clipped monthly from June through September, suggesting the response of plants to mowing will vary with environmental conditions (Kennett et al. 1992).

• Revegetation

In areas where residual plant species are absent, long-term control of spotted knapweed is unlikely, because desirable species are not available to occupy niches opened by the control procedure (James 1992, Sheley et al. 1996b). In these areas, establishing competitive plants is essential for the successful management of spotted knapweed and the restoration of desirable plant communities. Revegetation with aggressive species has been shown to inhibit the reinvasion of knapweeds (Hubbard 1975, Larson and McInnis 1989, Borman et al. 1991). However, there is a lack of data on reliable and affordable revegetation methods.

Typically, revegetation of spotted knapweed-infested rangeland involves late-fall cultivation, followed immediately by a dormant seeding of grass. Grass and knapweed emerge the next spring; however, knapweeds usually emerge first. At that time, glyphosate may be applied before grass seedlings emerge to control knapweed and cheatgrass (*Bromus tectorum*) seedlings. As long as there is adequate spring precipitation, both grass and knapweed seedlings survive. If grass seedlings survive until mid-summer, a reduced rate of 2,4-D or mowing is applied to weaken knapweed plants. Mowing and/or 2,4-D applications may need to be repeated. Grass seedling establishment is enhanced by increasing seeding rate (Velagala et al. 1997). Although the species most effective for revegetation of spotted knapweed infested rangeland depend on site condition, it could be assumed that a healthy, weed-resistant plant community consists of a diverse group of species which occupies most of the niches (Sheley et al. 1996b).

• Integrated Weed Management

The magnitude and complexity of the spotted knapweed problem indicates that successful management requires the adoption of integrated strategies. Integrated knapweed management involves the use of several techniques in a well-planned, coordinated, and ecologically-based strategy to maintain desired plant communities, or shift plant communities to those that are desired (Sheley et al. 1996b). Once land-use objectives and the desired plant community have been identified, inventory and mapping are the first phases of any integrated weed management program. The second phase includes prioritizing, choosing, and strategically implementing management techniques for a particular unit. The third phase is adopting proper grazing management practices as a portion of the integrated program. The integrated weed management program must fit into an overall range management plan.

On areas with a competitive grass stand, proper grazing management ensures that grasses remain strong and vigorous, thereby minimizing knapweed encroachment. Proper grazing is essential to maintaining competitive desirable plants. Sheley et al. (1997) found that a single moderate (<60%) grass defoliation did not enhance the invasion of diffuse knapweed into either a pristine bluebunch wheatgrass/needle-and-thread (*Stipa comata*) or a crested wheatgrass community. Spotted knapweed emergence and growth was enhanced by increasing Idaho fescue (*Festuca idahoensis*) defoliation over 60% (Jacobs and Sheley 1997). Increasing defoliation increased soil moisture availability to spotted knapweed. To minimize weed invasion, grazing systems should alter the season of use, rotate or combine livestock types and pastures, which allows grazed plants to recover before being regrazed, and promote litter accumulation for proper nutrient cycling (Sheley et al. 1996a, Sheley et al. 1997).

Literature Cited

Boggs, K.W. ,and J.M. Story. 1987. The population age structure of spotted knapweed (*Centaurea maculosa*) in Montana. Weed Sci. 35:194-98.

Borman, M.M., W.C. Krueger, and D.E. Johnson. 1991. Effects of established perennial grass on yields of associated annual weeds. J. Range Manage. 44:318-26.

Cooksey, D., and R. Sheley. 1996. Montana noxious weed survey and mapping system. Montana State Univ. Coop. Ext. Serv. MontGuide 96.

Davis, E.S. 1990. Spotted knapweed (*Centaurea maculosa* Lam.) seed longevity, chemical control and seed morphology. M.S. Thesis. Montana State Univ., Bozeman, MT.

Davis, E.S., P.K. Fay, T.K. Chincoine, and C.A. Lacey. 1993. Persistence of spotted knapweed (*Centaurea maculosa*) seed in soil. Weed Sci. 41:57-61.

Fletcher, R.A. 1961. A growth inhibitor found in *Centaurea* spp. M.S. Thesis, Dept. Plant Science, Univ. Of British Columbia, Vancover, B.C.

Fletcher, R.A., and A.J. Renney. 1963. A growth inhibitor found in *Centaurea* spp. Can. J. Plant Sci. 43:475-81.

Forcella, F., and S.J. Harvey. 1980. New and Exotic Weeds of Montana. II: migration and distribution of 100 alien weeds in northwestern USA, 1881-1980. Montana State Univ. Herbarium, Bozeman, MT.

Ford, E.J. 1989. *Sclerotinia* as a mycoherbicide. *In* P.K. Fay and J.R. Lacey (eds.), Knapweed Symp. Proc. Montana State Univ. Bozeman, MT, 182-89.

Griffith, D., and J.R. Lacey. 1991. Economic evaluation of spotted knapweed (*Centaurea maculosa*) control using picloram. J. Range Manage. 44:42-44.

Groh, H. 1944. Canadian weed survey. 2nd Ann. Rep. Can. Dep. Agric.

Hakim, S.E.A. 1979. Range condition on the Threemile game range in western Montana. M.S. Thesis, Univ. of Montana, Missoula, MT.

Harrison, P., and R. Cranston. 1979. An economic evaluation of control methods for diffuse and spotted knapweed in western Canada. Can J. Plant Sci. 59:375-82.

Hubbard, W.A. 1975. Increased range forage production by reseeding and the chemical control of knapweed. J. Range Manage. 28:406-407.

Jacobs, J.S., and R.L. Sheley. 1997. Relationship among Idaho fescue, soil water, and spotted knapweed emergence and growth. J. Range Manage. 50:258-62.

Jacobs, J.S., R.L. Sheley, and B.D. Maxwell. 1996. Effect of *Sclerotinia sclerotiorum* on the interference between bluebunch wheatgrass (*Agropyron spicatum*) and spotted knapweed (*Centaurea maculosa*). Weed Technol. 10:13-21.

James, D. 1992. Some principles and practices of desert revegetation seeding. Arid Lands Newsletter. 32:22-27.

Kearing, S.A., and R.M. Nowierski. 1997. First report of stem and bud blight by *Pseudomonas syringae* pv. *Syringae* on spotted knapweed (*Centaurea maculosa* Lam.). Plant Dis. 81:113.

Kelsey, R.G., and L.J. Locken. 1987. Phytotoxic properties of cnicin, a sequiterpene lactone from *Centaurea maculosa* (spotted knapweed). J. Chem. Ecol. 13(1):19-33.

Kelsey, R.G., and L.J. Locken 1989. *In:* P.K. Fay and J.R. Lacey (eds.), Proc. Knapweed Symp. Montana State Univ., Bozeman, MT, 172-74.

Kelsey, R.G., and R.D. Mihalovich. 1987. Nutrient composition of spotted knapweed (*Centaurea maculosa*). J. Range Manage. 40(3):277-81.

Kennett, G.A., J.R. Lacey, C.A. Butt, K.M. Olson-Rutz, and M.R. Haferkamp. 1992. Effects of defoliation, shading and competition on spotted knapweed and bluebunch wheatgrass. J. Range Manage. 45:363-69.

Lacey, C.A., J.R. Lacey, P.K. Fay, J.M. Story, and D.L. Zamora. 1995. Controlling knapweed in Montana rangeland. Circular 311, Coop. Ext. Serv., Montana State Univ., Bozeman, MT.

Lacey, J.R., C.B. Marlow, and J.R. Lane. 1989. Influence of spotted knapweed (*Centaurea maculosa*) on surface water runoff and sediment yield. Weed Technol. 3:627-31.

Lacey, J., P. Husby, and G. Handl. 1990. Observations on spotted and diffuse knapweed invasion into ungrazed bunchgrass communities in western Montana. Rangelands. 12(1):30-32.

Larson. L.L., and M.L. McInnis. 1989. Impact of grass seedlings on establishment and density of diffuse knapweed and yellow starthistle. Northwest Sci. 63:162-66.

Muller, H., D. Schroeder, and A. Gassmann. 1988. *Agapeta zoegana* (L) (Lepidoptera: Cochylidae), a suitable prospect for biological control of spotted and diffuse knapweed, *Centaurea maculosa* Monnet De La Marck and *Centaurea diffusa* Monnet De La Marck (Compositae) in North America. Can. Ent. 120:109-24.

Myers, J.H., and D.E. Berube. 1983. Diffuse knapweed invasion into rangeland in the dry interior of British Columbia. Can. J. Plant Sci. 63:981-87.

Nolan, D.G., and M.K. Upadhyaya. 1988. Primary seed dormancy in diffuse and spotted knapweed. Can. J. Plant Sci. 68: 775-83.

Olson, B.E., R.T. Wallander, and J.R. Lacey. 1997. Effects of sheep grazing on a spotted knapweed-infested Idaho fescue community. J. Range Manage. 50:386-90.

Popova, A. Ya. 1960. *Centaurea diffusa* Lam., a steppe pasture weed in the Crimea (English Transl.) Bot Zh. (Moscow) 45:1207-13.

Rees, N.E., P.C. Quimby, Jr., G.L. Piper, E.M. Coombs, C.E. Turner, N.R. Spencer, and L.V. Knutson. 1996. Biological control of weeds in the west. West. Soc. of Weed Sci., USDA Agric. Res. Serv., Montana Dept. Agric., Montana State Univ. Bozeman, MT.

Renney, A.J., and E.C. Hughes. 1969. Control of knapweed *Centaurea* species in British Columbia with Tordon herbicides. Down to Earth 24:6-8.

Roché, Jr., B.F., and C.J. Talbott. 1986. The collection history of *Centaurea* found in Washington State. Agric. Res. Center Res. Bull. XB0978. Washington State Univ. Coop. Ext., Pullman, WA.

Sheley, R.L., and J.S. Jacobs. 1997. Response of spotted knapweed and grass to picloram and fertilizer combinations. J. Range Manage. 50:263-67

Sheley, R.L., and L.L. Larson. 1996. Emergence date effects on resource partitioning between diffuse knapweed seedlings. J. Range Manage. 49(3):241-44.

Sheley, R.L., and B.F. Roché, Jr. 1982. Rehabilitation of spotted knapweed infested rangeland in northeastern Washington. Abstr. of papers, W. Soc. Weed Sci., Denver, CO.

Sheley, R., M. Manoukian, and G. Marks. 1996a. Preventing noxious weed invasion. Rangelands 18:100-101.

Sheley, R.L., T.J. Svejcar, and B.D. Maxwell. 1996b. A theoretical framework for developing successional weed management strategies on rangeland. Weed Technol. 10:712-20.

Sheley, R.L., B.E. Olson, and L.L. Larson. 1997. Effect of weed seed rate and grass defoliation level on diffuse knapweed. J. Range Manage. 50:39-43.

Shirman, R. 1981. Seed production and spring seedling establishment of diffuse and spotted knapweed. J. Range Manage. 34:45-47.

Spoon, C.W., H.R. Bowles, and A. Kulla 1983. Noxious weeds on The Lolo National Forest. A situation analysis staff paper. USDA Forest Ser. North. Reg., Missoula, MT.

Story, J.M., K.W. Boggs, W.R. Good, and R.M. Nowierski. 1989. The seed moth, *Metzneria paucipuntella*: Its impact on spotted knapweed seed production and two seedhead flies, *Urophora* spp. *In:* P.K. Fay, and J.R. Lacey (eds.), Proc. Knapweed Symp., Montana State Univ., Bozeman, MT, 172-74.

Tyser, R.W., and C.H. Key. 1988. Spotted knapweed in natural area fescue grasslands: An ecological assessment. Northwest Sci. 62:981-87.

Velagala, R.P. 1996. Using seed rate and plant densities to enhance intermediate wheatgrass establishment in spotted knapweed dominated rangeland. M.S. Thesis. Montana State Univ., Bozeman, MT.

Velagala, R.P., R.L. Sheley, and J.S. Jacobs. 1997. Influence of density on intermediate wheatgrass and spotted knapweed interference. J. Range Manage. 50:523-29.

Wallander, R.T., B.E. Olson, and J.R. Lacey. 1995. Spotted knapweed seed viability after passing through sheep and mule deer. J. Range Manage. 48:145-49.

Watson, A.K., and A.J. Renney. 1974. The biology of Canadian weeds. 6. *Centaurea diffusa* and *C. maculosa*. Can. J. Plant Sci. 54:687-701.

Squarrose Knapweed

Cindy Talbott Roché

If a standard model knapweed were redesigned by an engineering department for the harsh climate of the Great Basin of the Intermountain West, it would resemble squarrose knapweed (*Centaurea virgata* subsp. *squarrosa*). A long-lived perennial with deep roots and a stout crown, squarrose knapweed can endure drought at either temperature extreme, and does not depend on annual seed production or frequent seedling success to maintain its populations. Rosettes may grow slowly for years before flowering, constituting a vegetative equivalent of a seed bank. In addition, seeds remain inside the protective cover of the heads when they fall from the plant. These bur-like heads link seed dispersal with animal movement, thus increasing the probability of landing at a site disturbed by grazing or trampling. An ancient adaptation to animal-based transportation and seasonal migrations of herders and flocks, this dissemination strategy performs just as well today, using vehicle tires for access to disturbed roadsides and off-road tracks.

Identification

Squarrose knapweed is a member of the thistle tribe (Cynareae) in the sunflower family (Asteraceae). Its woody crown consists of one or more clusters of rosette leaves produced on a stout taproot. Several to many profusely branched stems grow 1 to 3 ft (0.3 to 0.9 m) tall from each crown. The stalked, deeply lobed basal leaves often wither by flowering time. Stem leaves are not stalked, and have fewer lobes progressively up the stems. Uppermost leaves are bract-like. Flower heads are borne singly or in pairs at the tips of the branches. The heads are smaller than other knapweeds in the western United States, 1/4- to 3/8-inch (6 to 10 mm) long and 3/16-inch (5 mm) wide, each containing only four to eight rose-purple or pink flowers. On the bracts that surround the flower head, the terminal spine is longer and stouter than are the four to six pairs of lateral spines. The tip of the bract usually spreads outward or curves backward toward the base.

The shapes of the head and bracts are somewhat similar to those of diffuse knapweed, but squarrose knapweed heads are a more slender urn shape. The heads are deciduous at maturity by the development of a well-defined abscission layer at the base of the head. Heads normally contain one to four seeds, but empty seedheads are common. Seeds are $^3/_{16}$- to $^1/_4$-inch (3 to 6 mm) long, including the whitish plume, which varies from one third the length of the seed body to entirely absent. Seeds are golden to dark brown with faint linear stripes and an oblique scar where they detach from the head.

References containing additional photographs, line drawings, or descriptions of squarrose knapweed include Abrams and Ferris (1960), Munz and Keck (1973), Holmgren and Anderson (1976), Welsh et al. (1987), Whitson et al. (1991), Roché and Roché (1991, 1993), and Rosenthal and Rees (1996).

Origin, History, and Distribution

Squarrose knapweed is native to Bulgaria, Lebanon, Anti-Lebanon, Transcaucasia, northern Iraq, Iran, Afghanistan, and Turkestan (Wagenitz 1975). In *The Flora of Turkey*, the U.S. specimens key to subspecies *squarrosa* (Group A) of *Centaurea virgata*, which is found mainly in Inner Anatolia (Wagenitz 1975). Much of the plateau of Inner Anatolia lies between 2,500 and 3,300 feet (750 to 1,000 m) elevation, falling to a large salt lake in the center (Davis 1965). The climate in that region is harsh, characterized by erratic precipitation, temperature extremes, wind, and devastating hail storms (Davis 1965). Precipitation falls predominantly as snow in winter and spring; summers are dry with very low humidity. Winter temperatures are lower than Mediterranean climates and, in summer, temperatures soar during the day and drop drastically at night.

Although squarrose knapweed was first collected in Big Valley, Lassen County, California, in July and August 1950 (Howell 1959), it was actually noted between 1934 and 1937 by the lessee who pastured his sheep at the Kramer ranch (Bellue 1952). By 1950, squarrose knapweed extended about 100 yards from both sides of the road into summer fallow and grain fields (Bellue 1952). This land was also used for overnight camping by several sheep operators who trailed their bands from the Sacramento Valley to higher ranges in the summer and returned them in the fall. By 1952, the largest known infestation covered an area about one mile wide by three miles long (1.6 by 4.8 km), running from the top of Big Valley Mountain down onto the valley floor, and extending over several ranches with stringers along highways to Pittville and into Shasta County (Bellue 1952). Squarrose knapweed was first documented in Siskiyou County around 1969, and by 1988 had spread to approximately 300 acres (120 ha) surrounding Hawkinsville, northwest of Yreka (Ed Hale, Ag. Commissioner, personal communication). By 1988, infestations in the three northern California counties were estimated as follows: Lassen, 800 to 1,000 acres (300 to 400 ha) in the northwestern part of the county (Big Valley and Big Valley Mountain); Shasta, 200 acres (80 ha) in the eastern part of the county; Modoc, 5 acres (2 ha) along roadsides and railroad rights-of-way (Roché and Roché 1989). The 1994 distribution of squarrose knapweed in California was mapped in seven counties: Del Norte, Lassen, Modoc, Plumas, Shasta, Siskiyou and Trinity (Barbe 1994).

363

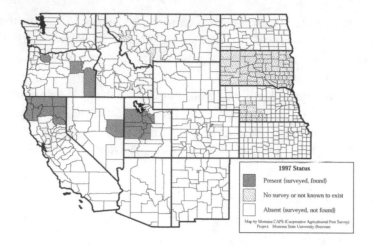

Distribution of squarrose knapweed by county in the western United States. Data were collected by surveying various weed authorities in each state.

In Utah, squarrose knapweed was first collected from Tintic Junction, Juab County, in August 1954 (Howell 1959). Minutes of a meeting held in Nephi, Utah, in November 1954, indicate that it was seen near the grain elevators in 1928, and that a seed company employee in Delta saw it in 1938 (Roché and Roché 1989). In 1954 it grew in varying densities over an area of about five square miles of depleted rangeland west of Tintic in Juab County, Utah (Tingey 1960). By 1960 it had spread along the highway from Eureka for about 7 miles (12 km) into Tooele County and along the foothills into Utah County as far as Elberta along Highways 50 and 6, and along the cattle trails over the Tintic Mountains (Tingey 1960). It had also appeared along Highways 50 and 6 from Tintic to Jericho, spreading out through the valley for about 30 miles (50 km). Scattered plants infested 400 to 500 acres (160 to 200 ha) east of the Star Ranch in northeastern Juab County, and a small patch grew along Highway 91 south of Santaquin (Tingey 1960). In 1989 the core of the Utah squarrose knapweed population was estimated at 10,000 acres (4,000 ha) in five counties: Juab, Tooele, Millard, Utah and Sanpete. Scattered plants had been found over 37,000 acres (15,000 ha) of Bureau of Land Management land west of Tintic Junction (Roché and Roché 1989).

In Oregon, 1988 brought the first report of squarrose knapweed, a 200-acre (80-ha) infestation within a total area of 800 acres (300 ha) near Long Creek, Grant County (Roché and Roché 1989). Control efforts reduced it to 25 acres (10 ha) by 1993 (Dennis Isaacson, Oregon Dept. Agric, personal communication). In 1991, an infestation less than 0.5 acre (1 ha) in size was found between Highway 20 and the Malheur River in Malheur County 38 miles (63 km) west of Vale (Roché 1992). A single squarrose knapweed plant was found intermingled with diffuse knapweed in Clackamas County in 1992 on the Clackamas Ranger District, Mt. Hood National Forest (D. Isaacson, personal communication).

It has been seen in White Pine County, Nevada (Pat Fosse 1998, personal communication). Squarrose knapweed has not been reported in Colorado, Idaho, Montana, or Washington.

Potential Invasion

In Utah, most squarrose knapweed grows on big sagebrush (*Artemisia tridentata*)-bunchgrass rangeland, but it also extends up into the juniper-dominated rangeland and down into the salt desert shrub range, particularly in sandy or gravelly washes. However, it also moves out of the washes into healthy black sagebrush types, starting on south-facing slopes (Pat Fosse 1998, personal communication). New infestations often establish on dry, rocky, south-facing slopes which naturally have less vegetative cover. Squarrose knapweed competes with crested wheatgrass (*Agropyron cristatum*) in rangeland seedings and has begun to invade fairly pristine mountain brush types (Pat Fosse 1998, personal communication). In northern California, squarrose knapweed grows on dry rocky sites of degraded juniper-shrub savanna with scattered western juniper (*Juniperus occidentalis*) and ponderosa pine (*Pinus ponderosa*) and chaparral-type understory (Roché and Burrill 1992). In Oregon, it has invaded juniper-Idaho fescue (*Festuca idahoensis*) rangeland and big sagebrush-bunchgrass rangeland with cheatgrass (*Bromus tectorum*).

In the Great Basin and intermountain foothills, the sagebrush and juniper range types appear to be the most susceptible to invasion by squarrose knapweed. Although squarrose knapweed is credited with crowding out most desirable vegetation in Utah (Rosenthal and Rees 1996), by the time squarrose knapweed was discovered, much of the rangeland in these vegetation types in Juab and Utah counties was "greatly misused in the past and in poor condition" (Stoddart 1945). The sagebrush type was "made up chiefly of common sagebrush with a little rabbitbrush (*Chrysothamnus*) and mixed grasses, mostly cheatgrass" (Stoddart 1945). Juniper rangelands were characterized by *Juniperus utahensis* with a sparse understory of sagebrush, Russian thistle (*Salsola kali*), and cheatgrass. In other areas, perennial vegetation had been removed entirely, including the sagebrush or juniper overstory. These included areas that had been plowed for dryland wheat production, then abandoned after two or three years; cheatgrass areas that were repeatedly burned; and sheep trails traveled by 100,000 to 150,000 sheep twice a year (Stoddart 1945). Since that time, many of the more productive sites (deeper soils) have been seeded with crested wheatgrass. Squarrose knapweed has invaded or persisted in some crested wheatgrass seedings.

Wildfires create the ideal conditions for rapid expansion. Squarrose knapweed's fire and drought tolerance, excellent seed dispersal, and rapid response to soil resources released by fire nearly guarantee spread into burned closed-canopy juniper sites with little understory.

Impacts

A long-lived perennial, squarrose knapweed appears better adapted than diffuse knapweed to the harsh climate of the shrub steppe rangeland in the Great Basin and high desert of eastern Oregon. Like the other knapweeds, squarrose knapweed competes with forage species on rangeland. In the rosette stage, its palatability and nutritive value may be equivalent to diffuse or spotted knapweed. As in the other knapweeds, both are much lower in reproductive-stage plants. Squarrose knapweed rosettes are grazed by sheep during late winter and spring (Roché et al. 1992). Cattle also graze rosettes, especially in years of below-normal rainfall when cheatgrass and other annual forage is in short supply.

Biology and Ecology

Squarrose knapweed is a long-lived perennial (Abrams and Ferris 1960, Wagenitz 1975). Although at least one source (Keffer 1978) lists it as a biennial, this is contradicted by field observations. Small rosettes of squarrose knapweed in Oregon and Utah had very large taproots and successive rows of weathered leaf bases, indicating that they were not seedlings. Under unfavorable conditions, plants appear to remain as taprooted rosettes for years before developing flowering stems. Bare, weathered flower stalks of the previous year often persist at the time of the next flowering. Although the potential lifespan of individual plants is not known, there is no indication that plants die after producing seed. Squarrose knapweed crowns branch under the soil surface, forming multiple rosettes from a single taproot. This morphology appears to be an adaptation to harsh growing conditions such as cold temperatures and drought. Favorable conditions for seedling establishment at relatively infrequent intervals, perhaps six or more years apart, appear adequate for maintenance and spread of squarrose knapweed populations (S. Dewey 1998, personal communication).

Squarrose knapweed flowers from June to August, then disperses seeds from August through the winter. The seed dissemination habit of squarrose knapweed is unique among adventive *Centaurea* species in the western United States. Historically, most movement of squarrose knapweed in the western United States has been associated with sheep (Bellue 1952, Tingey 1960, Roché and Roché 1989). It is ideally suited to this mode of transport, because the recurved spines of seedhead bracts, like those of burdock (*Arctium*) or cocklebur (*Xanthium*), perfectly complement the wool of sheep in a manner analogous to a Velcro® fastener. At fruiting time, the heads are closed (retaining the seeds) and deciduous; consequently, seeds are readily spread by animal wool, hair or fur (Wagenitz 1972). No reference has been found regarding the initial introduction of this species, but it is possible that seed was carried in wool, either on sheep or woolen products. Squarrose knapweed was among the 526 species introduced in France by seed cleaned out of fleece at Juvénal Gate, where imported wool was washed for 200 years, starting in 1686 (Thellung 1912).

At seed maturity, attachment of heads to the stems weakens along an abscission layer at the base of the head, so that slight motion of the plant causes heads to drop. Although many heads fall near the base of the parent plant, not all the heads drop during late summer and fall. Heads remaining on plants into the following spring greatly extend the distribution period. Dispersal by vehicles and trains appears increasingly important, judging by the expansion of squarrose knapweed along roads, railroads, and off-road vehicle trails.

Management

Because early detection is the first line of defense against new invaders, squarrose knapweed has a distinct advantage over survey crews. Since it so closely resembles diffuse knapweed, which is already widespread, careful observation is necessary to detect it. Thus, squarrose knapweed is probably more abundant in eastern Oregon, southern Idaho, and Nevada than has been reported. Priority survey sites should include livestock trails, recreational vehicle routes, and locations linked by current commerce with Utah and northern California.

• *Manual and Mechanical*

Squarrose knapweed may be eradicated when found as small infestations or individual plants by grubbing or digging the roots with a shovel. Stout taproots resprout when broken off, making hand-pulling ineffective. Tingey (1960) reported that squarrose knapweed forms adventitious buds well below the root crown. Tillage may be used in accessible pastures and fields. When cultivation and grubbing are used for control, the root should be cut at least 8 inches (20 cm) below the soil surface, to minimize forming new shoots. When dislodged by a single disking, rosettes continue to grow if they are attached to a piece of root that contacts the soil, requiring repeated cultivation.

• *Chemical*

Several herbicides are registered for control of knapweeds on rangeland, with varying degrees of residual activity for control of later germinants. For residual herbicides, recommendations are similar to those for spotted knapweed: 0.25 to 0.5 lb active ingredient (ai) per acre (0.28 to 0.56 kg/ha) of picloram or 0.5 lb ai/ac (0.56 kg/ha) of clopyralid. Dicamba at 2 lb ai/ac (2.24 kg ai/ha) was also effective, but 2,4-D did not give satisfactory control of squarrose knapweed in Utah (S. Dewey 1998, personal communication). However, in special situations in which a nonresidual chemical is needed, 2,4-D may be preferred, especially if seed prevention is the objective. Under Utah conditions (6 to 10 inches [15 to 25 cm] annual precipitation, occurring in winter and early spring, and high light intensity in summer), effectiveness of picloram appears to be maximized by application of the chemical directly to newly resprouted rosette leaves and surrounding bare soil, at a time when seasonal precipitation incorporates the herbicide into the soil. Under these conditions, which often occur

367

the first fall after a wildfire, essentially complete control has been achieved at the lower rate (within the recommended range) (S. Dewey 1998, personal communication). Effectiveness is hindered by factors that interfere with foliar or root uptake of the herbicide, including interception of picloram by non-target species and long, sunny periods prior to rainfall. Specific recommendations vary by site. Follow label directions, consult the state Extension weed specialist, and in Idaho, Oregon and Washington, refer to the current *Pacific Northwest Weed Control Handbook* (William et al. 1997).

• Biological Control

Six insects introduced for biological control of diffuse and spotted knapweed have accepted squarrose knapweed as a host plant. The gall-forming flies *Urophora affinis* and *U. quadrifasciata* are widespread in Utah and California, and have reduced seed production levels (Roché and Roché 1989). At 13 sites in Utah, 40% of the seedheads were attacked in 1996, reducing seed production by 70% in attacked heads (Ted Evans 1998, personal communication). *Bangasternus fausti*, a seed weevil, *Sphenoptera jugoslavica*, a root-feeding beetle, and *Cyphocleonus achates*, a root-feeding weevil, were established by the same date in Utah (Ted Evans 1998, personal communication). *Chaetorellia acrolophi*, a seedhead fly, reportedly feeds on squarrose knapweed, but so far is established only on other knapweeds in Montana and Oregon (Rees et al. 1996). In California, *Cyphocleonus achates* and *Bangasternus fausti* were released in 1995 and 1996, respectively, but were not recovered as of December 1996 (Woods et al. 1996). *Pterolonche inspersa*, *Agapeta zoegana*, a root-feeding moth, and *Larinus minutus*, a seedhead weevil, have been released but not recovered in Utah (Ted Evans 1998, personal communication).

• Integrated Management

The Bureau of Land Management (BLM) integrated management plan for squarrose knapweed in Utah includes seven goals: prevention and detection; education and awareness; inventory; planning; integrated weed control; coordination; and monitoring, evaluation, and research (Pat Fosse 1998, personal communication). An inventory of the infestation is helpful in planning for containment and local eradications. Perimeters, outlying or small infestations, and transportation routes receive priority for herbicide treatment. In eradication areas and along transportation routes, surviving plants and new seedlings should be controlled annually until no additional plants can be found. It is not known how long seeds remain viable in the soil, but seeds remaining in heads that become buried probably last longer than unprotected seeds. While long persistence is not indicated in either case (Tingey 1960), the problem of reintroduction requires constant vigilance.

Equipment used in construction and fire projects should be thoroughly cleaned before being moved from infested to weed-free areas. Preventing seed production by squarrose knapweed along transportation corridors and livestock trailing routes, in campgrounds, parking areas, and gravel pits or stockpiles, is far more cost-effective than attempting to remove weed seeds from every vehicle, load of gravel, and head of livestock moving to uninfested areas. Grazing plans should consider the potential for movement of squarrose knapweed seeds that cling to animal coats or lodge in mud on hooves. As sheep graze squarrose knapweed rosettes in late winter and early spring, they pick up knapweed seedheads in their wool, particularly on their heads (Roché et al. 1992). Knapweed heads in the wool are removed when the sheep are sheared, and recently-sheared sheep should gather fewer knapweed heads.

Large rangeland infestations may be managed with a combined system, using herbicides, revegetation with perennial forage species, and improved grazing management. Wildfires in Utah have given land managers the opportunity to rehabilitate squarrose knapweed-infested rangeland. By fall, squarrose knapweed rosettes have resprouted from the roots, constituting essentially the only green vegetation, surrounded by bare soil, in the areas scorched by intense summer fires. A window for seeding perennial grass is opened because the fire has consumed most of the surface seedbank of weedy species. Picloram or clopyralid is applied to kill the knapweed, and the desired grass species are broadcast-seeded two to four weeks later. Then the area is chained to incorporate the seed in the surface soil. Seeding is delayed to minimize contact with the herbicide and to prevent fall germination. The objective is to take advantage of winter freezing and thawing to further incorporate the seed and avoid frost-heaving damage to newly germinated seedlings. Early spring germination maximizes seedling establishment prior to summer drought.

Similar results may be achieved with prescribed fire, if sufficient fuel is present to carry the fire for relatively uniform site preparation. In accessible areas, a rangeland drill is used in place of chaining to incorporate the seed. In areas where shrub removal is not possible or desired, animal trampling following broadcast seeding may be applicable. Herbicides are more effective when followed by sufficient precipitation to carry them into the soil, either in the fall or early spring (S. Dewey 1998, personal communication).

Literature Cited

Abrams, L., and R.S. Ferris. 1960. Illustrated Flora of the Pacific States. Vol. IV. Stanford Univ. Press, Stanford, CA.

Barbe, D. 1994. Noxious Weeds of California I. Distribution Maps. State of Calif. Dept. of Food and Agric., Sacramento, CA.

Bellue, M.K. 1952. Virgate star thistle, *Centaurea virgata* var. *squarrosa* (Willd.) Boiss., *in* California. Calif. Dept. Agric. Bull. 41:61-63.

Davis, P.H. (ed.). 1965. Flora of Turkey and the East Aegean Islands, Vol. I. Edinburgh Univ. Press, Edinburgh, 1-7.

Holmgren, A.H., and B.A. Anderson. 1976. Weeds of Utah. Utah Agric Expt. Sta. Special Rept. 21. Utah State Univ., Logan, UT.

Howell, J.T. 1959. Distributional data on weedy thistles in western North America. Leafl. West. Bot. 9:17-32.

Keffer, M. 1978. Squarrose knapweed (*Centaurea squarrosa*). Calif. Dept. Food Agric. Detection Manual 6:33.

Munz, P.A., and D.D. Keck. 1973. A California Flora. Univ. of Calif. Press, Berkeley, CA.

Rees, N.E., P.C. Quimby, Jr., G.L. Piper, E.M. Coombs, C.E. Turner, N.R. Spencer, and L.V. Knutson (eds.). 1996. Biological control of weeds in the West. Western Soc. Weed Sci., USDA Agric. Res. Serv., Montana Dept. Agric., and Montana State Univ., Bozeman, MT.

Roché, B.F., Jr., and C.T. Roché. 1991. Identification, introduction, distribution, ecology and economics of *Centaurea* species. *In:* L.F. James, J.O. Evans, M.H. Ralphs and R.D. Child (eds.), Noxious Range Weeds. Westview Press, Boulder, CO, 274-91.

Roché, B.F., Jr., C. Roché, and A. Rasmussen. 1991. Squarrose knapweed: clinging to the wool. Knapweed Newsletter. 5(3):2.

Roché, C. 1989. Lesser known knapweeds: potential threats? *In:* Proc. Knapweed Symposium, April 4-5, 1989, Soil and Plant Science Dept. and Ext. Ser., Montana State Univ., Bozeman, MT, 47-53.

Roché, C. 1992a. *Centaurea virgata* Lam. subsp. *squarrosa* Gugl. Noteworthy Collections. Madroño 39(3):242-43.

Roché, C. 1992b. Knapweeds on the Move: Short-fringed Knapweed and Squarrose Knapweed. Knapweed Newsletter 6(1):1.

Roché, C., and L.C. Burrill. 1992. Squarrose Knapweed (*Centaurea virgata* ssp. *squarrosa*). Pacific Northwest Coop. Ext. Bull. PNW422, Oregon State Univ., Corvallis, OR.

Roché, C.T., and B.F. Roché, Jr. 1989. Introductory notes on squarrose knapweed (*Centaurea virgata* Lam. ssp. *squarrosa* Gugl.). Northwest Sci. 63:246-52.

Roché, C.T. and B.F. Roché, Jr. 1993. Identification of Knapweeds and Starthistles in the Pacific Northwest. Pac. Northwest Ext. Pub. PNW 432. Washington State Univ, Pullman, WA.

Roché, C.T., B.F. Roché, Jr., and G.A. Rasmussen. 1992. Dispersal of squarrose knapweed (*Centaurea virgata* ssp. *squarrosa*) capitula by sheep on rangeland in Juab County, Utah. Great Basin Naturalist 52(2):185-88.

Rosenthal, S.S., and N.E. Rees. 1996. Squarrose knapweed. *In:* N.E. Rees, P.C. Quimby, Jr., G.L. Piper, E.M. Coombs, C.E. Turner, N.R. Spencer, and L.V. Knutson (eds.), Biological control of weeds in the West. West. Soc. Weed Sci., USDA Agric. Res. Serv., Montana Dept. Agric., and Montana State Univ., Bozeman, MT.

Stoddart, L.A. 1945. Range lands of Utah county and their utilization. Bull. 317, Agric. Expt. Sta. Utah State Agric. College, Logan, UT.

Thellung, A. 1912. La flore adventice de Montpellier. Memoires de la Société Naturelles et Mathématiques de Cherbourg 38:57-728.

Tingey, D.C. 1960. Control of squarrose knapweed. Utah State Univ. Exp. Sta. Bull. No. 432.

Wagenitz, G. 1972. Beiträge zur kenntnis der gattung *Centaurea* L. Willdenowia 6:479-507.

Wagenitz G. 1975. *Centaurea* L. *In* P.H. Davis (ed.), Flora of Turkey and the East Aegean Islands. Vol. 5, Univ. Press, Edinburgh, 465-85.

Welsh, S.L., N.D. Atwood, S. Goodrich, and L.C. Higgins (eds.), 1987. A Utah Flora. Great Basin Naturalist Memoirs No. 9, Brigham Young University, Provo, UT.

Whitson, T.D., L.C. Burrill, S.A. Dewey, D.W. Cudney, B.E. Nelson, R.D. Lee, and R. Parker. 1991. Weeds of the West. Western Soc. Weed Sci. and West. U.S. Land Grant Universities Coop. Ext. Serv. 96-97.

William, R.D., D. Ball, T.L. Miller, R. Parker, J.P. Yenish, R.H. Callihan, C. Eberlein, G.A. Lee, and D.W. Morishita. 1997. Pacific Northwest Weed Control Handbook. Oregon St. Univ., Corvallis, OR.

Woods, D.M., B. Villegas, and D.B. Joley. 1996. Attempts to establish insects for the biological control of squarrose knapweed, *Centaurea squarrosa. In:* D. M. Woods (ed.), Biological Control Program Annual Summary, 1996. Calif. Dept. Food and Agric., Div. of Plant Industry, Sacramento, CA.

St. Johnswort

Gary L. Piper

St. Johnswort (*Hypericum perforatum*), also commonly referred to as goatweed or Klamath weed, is a member of the family Clusiaceae (= Guttifereae, Hypericaceae). This Eurasian plant has been accidentally or intentionally introduced throughout the world, and has become an important pest in many temperate regions. In the western United States it has become a pernicious weed of grazing lands, where it displaces desirable wildlife and livestock forage plants and may poison certain livestock that feed upon it.

Identification

St. Johnswort is a multistemmed, herbaceous perennial growing to a height of 1 to 4 ft (0.3 to 1.2 m). The smooth, reddish, upright flowering stems are characterized by two opposite longitudinal ridges and black glands. They are semi-woody near the base and much branched near the top. A number of non-flowering, prostrate, densely-leaved stems up to 1 ft (0.3 m) in length also develop from the root crown. Leaves are sessile, opposite, entire, linear-oblong with inrolled edges, and $^3/_8$ to 1 inch (10 to 25 mm) long. They are dark green above and light green below, and dotted with translucent and black glands (Fields et al. 1990). The inflorescence is an open, flat-topped, terminal cyme. Flowers are $^1/_2$ to 1 inch (1 to 2.5 cm) in diameter, and have five golden yellow petals bearing black glands along the margins, and numerous stamens. The fruit is an egg-shaped, three-valved, $^3/_{16}$ to $^7/_{16}$ inch (4 to 10 mm) long capsule that bursts at maturity. Upon opening, the capsule releases numerous seeds. They are $^3/_{64}$ inch (1 mm) long, somewhat cylindrical and slightly pointed at the ends, coarsely pitted, and shiny dark brown. The root system consists of a woody taproot that may penetrate the soil to a depth of 4 to 5 feet (1.2 to 1.5 m), and lateral roots that grow 2 to 3 inches (5 to 8 cm) beneath the soil surface and may extend 3 feet (0.9 m) (Tisdale et al. 1959). The lateral roots produce vegetative buds from which new crowns develop (Campbell and Delfosse 1984). St. Johnswort seedlings at the two- to four-leaf stage have rounded, smooth, sessile leaves. The black glands become evident at the six-leaf stage (Crompton et al. 1988). The seedlings are very small, grow slowly (Campbell 1985), and compete poorly with other vegetation (Moore and Cashmore 1942). Seedlings do not emerge from seeds found at soil depths greater than ³/₁₆ inch (5 mm) (Campbell 1985).

Origin, History, and Distribution

St. Johnswort is native to Europe, North Africa, and parts of Asia, including northern India, China, and Japan (Harris and Peschken 1971). It has been introduced into most continents, and is listed as a weed in 21 countries, including Australia, Canada, Chile, New Zealand, South Africa, and the United States (Goeden 1978, Holm et al. 1979). The major reason for the plant's introduction into foreign countries was its cultivation for medicinal purposes (Mitich 1994) or ornamental value (Campbell and Delfosse 1984). In Europe, St. Johnswort has long been valued as an herb, and is still widely used by folk medicine practitioners (Coon 1974, LeStrange 1977). Extracts of the plant have been administered for the treatment of mania, hysteria, hypochondriasis, depression, dysentery, jaundice, menorrhagia, and a host of other afflictions (Millspaugh 1974, Mitich 1994).

In the United States, St. Johnswort is distributed from Minnesota south to central Texas and east to the Atlantic Coast. It also occurs in northern California, Idaho, Oregon, Montana, and Washington (Reed 1970). The plant was first reported in the United States in 1793 in Pennsylvania (Sampson and Parker 1930). According to Campbell and Delfosse (1984), it was carried to Oregon between 1840 and 1850, then subsequently to California in the early 1900s. By 1945, the weed infested about 2.5 million acres (1 million ha) in California, and about 1.2 million acres (500,000 ha) in Idaho, Oregon, and Washington (Tisdale 1976). Following a successful biological control agent introduction program begun more than 50 years ago, infestation levels have been reduced by as much as 99% in many areas of the western United States (McCaffrey et al. 1995). However, in other areas, even where the

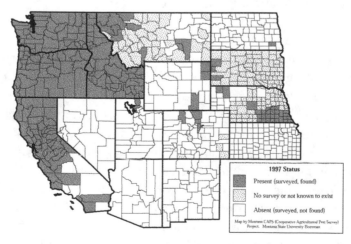

1997 Status
Present (surveyed, found)
No survey or not known to exist
Absent (surveyed, not found)
Map by Montana CAPS (Cooperative Agricultural Pest Survey) Project. Montana State University Bozeman

Distribution of St. Johnswort by county in the western United States. Data were collected by surveying various weed authorities in each state.

biological control organisms are present, many St. Johnswort populations are still increasing in size, while others have remained relatively static. Unfortunately, a reliable published estimate of the amount of land presently infested by St. Johnswort is not available.

Potential Invasion

St. Johnswort can become established in either highly degraded or pristine rangelands. In the western United States, the weed characteristically occupies low elevation sites where the annual precipitation is between 15 and 30 inches (38 and 76 cm) (Tisdale et al. 1959). Any soil disturbance, as typified by overgrazing of St. Johnswort-infested rangeland and pasture, favors its continued survival because of the reduction or elimination of competitive plant species, especially annual grasses. The plant's deeply penetrating taproot makes it a formidable competitor with other vegetation, particularly shallow-rooted species, for nutrients and water. St. Johnswort prefers sunny exposures (south- and southeast-facing slopes) and well-drained gravelly or sandy soils, but it also grows in heavier soils. It does poorly where the soil remains continually moist or in heavily shaded areas.

Impacts

• *Ecological and Environmental*

St. Johnswort can be a serious problem in rangeland and pastures where dense stands, through their displacement of valued forage and indigenous plant species, can greatly depreciate livestock and wildlife carrying capacities, and endanger the biological diversity of these grazing lands. The weed also infests forest clearings, transportation rights-of-way, and neglected lands, frequently to the extent that management in such locations becomes necessary. It can also become a problem in orchards, bush fruits, and conifer plantations (Crompton et al. 1988).

• *Economic*

St. Johnswort is usually avoided by livestock, and is consumed only when desirable forage is scarce (Kingsbury 1964). However, if ingested in sufficient quantity, the plant can be toxic to livestock, through the occurrence of the photodynamic fluorescent pigment hypericin, a naphthodianthrone derivative (Jensen et al. 1995). Hypericin, produced by glands found in the stems, leaves, flowers, stamens, and fruits, causes severe dermatitis in animals upon their exposure to intense sunlight (Giese 1980). St. Johnswort is considered a primary photosensitizer, meaning that the phototoxic substance comes directly from the plant, is unaltered by ingestion or digestion, not removed by the liver, and absorbed into the bloodstream (Kingsbury 1964). All growth stages are toxic, including dried plants in hay (James et al. 1980). The greatest toxic effects are expressed during flowering (Dodd 1920). The minimum toxic dose of foliage for cattle and sheep is about 1% and 4% of their body weight, respectively (Jessup 1970).

Classical symptoms of St. Johnswort poisoning or hypericism have been reported in cattle, horses, sheep, and goats, with goats being most resistant to the toxicant (Kingsbury 1964). Symptoms usually become detectable 2 to 21 days following ingestion of the weed (James et al. 1980). Light-haired or unpigmented skin areas such as the mouth, nose, ears, and hooves are most sensitive to the chemical. When hypericin is in the circulatory system, the permeability of the cell membrane is altered in these unpigmented skin areas. This cell damage leads to blistering, and causes such intensive itching that the animals often rub the affected areas raw. Livestock suffering from hypericism usually lose weight, are difficult to manage, and possess reduced market value. Once consumption of the weed is halted, affected animals usually fully recover within three to six weeks (Dodd 1920). Livestock rarely die directly from St. Johnswort ingestion; however, the aftereffects of hypericism, such as blindness or swelling and soreness of the mouth, may prevent affected animals from foraging and drinking, and thereby contribute to death by dehydration and/or starvation (Kingsbury 1964, Giese 1980).

Biology and Ecology

St. Johnswort seeds germinate during the warm summer months, and seedlings may require several years to reach reproductive maturity (Parsons 1981). The seedlings are not strong competitors with other vegetation for light, nutrients, space, and moisture, and may exhibit high mortality under stress conditions (Campbell and Delfosse 1984). A mature plant may consist of one or more woody crowns to which are attached vertical and lateral roots (Clark 1953). Large crowns may produce up to 30 erect, reddish, flowering stems annually (Campbell and Delfosse 1984). Crowns are usually well spaced, ranging from 12 to 37 per m^2 (Clark 1953). During spring and autumn, new crowns may be produced from lateral root buds. These crowns may later become independent through the decay of connecting lateral roots.

Following senescence of the aerial stems in mid- to late summer, fall regrowth occurs in response to increased soil moisture. This regrowth consists of numerous branched, more or less prostrate, spindly, non-flowering stems. These stems continue to grow during the winter and early spring months. They do not persist, dying by the end of May (Wilson 1943). Some plants begin to bolt during early March. By April, most two-year-old or older plants have produced floral shoots. The peak flowering period is reached in mid- to late June, with flowers being present from May through September. Flowering duration is determined by soil moisture (Wilson 1943). By midsummer most of the flowers have withered, and the developing seed capsules become moist, green, and sticky (Campbell and Delfosse 1984). By late autumn, each capsule contains 400 to 500 ripe seeds (Crompton et al. 1988). An average-sized plant produces between 15,000 and 33,000 seeds annually (Tisdale et al. 1959, Parsons 1981, Crompton et al. 1988). Seeds are disseminated short distances by the wind (Tisdale et al. 1959). Long-distance seed dispersal mechanisms include adherence to animals (facilitated by a gelatinous seed coat), animal ingestion and

375

subsequent deposition in feces, water movement, and through the activities of humans (Campbell and Delfosse 1984). The seeds apparently require a four- to six-month after-ripening period before germination occurs (Cashmore 1939, Campbell 1985). Longevity of viable seeds in the soil may range from six (Clark 1953) to 10 years (Bellue 1945). In late summer, the flowering stems turn reddish-brown and die, frequently remaining attached to the root crown for one or more years.

Management

• *Prevention*

A preventive weed strategy should be developed first. Prevention can be viewed as the process of forestalling the contamination of an area by a noxious weed (Piper 1992). Prevention includes early-detection procedures and measures taken to alleviate the conditions that cause and foster the spread of undesirable plants. Even though St. Johnswort has a widespread distribution, there are still numerous areas where it has not yet invaded, and other areas where it occurs at very low densities. Preventing the further introduction and spread of the plant to uninfested areas is much easier, more environmentally desirable, and more cost-effective than is the subsequent management of large-scale infestations.

Prevention requires education of resource managers and users concerning 1) the identification of invasive species such as St. Johnswort; 2) the detrimental environmental impacts associated with their spread; 3) the importance of curtailment of weed development in nearby wastelands or along transportation and utility line corridors; 4) the need for inspection of and/or cleaning of machinery and vehicles prior to their movement from weed-infested to uninfested sites; and 5) the revegetation of disturbed soils with site-appropriate replacement plant species to inhibit St. Johnswort entry. Upon discovery, pioneer plants should be eliminated by hand extraction or herbicides.

• *Cultural Control*

Practices that encourage residual competitive species or that establish competitive, well-adapted, palatable grasses or legumes at a St. Johnswort-infested site should also be part of any management effort. If the weed is suppressed by one or more methods but its ecological niche remains unfilled, reinvasion by St. Johnswort or invasion by other undesirable species will eventually occur (Campbell and Delfosse 1984, Campbell and McCaffrey 1991). It is imperative to maintain competitive, closed-canopy, beneficial plant communities by using site-specific range management recommendations, and implementing ecologically-based grazing management programs. In Australia, Campbell and Delfosse (1984) noted that a program of cultivation, fertilization, and reseeding with a competitive perennial grass controlled St. Johnswort populations in pastures and arable rangeland over a two- to five-year period.

The choice of plant species to be seeded should reflect site conditions, management, and future use. The Natural Resources Conservation Service, Cooperative Extension Service, or other technical assistance group personnel can evaluate your management unit and recommend appropriate plant species for revegetation purposes. Before embarking upon a large-scale vegetation enhancement program, small-scale plantings, evaluated for several years, may be the best means to determine which plant species are most competitive with St. Johnswort under prevailing climatic and land use conditions.

St. Johnswort does not grow well in dense shade, and therefore afforestation has been used as a control measure in some areas. Unfortunately, this method does not completely eliminate the weed (Wilson 1943). The plant will grow along roadways or waterways in forested areas, and when thinning or logging operations are conducted, will often invade these open, disturbed sites (Campbell and Delfosse 1984).

• Physical Management

Several physical methods have been used for St. Johnswort management. Hand-pulling or digging of young, isolated plants can often be effective, but is not considered practical for large populations of established, deeply rooted plants. The lateral roots of older plants are frequently left in the soil during extraction and can give rise to new plants (Sampson and Parker 1930). It is highly unlikely that a single pulling will result in complete removal of the weed from a site, particularly if the infestation is comprised of different-aged plants. Pulled or dug plants should be removed from the area and destroyed to prevent possible vegetative regrowth and/or seed dissemination. St. Johnswort can often be effectively managed by repeated tillage when found in intensively cropped situations (Crompton et al. 1988).

Cutting or mowing are ineffective as management methods; however, they may help diminish spread of the plant if performed prior to seed formation. Two or more cuttings may be necessary during the growing season. These techniques may not be feasible in many sites occupied by St. Johnswort because of inaccessible terrain.

Burning actually has increased the density and vigor of St. Johnswort stands (Sampson and Parker 1930, Campbell and Delfosse 1984). This increase may be attributable to increased germination of seeds that have been exposed to intense heat (Sampson and Parker 1930), and to vegetative regrowth from the fire-resistant root crowns and lateral root buds (Crompton et al. 1988). Repeated burning also diminishes decaying organic matter and potential nutrient levels in the soil, and may lower population densities of fire-intolerant competing plants, thus favoring the continued domination of a site by St. Johnswort and other noxious weeds.

• Chemical Management

Chemical suppression of small infestations of St. Johnswort can readily be achieved by using various herbicides. In noncropland sites, foliar applications of 2,4-D at a 2 lb acid equivalent per acre (ae/ac) (2.24 kg ae/ha) rate, will kill the plant in the seedling and pre-flowering stages (William et al. 1996). Ester formulations of phenoxy herbicides may be more effective than amine or salt formulations (Crompton et al. 1988). Spring applications of picloram (0.125 to 1.5 lb active ingredient per acre [ai/ac] [0.14 to1.68 kg ai/ha]) or glyphosate (0.187 to 0.375 lb ae/ac [0.21 to 0.42 kg ae/ha]) are recommended for the suppression of the weed in pasture, rangeland, and noncropland sites. Postemergence applications of metsulfuron at a rate of 0.04 lb ai/ac (0.045 kg ai/ha) also effectively control the plant (William et al. 1996). Herbicides are often too costly to be of practical value as a management tool for extensive infestations of St. Johnswort (James et al. 1980, Campbell and Delfosse 1984). Herbicides should be employed only in situations where chemical control is economically justifiable, and preferably in concert with other management practices. If herbicides are used, the best suppression is obtained when the timing of the application is synchronized with the susceptible life stage(s) of the weed. Repeated applications are often required to achieve adequate management (Campbell and Delfosse 1984).

• Biological Control

Biological control of St. Johnswort was initiated in Australia, where the invasive plant had become a serious pest by the 1940s. Biological control agents were identified in Europe, and several insects were subsequently evaluated, introduced, and established (Holloway 1964, Goeden 1978). These included the foliage-consuming beetles *Chrysolina hyperici* and *C. quadrigemina*, the root-boring beetle *Agrilus hyperici*, and the leaf bud gall-forming midge *Zeuxidiplosis giardi*. The spectacular success of these insects in Australia encouraged researchers in California to investigate this means of plant population suppression. From 1945 to 1946, shipments of both *Chrysolina* spp. were obtained from Australia and released in the state. Within two years, both species had become well established. The releases constituted the first attempt at control of a weed species by the intentional introduction of insects into North America (Holloway 1957). In 1950, *A. hyperici* and *Z. giardi* were imported from southern France to supplement the control provided by the beetles (Holloway and Huffaker 1953). Subsequent releases of all four St. Johnswort bioagents were made by collaborators throughout the western United States with progeny from established colonies in California. Also, a St. Johnswort foliage- and flower-feeding moth, *Aplocera* (=*Anaitis*) *plagiata*, has been recently released and established in the northwestern United States (McCaffrey et al. 1995). Excellent reviews of the biological control of this weed have been provided by Goeden (1978), Rosenthal et al. (1984), and McCaffrey et al. (1995).

• *Integrated Management*

Integrated pest management (IPM) is the deliberate selection, integration, and implementation of pest suppressive measures on the basis of predicted economic, ecological, and sociological consequences (Ennis 1977, Akobundu 1987, Piper 1992). The IPM concept implies a manipulation of the ecosystem so that noxious plant species are maintained at noninjurious population levels. An effective St. Johnswort suppression program requires planning, use of appropriate management methods, monitoring/evaluation, and persistence. One must first of all determine whether the weed should be classified as either a new or established invader (Hoglund et al. 1991) because the management approaches employed will differ. For example, in a new invader scenario, emphasis should be placed upon early treatment and prevention. Incipient infestations could effectively be reduced through the use of hand-pulling and spot treatment herbicide applications. Preventive methods, including revegetation practices, should then be invoked to deter further St. Johnswort invasion of the site. When dealing with an established invader population, be it small or large, the management approach may be more involved and take longer to implement. An established St. Johnswort population is comprised of outlier, perimeter, and core plants. Management activities are prioritized based on plant location within the infestation. The first priority is management of the outlier populations to restrict continued spread of the demarcated infestation, followed by plant suppression within the perimeter and core infestation zones, respectively. Physical extraction and herbicides can be used against outlier plants. A mixture of physical, cultural, chemical, and biological methods can be used to contain and eventually reduce perimeter and core population plants.

The design of a specific IPM plan is dependent upon plant invasive status and infestation density, the nature of the resource to be protected, economic, labor, and other constraints. Consequently, it is not possible or realistic to detail procedures to be followed in every potential management situation. The decision to use a combination of management methods must be based upon an assessment of plant developmental status, a characterization of sites infested or susceptible to infestation, method use constraints, and additional considerations. Emphasis should be placed on the use of methods that interrupt St. Johnswort seed dispersal and longevity, and that minimize habitat perturbations contributing to the weed's occupancy of the site. Plan development specifics, described in Hoglund et al. (1991), Hoglund (1992), and Piper (1992), can be followed to design site-specific integrated management programs for St. Johnswort or other weeds.

Literature Cited

Akobundu, I.O. 1987. Weed science in the tropics: principles and practices. John Wiley and Sons, New York.

Bellue, M.K. 1945. Weed seed handbook, series III. Calif. Dept. Agric. Bull. 34:116-23.

Campbell, C.L., and J.P. McCaffrey. 1991. Population trends, seasonal phenology, and impact of *Chrysolina quadrigemina, C. hyperici* (Coleoptera: Chrysomelidae), and *Agrilus hyperici* (Coleoptera: Buprestidae) associated with *Hypericum perforatum* in northern Idaho. Environ. Entomol. 20:303-15.

Campbell, M.H. 1985. Germination, emergence and seedling growth of *Hypericum perforatum* L. Wced Res. 25:259-66.

Campbell, M.H., and E.S. Delfosse. 1984. The biology of Australian weeds. 13. *Hypericum perforatum.* J. Aust. Inst. Agric. Sci. 50:63-73.

Cashmore, A.B. 1939. A note on the germination of St. John's wort seed. J. Counc. Sci. Ind. Res. Aust. 12:181-82.

Clark, N. 1953. The biology of *Hypericum perforatum* L. var. *angustifolium* D.C. (St. Johnswort) in the Owens Valley, Victoria, with particular reference to entomological control. Aust. J. Bot. 1:95-120.

Coon, N. 1974. Dictionary of useful plants. Rodale Press, Emmaus, PA.

Crompton, C.W., I.V. Hall, K.I.N. Jensen, and P.D. Hildebrand. 1988. The biology of weeds. 83. *Hypericum perforatum* L. Can. J. Plant Sci. 68:149-62.

Dodd, S. 1920. St. John's wort and its effects on livestock. Agric. Gaz. N. S. W. 31:265-72.

Ennis, Jr., W.B. 1977. Integration of weed control technologies. *In:* Fryer, J.D. and S. Matsunaka (eds.), Integrated control of weeds. Univ. of Tokyo Press, Tokyo, Japan, 229-43.

Fields, P.G., J.T. Arnason, and R.G. Fulcher. 1990. The spectral properties of *Hypericum perforatum* leaves: the implications for its photoactivated defences. Can. J. Bot. 68:1166-70.

Giese, A.C. 1980. Hypericism. Photochem. Photobiol. Rev. 5:229-55.

Goeden, R.D. 1978. Biological control of weeds, Hypericaceae: St. Johnswort, Klamath weed (*Hypericum perforatum* Linnaeus). *In:* Clausen, C.P., Introduced parasites and predators of arthropod pests and weeds: a world review. USDA Agric. Handbk. No. 480, 387-93.

Harris, P., and D.P. Peschken. 1971. Biological control programmes against insects and weeds in Canada, 1959-68. Part II: weeds. 32. *Hypericum perforatum* L., St. John's wort (Hypericaceae). CIBC Tech. Comm. No. 4: 89-94.

Hoglund, G.E. 1992. Design and implementation of integrated weed management on public lands. *In:* Proc. Oregon Interagency Noxious Weed Symp., 3 & 4 December 1991, Corvallis, OR. Oregon Dept. Agric., Salem, OR, 26-30.

Hoglund, G.E., J. Stiverson, H. Knorr, and J. Stiverson. 1991. Integrated weed management: a guide for design and implementation. USDA-USFS Okanogan National Forest, Okanogan, WA.

Holloway, J.K. 1957. Weed control by insect. Sci. Amer. 197:56-62.

Holloway, J.K. 1964. Projects in biological control of weeds. *In:* DeBach, P., Biological control of insect pests and weeds. Reinhold Publ. Corp., New York, 650-70.

Holloway, J.K., and C.B. Huffaker. 1953. Establishment of a root borer and a gall fly for control of Klamath weed. J. Econ. Entomol. 46:65-67.

Holm, L., J.V. Pancho, J.P. Herberger, and D.L. Plucknett. 1979. A geographical atlas of world weeds. John Wiley and Sons, New York.

James, L.F., R.F. Keeler, A.E. Johnson, M.C. Williams, E.H. Cronin, and J.D. Olson. 1980. Plants poisonous to livestock in the western states. USDA Agric. Info. Bull. No. 415.

Jensen, K.I.N., S.O. Gaul, E.G. Specht, and D.J. Doohan. 1995. Hypericin content of Nova Scotia biotypes of *Hypericum perforatum* L. Can. J. Plant Sci. 75: 923-26.

Jessup, C.T. 1970. St. John's wort in New Zealand. Tussock Grassld. Mt. Lands. Inst. Rev. 20: 76-83.

Kingsbury, J.M. 1964. Poisonous plants of the United States and Canada. Prentice-Hall, Inc., Englewood Cliffs, NJ.

LeStrange, R. 1977. A history of herbal plants. Angus and Robertson, Publ., London, England.

McCaffrey, J.P., C.L. Campbell, and L.A. Andres. 1995. St. Johnswort, *Hypericum perforatum* L. (Hypericaceae). *In:* Nechols, J. R., L. A. Andres, J. W. Beardsley, R.D. Goeden, and C.G. Jackson (eds.), Biological control in the western United States: accomplishments and benefits of regional research project W-84, 1964-1989. Univ. Calif. Div. Agric. Nat. Res. Pub. 3361. Oakland, CA, 281-85.

Millspaugh, C.F. 1974. American medicinal plants. Dover Publ., New York.

Mitich, L.W. 1994. Common St. Johnswort. Weed Technol. 8:658-61.

Moore, R.M., and A.B. Cashmore. 1942. The control of St. John's wort by competing pasture plants. C.S.I.R. Bull. No. 151.

Parsons, W.T. 1981. Noxious weeds of Victoria. Inkata Press, Melbourne, Australia.

Piper, G.L. 1992. Principles of integrated noxious weed management. *In* Proc. Oregon Interagency Noxious Weed Symp., 3 & 4 December 1991, Corvallis, OR. Oregon Dept. Agric., Salem, OR, 10-13.

Reed, C.F. 1970. Selected weeds of the United States. U.S. Dept. Agric. Handbk. 366.

Rosenthal, S.S., D.M. Maddox, and K. Brunetti. 1984. Biological methods of weed control. California Weed Conf. Monogr. No. 1. Thomson Publ., Fresno, CA.

Sampson, A.W., and K.W. Parker. 1930. St. Johns-wort on rangelands of California. Calif. Agric. Exp. Sta. Bull. 503.

Tisdale, E.W. 1976. Vegetational responses following biological control of *Hypericum perforatum* in Idaho. Northwest Sci. 50:61-75.

Tisdale, E.W., M. Hironaka, and W.L. Pringle. 1959. Observations on the autecology of *Hypericum perforatum*. Ecology 40:54-62.

William, R.D., D.Ball, T.L. Miller, R. Parker, K. Al-Khatib, R.H. Callihan, C. Eberlein, and D.W. Morishita. 1996. Pacific Northwest Weed Control Handbook. Washington State Univ. Coop. Ext. Serv., Pullman, WA.

Wilson, F. 1943. The entomological control of St. John's wort (*Hypericum perforatum* L.) with particular reference to the insect enemies of the weed in southern France. Aust. Counc. Sci. Ind. Res. Bull. 169.

Sulfur Cinquefoil

Peter Rice

A member of the rose family and closely related to strawberries, sulfur cinquefoil (*Potentilla recta*) is a long-lived perennial that has become one of the most serious invaders of the Northern Rockies. At least 29 species of herbaceous cinquefoils (*Potentilla*) are found in the Columbia River Basin of the Northwest. The introduced sulfur cinquefoil is sometimes confused with native northwest cinquefoil (*Potentilla gracilis*) which grows at the same low and mid-elevations. The misidentification of sulfur cinquefoil as a native variety has contributed to the unchecked expansion of this introduced species.

Identification

Prior to flowering, sulfur cinquefoil might be mistaken for marijuana because the leaves are composed of five to seven leaflets attached in a palmate pattern to a central leafstalk. The leaflets are toothed about halfway to the midvein. Up to eight leaves grow along the length of the upright stem, but only a few leaves are attached to the base of the stem. The length of the leafstalk and size of the leaflets decrease up the stem until the leaves are directly attached to the stem near its top. The erect stems are single to several, 12 to 28 inches (30 to 71 cm) tall, with few (if any) slender branches. Slender, pointed hairs project outward at right angles to the stem and leafstalks. These hairs are longer than the diameter of the stem or leafstalk to which they are attached. The longest hairs may be $^1/_4$ inch (0.635 cm) long.

The many-flowered inflorescence is elevated above most of the leaves. Five deeply-notched, pale yellow petals are slightly longer than the five enclosing green sepals and five small bracts. Individual flowers are $^5/_8$ to 1 inch (1.5 to 2.5 cm) wide and form an open, flaring cup. There are 25 to 30 stamens and numerous pistils.

The small, comma-shaped seeds are slightly flattened, brownish-purple, and covered with a net-like pattern of veins.

The plant has a single taproot, and may have several shallow, spreading branch roots, but no rhizomes.

Northwest cinquefoil (*Potentilla gracilis* Dougl.)	Sulfur cinquefoil (*Potentilla recta* L.)
Short, spreading hairs on leafstalk and stem	Longer hairs perpendicular to leafstalk and stem
Few stem leaves; mostly basal leaves	Numerous stem leaves; fewer basal leaves
Smooth seed coat	Net-like pattern on seed coat
Most have a dense, woolly underleaf	Sparse, stiff hairs; both sides of leaf are similar
Short rhizomes	Woody taproot with short branch roots
Flowers brighter yellow	Flowers paler yellow
Leaves are green to gray	Leaves are more yellowish
About 20 stamens	25 or more stamens
Leaflet serrations sometimes deep	Leaflet serrations halfway to midvein

Sulfur cinquefoil is sometimes confused with the widespread native northwest cinquefoil (*Potentilla gracilis*). Northwest cinquefoil is seldom weedy, but sometimes reaches locally heavy cover values in high elevation or subalpine sagebrush-bunchgrass rangelands. Further confusing identification, seven varieties of *Potentilla gracilis* have been recognized (Hitchcock and Cronquist 1973). The following list of contrasting characteristics are suggested to help separate sulfur cinquefoil from other species. Because of the variability of the native species, several specimens should be thoroughly examined.

Origin, History, and Distribution

Sulfur cinquefoil is native to the eastern Mediterranean region of Eurasia. The first collection in North America was made before 1900 in Ontario (Britton and Brown 1897). Sulfur cinquefoil had become a well-established weed in Canada, the northeastern United States, and the Great Lakes region by the 1950s (Werner and Soule 1976). Scattered populations were recorded in southern British Columbia. The earliest records of sulfur cinquefoil in the five-state Columbia River Basin are: Idaho in 1934, Washington in 1937, Montana and Wyoming in 1947, and Oregon in 1988.

By 1996, sulfur cinquefoil had spread to at least 30 counties in western Montana, including Glacier and Yellowstone national parks. Idaho reported infestations in 14 counties, and Wyoming in five. This weed has also been found in at least 12 counties in Washington and one in Oregon. This rapid spread over large geographic areas is similiar to the exponential spread pattern of spotted knapweed (*Centaurea maculosa*) and leafy spurge (*Euphorbia esula*), although sulfur cinquefoil was introduced several decades later.

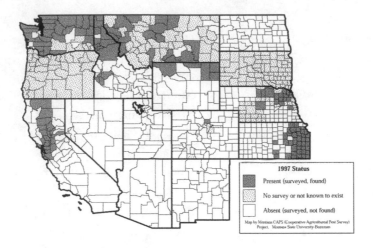

Distribution of sulfur cinquefoil by county in the western United States. Data were collected by surveying various weed authorities in each state.

Potential invasion

Sulfur cinquefoil has a wide ecological amplitude. Rice (1993) conducted a summary analysis of ecological and management data for 85 sulfur cinquefoil sites in Montana. Infestations were found at elevations as high as 6,580 feet (2,006 m). Conifer, grassland, shrubland, and seasonal wetland ecosystems are susceptible to invasion. The weed was found in 31 different habitat types (Pfister et al. 1977, Mueggler and Stewart 1978).

Sulfur cinquefoil does not seem to be limited by soil texture. It was found on sites with all soil textures except pure silt.

Some sulfur cinquefoil sites were relatively free of other noxious weeds, but the majority of the sites had one or more additional noxious weeds present. Spotted knapweed was most often associated with sulfur cinquefoil; the habitat requirements appear to be similar. Reports from land managers indicated that spotted knapweed is declining, while sulfur cinquefoil is increasing, on numerous sites. Preferential grazing of up to 30% of the spotted knapweed was often observed, with only trace utilization of sulfur cinquefoil. Sulfur cinquefoil was even found competing successfully with yellow starthistle (*Centaurea solstitialis*) and leafy spurge on several sites.

Roadsides, waste places, abandoned fields, clearcuts, and other disturbed sites are particularly susceptible to early colonization and rapid dominance by sulfur cinquefoil. However, sulfur cinquefoil is now successfully invading low-disturbance sites, including native plant communities that are remote from any apparent human disturbance. This weed is now common in natural grasslands, shrubby areas, and open canopy forests. Shading from a dense overstory prevents its establishment in mature forests, but sulfur cinquefoil can successfully occupy areas below natural gaps in the forest canopy.

Impacts

By 1996, one-quarter of the initial sulfur cinquefoil colonies evaluated by Rice (1993) in Montana had expanded to more than 100 acres (40 ha). Several colonies had expanded to more than 1,000 contiguous acres (404 ha). Although large infestations are not uncommon, the majority of the colonies are still small, half being less than 10 acres (4 ha).

Canopy cover is a useful measure of the severity of a weed infestation on individual sites and the ability of that weed to outcompete other plants. Sulfur cinquefoil often becomes a significant component of the plant community and dominates many sites.

In spite of its abundance, sulfur cinquefoil is avoided by most grazing animals. Utilization was less than 1% on 98% of the sites. This trace grazing usually consists of removal of the bud and flower tops from a limited number of plants. Intensive grazing management can increase utilization above 5%, but sulfur cinquefoil appears to be one of the last plants selected. The low preference is believed to be a result of a high concentration of phenolic tannins in the leaves and stems.

Biology and Ecology

Sulfur cinquefoil is a perennial. It is one of the first plants to emerge in the spring, and one of the fastest plants to "green-up" in the fall in response to late summer/early fall rains. It continues to grow until freezing temperatures are sustained.

Rice (1993) recorded growth stages in 1991 and 1992. In Montana, the first basal leaves of sulfur cinquefoil emerge in early to mid-March; the basal rosette is fully formed in April. In May, the plants bolt and form buds which bloom in June. (Flower production can continue into the fall if sufficient moisture is available.) Seeds are set in July, and by late July and early August the seeds are being dispersed. Seeds remain viable in the soil for at least three years. The sulfur cinquefoil leaves deteriorate (senesce) in August, but new, green basal leaves appear with fall rains in September and October. The plants stop growing after an extended freeze.

Management

Sulfur cinquefoil is rapidly increasing its geographic distribution. The number of new colonies is increasing exponentially. Many of these infestations are reaching environmentally severe sizes and densities.

• Identification

Correct identification is the first step in controlling this noxious weed. Initial recognition can be difficult because of the large number of cinquefoils in the Northwest. The three main identifying characteristics of sulfur cinquefoil are: 1) long, right-angled hairs on the leafstalks and stems; 2) many stem leaves, but few basal leaves; and 3) a seed coat with a net-like pattern.

• *Biological Control*

Biological control of sulfur cinquefoil has been problematic because of the plant's close genetic relationship to strawberries. Natural enemies of sulfur cinquefoil have been collected and evaluated domestically and overseas.

Rice (1993) collected root- and crown-boring insects from sulfur cinquefoil plants on numerous sites in Montana. Three of the identified species are known to be strawberry pests. A bright orange-and-black rust fungus (*Phragmidium ivesiae*) was found on sulfur cinquefoil at 79% of the sampled sites. The potential management value of this rust fungus had not been evaluated as of 1996.

A foreign survey of natural enemies of sulfur cinquefoil in eastern Europe was initiated in 1992. The survey, funded by the State of Montana, is being conducted by the CAB International Institute of Biological Control. By 1996, two potential biological control agents were targeted for screening (Rees et al. 1996). Field releases of any such insects would occur a decade or more in the future.

• *Mechanical Control*

Hand-digging sulfur cinquefoil plants can be effective in small infestations. The topgrowth dies back each winter, and the plant regrows from the root crown in the spring. Digging tools can be easily slipped under the root crown to remove it.

Mowing is not effective. The plants respond by developing heavier rootstocks and increasing their vegetation near ground level.

• *Chemical Control*

Selective herbicides are the most effective tool for controlling larger populations of sulfur cinquefoil at this time. Picloram (0.25 lb. acid equivalent picloram/acre) applied in the fall or spring up to late bud stage will provide several years of control. The ability of sulfur cinquefoil to green-up in response to late summer and fall rains increases the potential effectiveness of fall picloram treatment. Spring application (rosette through bud) of 2,4-D ester (2 lbs. ae/acre) also provides good control, but without the multi-year residual activity obtained from picloram. 2,4-D ester may be a better choice where the potential for water contamination is significant. On more typical dryland sites, picloram is preferred because the residual activity will suppress re-establishment from seeds in the soil bank.

Herbicide trials for sulfur cinquefoil management were initiated in 1991 and 1992. A variety of chemicals at different rates and timings are being tested (Duncan 1993). The efficacy of 2,4-D amine was less consistent than the 2,4-D ester formulation. A mix of dicamba and 2,4-D amine (1 qt/acre + 1 qt/acre) applied at the rosette stage had a 97% efficacy similar to 2,4-D ester one year after application. However, the efficacy of the dicamba + 2,4-D amine mix declined when applied at later growth stages. The dicamba + 2,4-D treatment is considerably more expensive than 2,4-D ester alone or with picloram. Picloram at 0.25 lb. ae/ac still appears to be the most consistently effective chemical prescription from the rosette stage through fall applications, including the flowering period (Duncan 1993).

Sulfur cinquefoil is not sensitive to clopyralid. Sulfur cinquefoil and spotted knapweed often grow in association. Application of clopyralid to these mixed stands will depress the spotted knapweed without harming the sulfur cinquefoil, thereby giving it more room to grow and spread.

• Grazing

Most livestock grazing practices accelerate the dominance of sulfur cinquefoil over grasses and forbs, including several noxious weeds. Sulfur cinquefoil is unpalatable to most livestock, possibly because of a high tannin content. Some utilization has been observed under intensive grazing in confined pastures. On open range or at low stocking rates, most livestock prefer spotted knapweed over sulfur cinquefoil. Animals will graze off the spotted knapweed flowering tops while completely avoiding sulfur cinquefoil. Spotted knapweed seed production is lowered relative to sulfur cinquefoil, and because sulfur cinquefoil is a long-lived perennial while spotted knapweed is a short-lived perennial, the population dynamics favor the replacement of spotted knapweed with sulfur cinquefoil. The abundance of native forbs and grasses continues to decrease, regardless of which weed is most successful.

Goats are the only animals that have been reported to select for sulfur cinquefoil.

• Monitoring

Sulfur cinquefoil seeds remain viable in the soil for at least three years. It is necessary to appraise treated sites in subsequent years until the seed bank is depleted. Systematic re-treatments should be planned if eradication is the management goal. Annual monitoring is particularly necessary if hand-pulling techniques are used.

• Realistic Expectations

Early detection of new colonies and an aggressive chemical control program with eradication as the goal is a feasible management strategy for areas outside the zone of a major infestation. Integrated weed management techniques will have to be developed to maintain sulfur cinquefoil at environmentally and economically acceptable levels within major infestation zones.

Literature Cited

Batra, S.W.T. 1979. Insects associated with weeds in the Northeastern United States. II Cinquefoils, *Potentilla norvegica*, and *P. recta* (Rosaceae). New York Entomological Soc. 87(3):216-22.

Britton, N.L., and H.A. Brown. 1897. An illustrated flora of the northern USA and Canada. Vol. II. Scribner, New York.

Duncan, C.L. 1993. Chemical control of sulfur cinquefoil on range and pasture. *In:* Proc. Montana Weed Control Assoc. Annual Meeting, Nov. 17-19, 1993, Billings, MT. MWCA, Clancy, MT, 11-13.

Hitchcock, C.L., and A. Cronquist. 1973. Flora of the Pacific Northwest. Univ. Wash. Press, Seattle, WA.

Mueggler, W.F., and Stewart, W.L. 1978. Grassland and shrubland habitat types of western Montana. USDA Forest Service Gen. Tech. Rep. INT-66. 154 p. Intermountain Forest and Range Exp. Sta., Ogden, UT.

Pfister, R.D., B.L. Kovalchick, S.F. Arno, and R.C. Preby. 1977. Forest habitat types of Montana. USDA Forest Service Gen. Tech. Rep. INT-34. Intermountain Forest and Range Exp. Sta., Ogden, UT.

Rees, N.E., P.C. Quimby, Jr., G.L. Piper, E.M. Coombs, C.E. Turner, N.R. Spencer, and L.V. Knutson (eds.). 1996. Biological Control of Weeds in the West. West. Soc. Weed Sci, USDA-ARS, Montana St. Univ., Bozeman, MT.

Rice, P.M. 1993. Distribution and ecology of sulfur cinquefoil in Montana, Idaho and Wyoming. Final Report, Montana Noxious Weed Trust Fund Project, Montana Dept. Agric., Helena, MT.

Rice, P.M., C.A. Lacey, J.R. Lacey, and R. Johnson. 1994. Sulfur cinquefoil: Biology, ecology and management in pasture and rangeland. MSU Extension Svc, EB 109, Rev. March 1994. Montana State Univ., Bozeman, MT.

Werner, P.A., and J.D. Soule. 1976. The biology of Canadian weeds. 18. *Potentilla recta* L., *P. norvegica* L., and *P. argentea* L. Can. J. Plant Sci. 56:591-603.

Tansy Ragwort

Eric M. Coombs, Peter B. McEvoy, and Charles E. Turner

Tansy ragwort (*Senecio jacobaea*) is a member of the sunflower family (Asteracaeae). It is a poisonous non-indigenous weed responsible for the deaths of thousands of livestock in the Pacific Northwest. Ragwort is a major pest in millions of acres west of the Cascade Mountains in Oregon and Washington, and in northern California. Infestations at inland sites and in states not normally considered as susceptible, such as Montana, have become an increasing problem. Disturbances in natural plant communities, such as logging, road construction, and overgrazing, provide potential tansy ragwort habitat. Ragwort infestations can range from isolated plants to linear colonies along roadsides, or dense infestations in pastures and clearcuts. Biological control of tansy ragwort has been heralded as one of the major regional success stories, saving over $5 million per year in Oregon alone. Integrated vegetation management methods can provide effective control of tansy ragwort where sound range management principles are implemented.

Identification

Tansy ragwort is an introduced biennial or short-lived perennial. Ragwort seedlings, $^1/_2$ to 2 inches (1.5 to 5 cm) in diameter, consist of two light green cotyledons, with green to dark green leaves, slightly lobed with ruffled edges, turned slightly under, and older petioles purplish. Young leaves are often covered with a web-like pubescence. Seedlings become rosettes during the first growing season. The leaves are generally 2 to 10 inches (5 to 26 cm) long, irregularly lobed, with the terminal lobe larger than lateral lobes. The compound leaflets are also lobed. The leaves, generally 10 to 20 in number, attach directly to the main stalk during the rosette stage. When the leaves are crushed they give off a rank odor, hence one of the colloquial names, stinking willy. Flowering plants are generally 1 to 3 feet (0.3 to 1.0 m) tall, up to 5 feet (1.7 m) in shaded areas. The leaves on flowering plants are compound and varied, most deeply lobed, becoming smaller closer to the flowers. The stems and leaf stalks are often purplish. The daisy-like flowering heads are about $^1/_2$ inch (1.6 cm) across, yellow, with 10 to 16 (usually 13) ray flowers. Many flowering heads can be produced on one to 15 clusters, depending on the size of the plant. Flowering may occur from July through October.

Tansy ragwort is somewhat similar in appearance to common tansy (*Tanacetum vulgare*), common groundsel (*Senecio vulgaris*), and woodland groundsel (*Senecio sylvaticus*) (Coombs et al. 1997). Common tansy, a perennial, lacks ray flowers, grows in thick stands, and may be 4 to 5 feet (1.2 to 1.5 m) tall. Common groundsel is an annual, 4 to 18 inches (10 to 47 cm) tall, leaves are attached to the stem with no leaf stalk, and the flowers do not open fully. Woodland groundsel is an annual, 1 to 3 feet (0.3 to 1 m) tall, leaves are woolly and lack the terminal lobe of tansy ragwort, flowers are small and inconspicuous. Some native species may appear similar, and should be identified using a local field guide to flowering plants.

Origin, History, and Distribution

Tansy ragwort is native to Europe and Asia Minor. It occurs in the more temperate areas from Great Britain east to Siberia. Meadows in oak and conifer woodlands, livestock pastures, and roadsides harbor the majority of ragwort in its native environment. Ragwort has been widely introduced throughout the world where livestock use and climates are similar to those in Europe. Infestations have been reported from the United States, Canada, New Zealand, Australia, and Argentina (Harper 1958). The major infestation or "core area" in the United States occurs in the Pacific Northwest in Oregon, Washington, and northern California, mostly west of the Cascade Mountains. Ragwort may have arrived in the United States as seed in ship ballast and/or in contaminated straw. The first record of ragwort in North America was in 1913 in British Columbia (Harris et al. 1978). Ragwort was reported in 1922 in Oregon along the Portland waterfront and now infests more than 3 million acres (1.2 million ha) (Isaacson and Schrumpf 1979). By the 1950s, ragwort was a serious economic problem for stockmen west of the Cascades (Snyder 1972).

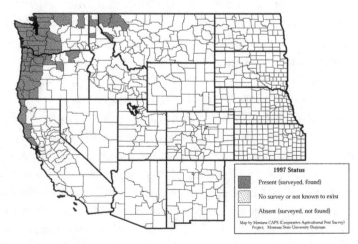

Distribution of tansy ragwort by county in the western United States. Data were collected by surveying various weed authorities in each state.

Tansy ragwort has essentially reached its biological potential west of the Cascade range (Isaacson et al. 1996). Infestations east of the Cascade Mountains have been reported in Oregon, Washington, and Idaho. A major infestation was recently detected in Montana: 1,500 acres (600 ha) spread over 150 square miles (390 sq. km) (Oliverez 1997). This alarming find indicates tansy ragwort can survive and increase in the intermountain zone, which had been thought unsusceptible. Ragwort will probably continue to spread into new areas within the northern part of the United States.

In the Pacific Northwest, ragwort is found from the upper beaches along the Pacific Ocean up to the 3,000-foot (900-m) level in the Cascade Mountains (Isaacson and Schrumpf 1979). At elevations less than 800 feet (240 m), most infestations occur in pastures and along roadsides. In western Oregon, clearcuts 10 years or older (30%) were three times as likely to be infested with ragwort when compared to clearcuts less than 10 years old (Isaacson et al. 1996). The worst infestations were in clearcuts that were more than 25 years old and grazed by cattle. In the 1960s, many pastures in Oregon that were once infested with St. Johnswort (*Hypericum perforatum*) became infested with tansy ragwort after the St. Johnswort was successfully controlled by biological agents.

Potential Invasion

Tansy ragwort has the potential to invade most areas in the United States that have a climate similar to that of central Europe. The Pacific Northwest west of the Cascade Mountains is especially vulnerable, and comprises the core area of ragwort infestations. Many satellite infestations have been traced to contaminated logging equipment or hay transported from the core area (Coombs et al. 1991, 1996b). In areas east of the Cascades, ragwort is generally found at disturbed sites in mountains where precipitation exceeds 16 to 20 inches (40 to 51 cm) per year. Open areas and south slopes are most vulnerable. In the western United States, Douglas fir (*Pseudotsuga menziesii*) is a general indicator species for potential tansy ragwort habitat.

Intact plant communities, sites dominated by sagebrush, desert and alkaline soils are generally resistant to ragwort infestations. The fact that tansy ragwort has been found in Idaho and Montana demonstrates the importance for survey and detection efforts throughout northwestern rangelands and forests.

Impacts

Tansy ragwort is a serious threat to livestock in pastures and on rangelands. Ragwort and several of its relatives contain pyrrolizidine alkaloids (Muth 1968). The pyrrolizidine compounds become toxic to cattle and horses after they are converted into pyrroles in the liver. The resulting damage to the liver is cumulative. When cattle have ingested 3% to 7% of their body weight in ragwort, they generally die of liver failure (Snyder 1972). The most toxic part of the plants are the leaves.

391

Concentrations of pyrrolizidine alkaloids average 0.18% of the weight of dry plants. Ragwort is mildly toxic to goats, and does not seem to affect sheep, which are sometimes used as a method of cultural or biologically based control (Sharrow and Mosher 1982). Poisoning usually occurs in the spring, when grazing animals cannot selectively eliminate ragwort leaves intermixed with grass. Ragwort retains its toxicity in hay and silage (Snyder 1972). Young animals and those weakened by illness are more susceptible to poisoning. There is evidence that suckling animals can acquire the toxins through contaminated milk (Snyder 1972).

An economic analysis of the impacts of tansy ragwort was conducted in Oregon (Radtke 1993, Coombs et al. 1996b). Annual economic losses due to ragwort were conservatively estimated at $5 million ($3.7 million from direct losses of livestock, $1.2 million from loss of pasture productivity, and $0.85 million from herbicide expense). In coastal counties in Oregon during the 1970s, some dairymen were forced out of business because of 2% to 10% annual herd losses caused by ragwort poisoning (Cannon, pers. comm., 1992). There were also significant opportunity costs for livestock production that did not occur because of ragwort infestations.

Along coastal areas in Oregon where ragwort infestations have been reduced by biological control agents, populations of hairy-stemmed checkered mallow (*Sidalcea hirtipes*), a rare native species, have increased by more than 40% (Gruber and Whytemare 1997). Infestations in pastures crowd out forage species, which accounted for 13% of the economic losses caused by ragwort in Oregon (Coombs et al. 1996b). At one clearcut near Mt. Hood, Oregon, tansy ragwort comprised 15% of the total plant cover. Severe infestations of ragwort reduce biological diversity, and may lead to increased soil erosion.

Some sociological impacts have resulted from ragwort infestations. Landowner relations have been strained because of untreated infestations of ragwort in forest management areas near livestock pastures. The chief incentive to control ragwort in infested clearcuts is not to speed forest regeneration, but to promote multiple-use policies, including grazing, and to foster "good neighbor" relations with ranchers in the valleys below.

A minor beneficial use of tansy ragwort is as an additional nectar source for the Oregon silverspot butterfly, *Speyeria zerene hippolyta* (Lepidoptera: Nymphalidae), which is federally listed as a threatened species (Stine 1982).

Biology and Ecology

Tansy ragwort is usually a biennial and occasionally a short-lived perennial. Seedlings that germinate in the spring generally form rosettes that bolt during the following spring and produce flowering plants. Some seedlings that germinate in the fall may bolt and flower the following growing season. The rosettes bolt in May or June, generally producing a main stem. Plants damaged early in the spring by cutting or grazing often produce several stems from the central root crown. Mature plants generally flower late in June through October, depending on available moisture, temperature, and elevation. Flowering plants damaged by mowing or grazing often regrow and reflower late in the growing season, or may overwinter and flower again the following growing season. The density of rosettes is generally several times higher than the density of mature plants produced at the same site.

Seeding occurs after the flowering period in the fall. Some large plants with multiple stems can produce more than 150,000 seeds. The central (disc) florets produce numerous light-colored seeds that have a small pappus to assist with short-range wind dispersal, and rows of trichomes that aid in animal transport (Baker-Kratz and Maguire 1984, McEvoy 1984a,b). The marginal (ray) florets produce darker, heavier, and less numerous seeds that lack dispersal structures and adhere longer to the seedhead following maturity. Seeds are small, about $^{13}/_{16}$ inch (2 mm) long, $^{5}/_{16}$ inch (0.7 mm) wide, light brown. Most seeds fall within 10 feet (3 m) of the parent plant (McEvoy 1984a, b). Some seeds may be blown further by strong winds, but most long-distance movement of seeds is by humans and animals.

Colonization and establishment can occur from propagules (seeds and vegetative buds) buried in the soil, and from seed transported to disturbed sites (McEvoy and Cox 1987). The size of the seed bank is positively correlated with the historical abundance of flowering plants (McEvoy et al. 1991). Viable seeds persist longer at increasing depths below the soil surface. Longevity estimates (time to reach 1% viability in the soil) indicate that ragwort seeds persist in the 0- to $^{3}/_{4}$-inch (0- to 2-cm) surface layer for at least four to five years, and at least 10 to 16 years when buried below $1^{1}/_{2}$ inches (4 cm) (Thompson and Makepeace 1983, McEvoy et al.1991). Seeds buried below $^{3}/_{4}$ inches (2 cm) do not normally receive the appropriate germination cues and persist in a forced state of dormancy.

The mortality rate of ragwort plants declines with the stage of development; Forbes (1977) estimated 57% die as seedlings, 29% as single rosettes, 6% as multiple rosettes, and the remaining 8% die after flowering. The death rate of juvenile plants is greatly increased by competition from perennial grasses and attacks by the ragwort flea beetle (*Longitarsus jacobaeae*), an introduced biological control agent (McEvoy et al. 1993)

Pristine plant communities are generally resistant to ragwort infestations. Micro-disturbances in the soil created by rodent burrows and the tracks of ungulates can provide sufficient habitat for ragwort to invade and maintain its presence in less disturbed plant communities. Logging, overgrazing, road construction, and fire create disturbances in plant communities where ragwort can dominate within 10 years.

Management

The management of tansy ragwort infestations, large and small, is especially important when ragwort occurs in new areas and where livestock are grazed (Bedell et al. 1981). Most landowners who have tansy ragwort want it eradicated. The reality is that ragwort persists in a region once large infestations are established. The most practical means of dealing with chronic infestations is through integrated vegetation management practices. For each infestation, the environment and land use practices should be considered to determine the best control strategy. In new, isolated infestations, prompt eradication of plants and propagules is imperative. The larger the treatment area, the greater the probability of ragwort plants being found for the next several years. In eastern Oregon, newly detected sites with more than 50 plants were found to have a 50% probability of requiring continuing treatment the following year. Rosettes should be treated with an appropriate herbicide, whereas bolting and flowering plants must be pulled. Recent infestations that have produced seeds may require intensive management practices for several years. Within large infestations, an integrated vegetation management approach can provide the best long-term options. Biological control is discussed in greater detail because it has been regionally effective.

• Prevention

Reducing or stopping the spread of ragwort seeds into uninfested areas is probably the most important and cost-effective control measure (Isaacson et al. 1996). In Oregon, use of ragwort-free hay has reduced the number of satellite infestations outside the western core infestation by 80% (Coombs et al. 1991, 1996b). Cleaning logging and construction equipment before leaving ragwort-infested areas is strongly recommended. A recently detected ragwort infestation in Montana was believed to be the result of contaminated logging equipment from western Oregon (Markin, personal communication 1997). In areas targeted for rehabilitation, e.g., after fires, logging, or construction, only certified weed-free seeds should be used for reseeding.

New infestations of ragwort outside of core areas should be targeted for eradication and prevented from going to seed. In computer simulations, the importance of detection and prompt, effective control of outlying infestations was demonstrated by Moody and Mack (1988) and Isaacson et al. (1996). Outlying sites that are treated must be marked or identified so they can be monitored for several years, even after no ragwort has been observed for two years.

• Mechanical or Physical Control

Mechanical or physical control of ragwort may be warranted in some situations, particularly for new infestations (Coombs et al. 1997). Single plants outside core infestations should be pulled before they produce seeds. Plants that are flowering or producing seeds should be bagged and burned to destroy the seeds. Pulling plants in areas where they have gone to seed may cause enough disturbance to create more habitat for seedlings. These sites should be reseeded with desirable species, then marked for future inspection and possible treatment.

Cutting or mowing bolting plants before flowering is generally ineffective, because they may produce smaller lateral shoots and flower later in the growing season (Bedell et al. 1981). The resulting smaller plants may be more difficult to cut. Cutting flowering plants late in the summer may reduce seed production, but the plants may survive through the winter to produce more robust plants the following year. Livestock should not be allowed to graze in fields where ragwort has been mowed because of the risk of poisoning. Mowed patches should be treated in the fall with a herbicide after re-sprouting occurs.

Annual tilling is effective in croplands to prevent ragwort from producing seeds (Coombs et al. 1997). Tilling also stimulates the germination of seeds, which if continued over time, can exhaust the seed bank in the soil.

• Cultural Control

Cultural control of ragwort is frequently integrated with other practices, particularly to enhance grazing. Competitive stands of high-yielding perennial grasses with subclover or white clover (*Trifolium repens*) generally exceed the production obtained by native or annual grasses (Bedell et al. 1981). Proper use of irrigation and fertilizers, including manure, are important to maintain vigorous forage stands. Nitrogen can be provided by clover, but phosphorus and sulfur should be provided by fertilizers. Soil test kits can be used to determine soil nutrient requirements.

Uniform grazing is important in maintaining a healthy stand of competing vegetation. Severe ragwort infestations in pastures can frequently be traced to poor grazing management (Bedell et al. 1981). Integrating sheep into the grazing system where ragwort is a severe problem is often beneficial (Sharrow and Mosher 1982). Sheep have been used on infested pastures to feed on ragwort before cattle are introduced (Sharrow and Mosher 1982). Tansy ragwort often increases after sheep are removed. Cattle and horses are usually poisoned by eating small ragwort plants that are mixed with grass. They rarely feed on mature plants, except when little other forage remains.

• Chemical Control

Chemical control can be effective in controlling ragwort infestations. Selection of the proper chemical, formulation, application time and method, consideration of environmental constraints, land use practices, and local regulations, etc., are all important factors to consider before purchasing a chemical. The following chemical guidelines are for general applications. Because local regulations frequently change, it is always best to consult local authorities and check with reputable pesticide dealers. Read all labels and comply with recommended application procedures before using any herbicides.

The following recommendations were taken from the current *Pacific Northwest Weed Control Handbook* (William et al. 1997). Livestock must be removed from

395

treatment areas and not allowed to return until label requirements have been met, generally after two weeks. The primary herbicides used to treat tansy ragwort infestations without clover are 2,4-D, picloram, dicamba, and metsulfuron. These herbicides are also commonly used against other broadleaf range weeds. Apply 2,4-D, (amine and LVE) at 1.0 to 2.0 lb acid equivalent (ae)/acre (1.12 to 2.24 kg/ha), before plants flower. Applications after spring growth begins are generally more effective. Fall application without a wetting agent helps minimize injury to clover. Picloram (Tordon®) at 0.25 lb ae/acre (0.28 kg/ha) or dicamba (Banvel®)at 1.0 lb ae/acre (1.12 kg/ha) can be used at the flowering stage. Because of its soil residual qualities, picloram can be used in the fall to control growing seedlings. Combination herbicides are also effective: 2,4-D + dicamba (Weedmaster®) at 2.0 quarts/acre (4.68 l/ha), or triclopyr + 2,4-D (Crossbow®) at 1.5 to 2.0 quarts/acre (3.51 to 4.68 l/ha). Metsulfuron at 0.6 oz active ingredient (ai)/acre (42 gm/ha), or Escort® 1.0 oz/acre (70 gm/ha) can be used against actively growing ragwort. These herbicides are selective when used properly, controlling ragwort and other broadleaf weeds while not harming grasses. Older infestations generally require several years of treatment integrated with intensive pasture management to promote perennial grasses.

• *Biological Control*

Biological control involves the use of host-specific natural enemies to damage and stress a target weed, so as to reduce its density below an economically damaging level. The most spectacular successes controlling ragwort using natural enemies have been in the Pacific Northwest region (Pemberton and Turner 1990, McEvoy et al. 1991, Turner and McEvoy 1995, and Coombs et al. 1996a,b). (For this discussion we do not include sheep as biological control agents, but rather as a form of cultural or biologically based control, because they are not host-specific to ragwort.) The level of suppression was highest when biological control agents were complemented with competitive vegetation (McEvoy et al. 1990 and 1993). Since 1959, three insects— the cinnabar moth, ragwort flea beetle, and ragwort seed-head fly—have been introduced into the western United States as biological control agents against tansy ragwort (Turner and McEvoy 1995, Coombs et al. 1996a, Rees et al. 1996).

The cinnabar moth *Tyria jacobaeae* (Lepidoptera: Arctiidae) was first released in California in 1959 (Hawkes 1968). Within a few years, the moth was widely distributed throughout major infestations in the Pacific Northwest. In the spring, the red and black adults emerge from overwintering pupae, mate, and females lay eggs on ragwort rosettes. The orange-and-black-banded larvae defoliate the bolting plant, often preventing flowering. After four to seven weeks, larvae pupate in soil debris. The cinnabar moth is univoltine, producing one generation per year. The cinnabar moth works best in large dense stands at lower elevations. Moths are redistributed by collecting several hundred larvae and transporting them in containers with plant material. Containers should be kept cool and dry, and transported as soon as possible to new sites. Areas above 3,000 feet (900 m), or that

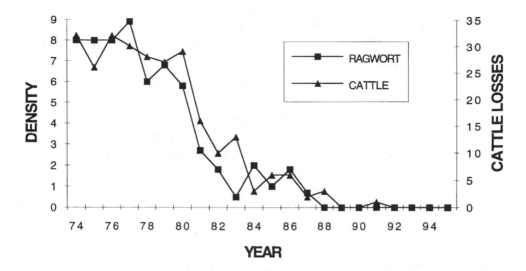

Figure 1. Changes in average flowering plant density of tansy ragwort at 11 Oregon Dept. of Agriculture study sites and the number of pyrrolizidine poisoning cases of cattle reported by the Oregon State Veterinary Diagnostic Laboratory. (In the early 1980s, ragwort populations declined significantly because of the widespread dissemination of the ragwort flea beetle and cinnabar moth).

have cool, rainy summers or very cold winters, are generally unsuitable for cinnabar moth survival. In some cases, after late summer rains, defoliated plants can reflower. Seeds produced by late-flowering plants have lower viability than those of early-flowering plants (McEvoy 1984b).

The ragwort flea beetle *Longitarsus jacobaeae* (Coleoptera: Chrysomelidae) is probably the most important agent in reducing stand density (McEvoy et al. 1993). The flea beetle was first released in California in 1969 (Frick 1970). It has been widely redistributed throughout the Pacific Northwest. Adults are small and golden-colored, and become most active in the fall after the first rains. Adults mate and lay eggs on rosettes and in the soil near the plants. Both the larvae and the adults attack the plant. The most serious damage to the plant is caused by the larvae that feed in the roots, often causing plant mortality. The adults chew small holes in the leaves, causing the characteristic "shot hole" appearance. Beetles are collected for redistribution in the fall using a motorized vacuum device. Adults are separated into containers containing several hundred beetles and plant material. Tissue should be added to provide sufficient surface area for the active adults to crawl on, otherwise they tend to wear themselves out by jumping around and disturbing each other. Releases should be made where there is at least one plant per square yard/meter.

In study sites in Oregon, flowering plant density has been reduced more than 90% within six years (Fig. 1) following flea beetle releases (McEvoy et al. 1991), with a corresponding decrease in reported cases of livestock poisoning (Coombs et al. 1996b). The flea beetle is the most effective agent for low-density infestations and areas close to the coast. At several study sites in Oregon, no mature ragwort plants were detected for several years after initial control by the flea beetle (McEvoy et al. 1991). Similar results of successful control have been documented in California (Pemberton and Turner 1990).

The ragwort seed-head fly (*Botanophila seneciella*), also known as *Pegohylemyia* and *Hylemyia* (Diptera: Anthomyiidae), was introduced into the United States in 1966 (Frick 1969). The univoltine seed fly appears similar to a housefly. Adults emerge in the spring, mate, and females lay eggs on developing flower buds. The larva hatches and tunnels into the involucre and feeds on the developing seeds and receptacle tissues. Attacked flowers often have a dark center which may later exude a clear frothy spittle. The mature larva exits the flower and pupates in the soil. The seed fly has been widely distributed (mostly on its own) throughout the range of ragwort in the Pacific Northwest, thus eliminating the need for further redistribution (Coombs et al. 1991). By itself, the seed fly is not an effective agent. Attack rates in Oregon rarely exceed 40% of the flowers. Generally, only the buds produced early in the flowering season are attacked. Seed flies have been found at inland ragwort infestations where the cinnabar moth and flea beetle are unable to establish.

The root-boring moth *Cochylis atricapitana* (Lepidoptera: Cochylidae) was introduced into Canada in 1990 (Julien 1992). Interest in importing the moth into the United States has been increasing, especially since more infestations are being detected at inland sites where the cinnabar moth and ragwort flea beetle do not occur. Additional host testing will be required before the moth is released in the United States.

Literature Cited

Baker-Kratz, A.L., and J.D. Maguire. 1984. Germination and dry-matter accumulation in dimorphic achenes of tansy ragwort (*Senecio jacobaea*). Weed Sci. 32:539-45.

Bedell, T., R.E. Whitesides, and R.B. Hawkes. 1981. Pasture management for control of tansy ragwort. Pacific Northwest Coop. Ext. Bull. PNW-120. Oregon State University, Corvallis, OR.

Cannon, L. (Personal communication 1992). Coos County Extension Service, Oregon State University, Corvallis, OR.

Coombs, E.M., T.E. Bedell, and P.B. McEvoy. 1991. Tansy Ragwort (*Senecio jacobaea*): Importance, Distribution, and Control in Oregon. *In*: Noxious Range Weeds, James L.F., Evans J.O., Ralphs M.H. & Child R.D. (eds). Westview Press, Inc., San Francisco, CA, ch. 42.

Coombs, E.M., P.B. McEvoy, and C.E. Turner. 1996a. Tansy ragwort. *In:* N.E. Rees, P.C. Quimby, Jr., G.L. Piper, C.E. Turner, E.M. Coombs, N.R. Spencer, and L.V. Knutson (eds). Biological control of rangeland weeds in the west. West. Soc. Weed Sci., USDA-ARS, Montana St. Univ., Bozeman, MT.

Coombs, E.M., H. Radtke, D.L. Isaacson, and S.P. Snyder. 1996b. Economic and regional benefits from the biological control of tansy ragwort, *Senecio jacobaea* (Asteracaeae), in Oregon. *In*: Proc. IX Internat. Symp. on Biol. Cont. of Weeds. V.C. Moran and J.H. Hoffmann (eds.). 19-26 January 1996, Stellenbosch, South Africa. University of Cape Town, 489-94.

Coombs, E.M., C. Mallory-Smith, L.C. Burrill, R.H. Callihan, R. Parker, and H. Radtke. 1997. Tansy ragwort *Senecio jacobaea* L. Oregon State Univ. Ext. Serv. (Revised) PNW 175.

Forbes, J.C. 1977. Population flux and mortality in a ragwort (*Senecio jacobaea*) infestation. Weed Res. 17:387-91.

Gruber, E., and A. Whytemare. 1997. The return of the native? *Sidalcea hirtipes* in coastal Oregon. *In*: T.N. Kaye, A. Liston, R.M. Love, D.L. Luoma, R.J. Meinke, and M.V. Wilson, (eds.). Conservation and management of native plants and fungi. Proc. Oregon Conference on the Conservation and Management of Native Vascular Plants, Bryophytes, and Fungi. Native Plant Soc. of Oregon, Corvallis, OR, 121-24.

Frick, K.E. 1969. Attempt to establish the ragwort seed fly in the United States. J. Econ. Entomol. 62:1135-38.

Frick, K.E. 1970. *Longitarsus jacobaeae* (Coleoptera: Chrysomelidae), a flea beetle for the biological control of tansy ragwort. 1. Host plant specificity studies. Annals Entomol. Soc. Amer. 63:284-96.

Harper, J.L. 1958. The ecology of ragwort *Senecio jacobaea* with especial reference to control. Herbage Abstr. 28:151-57.

Harris, P., A.T.S. Wilkinson, L.S. Thompson, and M.E. Neary. 1978. Reproductive biology of tansy ragwort climate and biological control by the cinnabar moth in Canada. *In*: Proc. IV International Biological Control of Weeds Symposium. T.E. Freeman (ed.), 30 August-2 September 1976. Univ. Florida, Gainesville, FL, 163-73.

Hawkes, R.B. 1968. The cinnabar moth *Tyria jacobaeae* for control of tansy ragwort. J. Econ. Entomol. 61:499-501.

Isaacson, D.L., and B.J. Schrumpf. 1979. Distribution of tansy ragwort in western Oregon. In: Proc. Symposium on pyrrolizidine (*Senecio*) alkaloids: toxicity, metabolism, and poisonous plant control measures. P.R. Cheeke (ed.). Oregon State Univ., February 23-24, 1979, Corvallis, OR, 163-69.

Isaacson, D.L., D.B. Sharratt, and E.M. Coombs 1996. Integrating biological control into management of invasive weed species. *In*: Proc. IX International Symposium on Biological Control of Weeds, J. H. Hoffmann and V.C. Moran (eds.). 21-26 January.1996. Capetown, South Africa.

Julien, M. 1992. Biological control of weeds, A world catalogue of agents and their target weeds. Third Edition. CAB International, CSIRO, Australia.

Markin, G.P. (Personal communication 1997). USDA Forest Service, Montana State University, Bozeman, MT.

McEvoy, P.B. 1984a. Dormancy and dispersal in dimorphic achenes of tansy ragwort *Senecio jacobaea* L. (Compositae). Oecologica (Berlin) 61:160-68.

McEvoy, P.B. 1984b. Seedling dispersion and persistence of ragwort *Senecio jacobaea* (Compositae) in a grassland dominated by perennial species. Oikos 42:138-43.

McEvoy, P.B., and C.S. Cox. 1987. Wind dispersal distances in dimorphic achenes of ragwort, *Senecio jacobaea*. Ecology 68:2006-15.

McEvoy, P.B., C.S. Cox, R.R. James, and N.T. Rudd. 1990. Ecological mechanisms underlying successful biological control: field experiments with ragwort *Senecio jacobaea*. *In*: E.S. Delfosse (ed.). Proc. VII International Symp. on Biol. Cont. of Weeds. 6-11 March 1988. Rome, Italy. Ministero dell' Agricoltura e delle Foreste, Rome/CSIRO, Melbourne, Australia, 55-66.

McEvoy, P.B., C.S. Cox, and E.M. Coombs. 1991. Successful biological control of ragwort. Ecol. App. 1:430-32.

McEvoy, P.B., N.T. Rudd, C.S. Cox, and M. Huso. 1993. Disturbance, competition and herbivory effects on ragwort *Senecio jacobaea* populations. Ecol. Monogr. 63: 55-75.

Moody, M.E., and R.N. Mack. 1988. Controlling the spread of plant invasions: the importance of nascent foci. J. Applied Ecol. 25:1009-21.

Muth, O.H. 1968. Tansy ragwort *Senecio jacobaea*, a potential menace to livestock. J. Amer. Vet. Med. Assoc. 153:310-12.

Oliverez, J. 1997. Tansy ragwort *Senecio jacobaea*. Briefing paper. Western Area Weed Council. Unpublished report. USDA Forest Service, Region 1, Reg. Office, Missoula, MT.

Pemberton, R.W., and C.E. Turner. 1990. Biological control of *Senecio jacobaea* in northern California, an enduring success. Entomophaga 35:71-77.

Radtke, H. 1993. An economic evaluation of biological control of tansy ragwort. Technical Report, Oregon Dept. of Agric., Salem, OR.

Rees, N.E., P.C. Quimby, Jr., G.L. Piper, E.M. Coombs, C.E. Turner, N.R. Spencer, and L.V. Knutson (eds.). 1996. Biological Control of Weeds in the West. West. Soc. Weed Sci., USDA-ARS, Montana St. Univ., Bozeman, MT.

Schroeder, D. (Personal communication 1992). International Institute of Biological Control, Delemont, Switzerland.

Sharrow, S.H., and W.D. Mosher. 1982. Sheep as a biological control agent for tansy ragwort. J. Range Manage. 35:480-82

Snyder, S.P. 1972. Livestock losses due to tansy ragwort poisoning. Oregon Agri-record, Oregon Dept. of Agric., Salem, OR. No. 225, 3-4.

Stine, P. 1982. Oregon silverspot butterfly recovery plan. U.S. Fish and Wildlife Service.

Thompson, A., and W. Makepeace. 1983. Longevity of buried ragwort (*Senecio jacobaea* L.) seed. New Zealand J. Exp. Agric. 11:89-90.

Turner, C. E., and P.B. McEvoy 1995. Tansy ragwort. *In:* J.R. Nechols, L.A. Andres, J.W. Beardsly, R.D. Goeden, and C.G. Jackson (eds.). Biological control in the western United States: Accomplishments and benefits of Regional Research Project W-84. Berkeley, Univ. CA., Div. Agric. and Nat. Res. Pub. No. 3361, 264-69.

William, R.D., D. Ball, T.L. Miller, R. Parker, K. Al-Khatib, R.H. Callihan, C. Eberlein, and D.W. Morishita. 1997. Pacific Northwest Weed Control Handbook. Extension Service, Agric. Comm., Oregon State University, Corvallis, OR, 331.

Whitetop

Roger Sheley and Jack Stivers

Whitetop (*Cardaria pubescens*), lens-podded whitetop (*C. chalapa*), and heart-podded whitetop or hoary cress (*C. draba*) are three closely related exotic noxious weeds. These weeds are well-adapted to moist habitats, especially sub-irrigated pastures, rangeland, roadsides, and ditch banks. Like many noxious weeds, these species dramatically reduce biodiversity and forage production, and displace native plant communities. They also have the potential to invade cropland.

Identification

The three *Cardaria* species are members of the mustard family (Brassicaceae). Flowers have four petals and six stamens. Individual white flowers ($1/8$ to $1/4$ inch [3 to 6 mm]) are borne on slender stalks that are about $1/2$ inch (1.25 cm) long. Dense blooming stands appear as a late-melting snowfield.

These species can be distinguished by the shape of their seed pods. Globe-podded whitetop has small, purplish, globe-shaped seed pods. The seed pod of lens-podded whitetop is flat and round. Heart-podded whitetop, also known as hoary cress, has a heart-shaped, broad, flat seed pod. In all species, the seed pod is tipped with a beak, which is rare in the mustard family. Each pod is oval, reddish brown, $1/2$-inch (1.25 cm) long, and contains two or four seeds.

In general, plants grow from 10 to 24 inches (25 to 60 cm) tall. They have both basal and stem leaves. Basal leaves taper to a short stalk that attaches to the crown near the ground. Stem leaves are grayish green, arrowhead-shaped, and have smooth or occasionally finely-toothed edges. All leaves have a covering of soft white hairs. They range from $3/4$ to 3 inches (1.8 to 7.5 cm) long, and are arranged alternately along the stem. The base of each leaf clasps around the stem at the point of attachment.

Origin, History, and Distribution

Whitetop is native to the former USSR, northern Iran, and Afghanistan, and can be found in Argentina and North America. Lens-podded whitetop is native to the former USSR, and is common in Pakistan, northern India, Afghanistan, Iran, Iraq, Syria, and Israel. It has been introduced into England, Argentina, and North America. Heart-podded whitetop is native to the Balkan Peninsula, Armenia, Turkey, Israel, Syria, Iraq and Iran. These species have been introduced onto every continent in the world (Mulligan and Findlay 1973).

These noxious weeds probably arrived in North America in contaminated alfalfa seed. In the United States, heart-podded whitetop was first collected at Long Island, New York, in 1862. Lens-podded whitetop and globe-podded whitetop were recorded at Chino, California, in 1918, and Ypsilanti, Michigan, in 1919, respectively (Miller and Callihan 1991). In the Pacific Northwest, all three of these species have been introduced into California, Washington, Oregon, Idaho, Wyoming, and Montana (Forcella and Harvey 1981). In California, more than 38,000 acres (15,300 ha) are reported to be infested with *Cardaria* spp. Heart-podded whitetop has been introduced in all but two of Montana's 56 counties, and infests about 32,000 acres (12,900 ha).

Potential Invasion

All three *Cardaria* species grow in open, unshaded areas. These weeds are generally found in disturbed areas dominated by other exotic species, such as smooth bromegrass (*Bromus inermus*) and Russian knapweed (*Centaurea repens*). Native species associated with these weeds are aster (*Aster ericoides*), common yarrow (*Achillea millifolium*), fringed sagebrush (*Artemisia frigida*), rose (*Rosa* spp.), and snowberry (*Symphoricarpos albus*) (Selleck 1964).

All three species grow under a variety of soil and range types. They require moderately moist sites (12 to 16 inches [30 to 40 cm] moisture annually) and are best adapted to alkaline soils that are wet during late spring. However, alkaline soils are not a requirement for invasion (Kiemnec and Larson 1990). Sites susceptible to invasion are mainly sub-irrigated pastures, rangeland, ditch banks, roadsides, and waste areas. Invasion of arid rangeland by whitetop is unlikely.

These weedy species are also well-adapted to cropland, especially under irrigation. Whitetop is a serious problem in small grains, stubble and fallow fields, alfalfa, peas, onions, sugar beets, other crops, and on Conservation Reserve Program (CRP) lands (Miller and Callihan 1991).

Impacts

• *Detrimental*

On rangeland and pastures, *Cardaria* spp. form dense monocultures, similar to other noxious weeds such as leafy spurge (*Euphorbia esula*). These species displace native plant species; reduce biodiversity, wildlife habitat, and forage production; and are a serious threat to the cattle industry. Whitetop contains glucosinolates which can be toxic to cattle (Selleck 1964). These weeds can also reduce yields in cropland by competing aggressively for soil moisture.

• *Beneficial*

These weeds provide nectar for honeybees. Seeds of whitetop can be used as a substitute for pepper.

Biology and Ecology

• *Life Cycle*

All three whitetop species are herbaceous, relatively long-lived, rhizomatous perennial weeds (Miller and Callihan 1991). They begin their life cycle in the fall with seed germination and seedling establishment. Seed leaves grow very rapidly, and lateral roots develop within three weeks. Seedlings overwinter as rosettes.

Established plants and first-year rosettes initiate growth very early each spring (Kiemnec and Larson 1990). These species frequently bloom in May, and begin producing seeds about a month later. After blooming, the plants continue to grow until frost. If conditions remain suitable, they will flower and produce a second crop of seeds late in the summer.

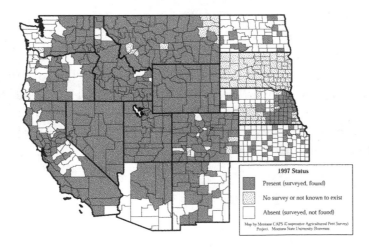

Distribution of whitetop by county in the western United States. Data were collected by surveying various weed authorities in each state.

• Seed Dispersal

Seeds are spread by wind, along waterways and irrigation systems, on vehicles and machinery (especially during cultivation), and in hay and crop seed (Corns and Frankton 1952). Small infestations spread by rhizomes, which are underground stems capable of producing shoots. A single plant can spread to an area 12 feet (3.6 m) in diameter in its first year of growth. Subsequent radial increases range from 2 to 5 feet (0.6 to 1.5 m) per year. Perimeter expansion appears to be related to the competitive ability of the vegetation surrounding a patch of whitetop.

• Roots

Roots grow very rapidly, reaching 12 to 30 feet (3.6 to 9 m) by the second or third growing season (Mulligan and Findlay 1973). The root system consists of a vertical taproot that develops several lateral roots. Lateral roots eventually turn down to become vertical roots, which often reach greater depths than the parent roots. In some cases, whitetop can place its roots directly into the water table. Both the vertical and lateral roots produce adventitious buds, which develop into rhizomes and shoots. The deep root system and the weeds' ability to reproduce vegetatively make these weeds very difficult to control.

• Shoots

The three whitetop species are capable of producing many shoots. Shoot development tends to occur at or just above the point where lateral roots bend down to become vertical. However, buds can form on any part of the root system. Under experimental conditions, without competition a single plant can produce more than 450 shoots in a single year. With competition, individual plants rarely exceed 50 shoots per year.

• Seed Production and Longevity

A single plant of these species can produce from 1,200 to 4,800 seeds each year. Of the three species, heart-podded whitetop is the most prolific seed producer. This species can produce up to 850 seeds per flowering stem.

Buried seeds remain viable for about three years in the soil. In one study, germination rates of one-year, two-year, and three-year-old heart-podded whitetop seeds were 84%, 31%, and 0%, respectively (Kiemnec and Larson 1990). Similar results were found for other whitetop species. Heart-podded whitetop is capable of germinating in salt-stressed environments.

Management

Successful management of whitetop requires the use of integrated weed management, which includes combining management strategies to prevent the movement of these weeds, containing existing infestations, and integrating control methods to reduce weed infestations to tolerable levels.

• *Prevention*

The most cost-effective method for managing these weeds is to prevent their encroachment. To prevent invasion, seed dispersal must be limited. Avoid driving vehicles and machinery through infested areas during the seeding period, and wash the undercarriage of vehicles and machinery before leaving infested areas. Livestock should not graze weed-infested areas during flowering and seed-set, or else they should be transported to a holding area for 10 to 14 days after grazing whitetop-dominated pastures, to allow time to digest and excrete the weed seeds.

In cropping systems, it is important to circumvent whitetop patches during cultivation to prevent spreading their roots throughout the field. Similarly, circumventing small weed infestations during harvesting will limit the spread of weed seeds in hay and in crop seed. Screening irrigation water before applying it to the field will minimize seeds spread through ditches and irrigation canals.

Recreationists, including campers, hikers, off-road vehicle enthusiasts, and horseback riders, spread weed seeds. They should be careful to brush and clean their equipment and animals, and place discards into a hot fire before leaving an area. Weed-seed-free forage should be used.

Proper livestock grazing is essential to maintain competitive grass plants that will slow or prevent whitetop encroachment. Grazing management plans should be developed for any management unit involved in a weed prevention program. A grazing management plan that alters the season of use, rotates livestock to allow plants to recover before being regrazed, and promotes litter accumulation, can minimize whitetop establishment and spread.

• *Containment*

An essential part of any integrated weed management program is the commitment to contain existing weed infestations. Generally, this includes a long-term program of annually spraying the perimeter of the infestation with herbicides.

• *Small-Scale Eradication*

Detecting satellite infestations and implementing eradication programs is an important step in integrated weed management. Local systematic surveys along roadways, railways, and waterways to detect weed infestations as early as possible are mandatory. Once an infestation is found, a small-scale eradication plan should be designed and implemented. The plan will consist of delimiting the infestation boundaries, applying control treatments, setting a control schedule, revegetating, planning follow-up monitoring, and estimating costs.

• *Mechanical Control*

Diligent digging can provide control of very small infestations. Hand-pulling of aboveground plant parts is ineffective (Corns and Frankton 1952). Successful digging requires complete plant removal within 10 days after weed emergence throughout the growing season for two to four years. Digging can be a useful method for controlling new introductions of whitetop in riparian areas and around the home. Removing whitetop is easiest when the soil is moist.

Successful eradication on arable land by cultivation is difficult because it requires exhausting the root reserves. Cultivation 6 inches (15 cm) deep must be repeated within 10 days of weed emergence throughout the growing season for two to four years.

Mowing to ground level during flowering reduces biomass and seed production, but does not provide long-term control (McInnis et al. 1990). However, preliminary research indicates that mowing combined with using herbicides can be effective.

• *Cultural Control*

Planting competitive legumes, such as alfalfa, can reduce whitetop in crop-pasture rotations. Extremely dense stands of legumes compete for soil moisture and shade the weeds (McInnis et al. 1993).

Preliminary evidence suggests that sheep will graze whitetop. However, a complete grazing management program has not been developed. Some generalizations from experience using sheep to graze leafy spurge and rush skeletonweed (*Chondrilla juncea*) may be useful. Continuous rather than rotational grazing generally produces the lowest weed densities and prevents flowering. Moderate grazing is as effective as heavy grazing in controlling most noxious weeds, because heavy use causes sheep to switch from the preferred forbs to grasses.

Flooding can control whitetop, but is of limited use because many areas are not suited for flooding (Corns and Frankton 1952). Flooding is useful in areas with heavy enough soil to hold water without excessive seepage, and where fertility loss following prolonged submersion and drainage is minimal. Successful flooding to control these weeds requires submersion from May to September in most cases.

• *Biological Control*

No natural enemies for use as biological controls are available for use in the United States as of 1998.

• *Chemical Control*

The whitetop species can be controlled using herbicides, although it is difficult. Successful control usually depends on an aggressive re-application program. On rangeland, roadsides, and waste areas, applications of metsulfuron at 0.12 to 0.45 oz. active ingredient per acre (8.4 to 31.5 g ai/ha) to actively growing rosettes early in the spring, to regrowth before bud stage, or to fall regrowth before the first killing frost, can be effective in managing whitetop (Whitson et al. 1989). Treatment after bloom is generally less effective. It is important to apply metsulfuron with at least 10 gallons of water per acre (125 l/ha) and to use a nonionic surfactant. Nitrogen fertilization will enhance the vigor of grasses and slow the rate of whitetop reinvasion.

Although 2,4-D is somewhat effective, picloram, a broad spectrum herbicide that provides excellent control of many rangeland weeds, is ineffective on whitetop species.

In cereal grains, metsulfuron applied at 0.06 oz ai/acre (4.2 g ai/ha) provides reasonable control of whitetop species. These weeds can be effectively controlled in stubble and fallow, but it is important to wait the period stated on the label before planting.

No herbicides are registered for use on alfalfa, peas, onions, and sugar beets that control whitetop. These weeds should be controlled during another part of the crop rotation.

• Integrated Weed Management

No single treatment provides effective, long-term control of the whitetop species. Managing these noxious weeds requires an integrated strategy. The first line of defense is to prevent introductions of these weeds. Systematic surveys, early detection, and implementation of an eradication program on small infestations is the second step. Once the weed becomes established, integrating various combinations of competitive plantings, crop rotations, sheep grazing, herbicides, and possibly the use of fertilizers, can reduce whitetop to manageable levels. A key component of any integrated weed management program is sustained effort, constant evaluation, and the adoption of improved strategies.

Literature Cited

Corns, W.G., and C. Frankton. 1952. Hoary cresses in Canada with particular reference to their distribution and control in Alberta. Sci. Agric. 32:484-95.

Forcella, F., and S.J. Harvey. 1981. New and exotic weeds of Montana. Part II. Montana Weed Survey. Montana State Univ., Bozeman, MT.

Kiemnec, G., and L. Larson. 1990. Germination of whitetop (*Cardaria draba* L.) and diffuse knapweed (*Centaurea diffusa* Lam.) as influenced by osmotic and salt stress. Northwest Sci. 64:107.

McInnis, M.L., L.L. Larson, and R.F. Miller. 1990. First year defoliation effects on whitetop. Northwest Sci. 64:107.

McInnis, M.L., L.L. Larson, and R.F. Miller. 1993. Nutrient composition of whitetop. J. Range Manage. 46:227-31.

Miller, T.W., and R.H. Callihan. 1991. Hoary cress and related whitetops. PNW Ext. Pub. 359:1-4.

Mulligan, G.A., and J.N. Findlay. 1973. The biology of Canadian weeds. 3. *Cardaria draba*, *C. chalepensis*, and *C. pubescens*. Can. J. Plant Sci. 54:149-60.

Selleck, G.W. 1964. An ecological study of lens and globe podded hoary cresses in Saskatchewan. Weeds 12:1-5.

Whitson, T.D., A. Mooney, and M. Griswold. 1989. Control of hoary cress (*Cardaria draba*) with various herbicides. W. Soc. Weed Sci., 39-41.

Yellow Starthistle

Roger L. Sheley, Larry L. Larson, and James S. Jacobs

Yellow starthistle (*Centaurea solstitialis*) is a member of the knapweed complex of weeds in the sunflower (Asteraceae) family. It is considered poor forage for all classes of livestock, and causes "chewing disease," a fatal neurological disorder, in horses. This noxious weed currently infests millions of acres of rangeland in California and the Pacific Northwest. Infestations range from scattered individuals to dense stands of yellow starthistle associated with annual grasses.

Identification

Yellow starthistle is a winter annual with bright yellow flowers. Flowerheads are single on the end of short branches. The flowerhead consists of many individual flowers and bracts. Each bract is armed with a stout, straw-colored spine 1 to 2 inches (2.5 to 5 cm) long. The spines radiate from the flowerhead in a star shape. Yellow starthistle produces a rosette of leaves that lie close to the ground. These basal leaves are deeply lobed, and generally less than 2 inches (5 cm) wide and 8 inches (20 cm) long. Stems range from 4 inches (10 cm) to 5 feet (1.5 m) tall, depending on the environmental conditions. Stem leaves are entire, linear, and have thin woolly hairs that persist through the growing season.

Origin, History, and Distribution

Yellow starthistle is native to the Mediterranean region of Europe (Roché and Talbott 1986). Seeds of this species are believed to have arrived in North America in alfalfa seeds shipped to California. Yellow starthistle seeds were found in the flora analyses of adobe brick from the post-Mission Period (after 1824) in California (Roché et al. 1986). Several herbarium specimens of this species were collected in California in the middle- and late 1800s.

In the Pacific Northwest, yellow starthistle was first reported near Bingen, Washington, around the turn of the century. Yellow starthistle began spreading into grasslands in the Pacific Northwest in the 1920s. As of the mid-1980s, infestations were estimated to be about 8 million acres (3.2 million ha) in California, 280,000

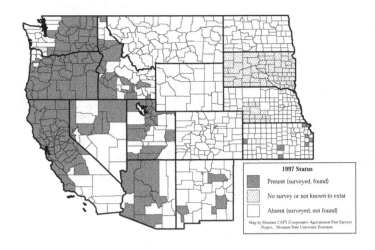

Distribution of yellow starthistle by county in the western United States. Data were collected by surveying various weed authorities in each state.

acres (113,000 ha) in Idaho, 135,000 acres (54,500 ha) in Oregon, and 148,000 acres (60,000 ha) in Washington (Maddox et al. 1985, Roché and Roché 1988, Callihan et al. 1989). Yellow starthistle appears to have reached its northern and southern boundaries, but continues to invade 7,000 to 20,000 acres (2,800 to 8,000 ha) of rangeland per year within these boundaries (Callihan et al. 1989).

Potential Invasion

Yellow starthistle has the ability to invade rangelands throughout the western United States. In the Pacific Northwest, the most susceptible rangelands are those with deep, loamy soils, south-facing slopes, and 12 to 25 inches (30.4 to 63.5 cm) of precipitation that peaks in winter or spring. Yellow starthistle favors sites originally dominated by perennial grasses, primarily bluebunch wheatgrass (*Agropyron spicatum*), Idaho fescue (*Festuca idahoensis*), and Sandberg bluegrass (*Poa secunda*). This weed does not appear to compete well with sagebrush (*Artemisia* spp.), but readily invades areas of soil or vegetation disturbance within sagebrush communities.

About 60% of the seeds produced by a yellow starthistle population actually survive the process of seed dispersal (Sheley and Larson 1994a). Birds, such as ring-necked pheasants, California quail, house finches, and American finches, feed on yellow starthistle seeds and have been implicated in both long- and short-distance dispersal (Roché 1991). Finches tend to shell seeds, leaving most of the consumed seeds non-viable. Quail and pheasants consume whole seeds, which may occasionally be passed through the digestive tract in a viable form. Humans, other animals, whirlwinds, and vehicles may transport yellow starthistle seeds.

409

Impacts

Yellow starthistle is a serious problem on rangeland, pastures,. roadsides, and wastelands. This species displaces native plant communities and reduces plant diversity. Many members of the knapweed family of plants accelerate soil erosion and surface runoff (Lacey et al. 1989). Yellow starthistle forms solid stands that drastically reduce forage production for livestock and wildlife.

Yellow starthistle is poisonous to horses, causing a nervous disorder called "chewing disease." Livestock can be injured if they are forced to feed on the spiny portions of yellow starthistle. However, in the pre-spiny stage, it can be grazed by cattle and sheep where quality forage is unavailable.

This weed can provide nectar for honeybees.

Biology and Ecology

• Seed Production

A winter annual, yellow starthistle depends on seed production for population survival. A single yellow starthistle plant has the potential to produce up to 150,000 seeds. In studies near Walla Walla, Washington, yellow starthistle produced 20 to 120 seeds per plant, depending upon the density of the plants and the amount of spring precipitation (Sheley and Larson 1994a).

Yellow starthistle plants produce two types of seeds: those with parachute-like plumes, and those without plumes. The majority of seeds are plumed, and are dispersed at maturity. Plumeless seeds are retained in the seedhead and dispersed at maturity between November and February. Studies of yellow starthistle seed dispersal suggest that more than 90% of the seeds fall within 2 feet of the parent plant (Roché 1991), creating a slow invasion front.

• Germination and Viability

In the field, most yellow starthistle seeds germinate in the fall. However, yellow starthistle germinates rapidly under a variety of conditions. Under optimum conditions (which are near 68° F [20° C] with unlimited moisture), plumed seeds germinate within 16 hours (Sheley et al. 1993). Germination rates approaching 75% for plumed yellow starthistle seeds can be attained within 48 hours under optimum conditions. In comparison, plumeless yellow starthistle germination was less than 44% under the same conditions. Under dry or salty conditions, yellow starthistle germination is reduced. The ability to germinate rapidly under favorable and unfavorable field conditions gives yellow starthistle an opportunity to occupy a site by capturing and using resources before neighboring species.

Approximately 95% of the seeds of yellow starthistle are viable, and 10% of the seeds can remain dormant for more than 10 years (Callihan et al. 1993). In heavily infested areas, the soil seed bank of yellow starthistle approaches 13% of total seed production; it primarily consists of plumeless seeds (Sheley and Larson 1994a). Seeds lying dormant in or on the soil create a difficult problem for land managers, because the seeds germinate and allow yellow starthistle to re-invade sites rapidly following most control applications.

• Seedlings

Yellow starthistle typically initiates growth in the fall following precipitation ($1/4$ to $1/2$ inch [0.6 to 1.3 cm]) (Sheley and Larson 1994a). If seeds are available, the number of yellow starthistle seedlings will continue to increase until the soil dries or gets too cold. In a study near Walla Walla, Washington, yellow starthistle seedling populations reached winter (mid-January) densities approaching 2,500 plants per square foot. Subsequent frost heaving reduced seedling populations by about 40%. Ninety-five percent of the seedlings observed in this study germinated in the fall and completed their life cycle as winter annuals. However, seedlings can emerge in the spring, and they can complete their life cycle in the same year or continue into the next growing season, depending upon growth conditions.

• Rosettes

The transition from seedling to rosette begins in March and continues into May. Yellow starthistle rosettes form as leaves emerging from the base of the plant. Rosettes generally have 6 to 28 leaves that range from 1 to 8 inches (2.5 to 20 cm) long. The rosette stage appears to be a very vulnerable time for yellow starthistle. In most years, 60% to 75% of yellow starthistle rosettes die by July, either from moisture stress or self-thinning.

• Adults

As yellow starthistle matures, a flower stalk elongates from the center of the rosette. Bright yellow flowers are surrounded with spines up to 2 inches (5 cm) long. Adult plants can die in stressful environments. Yellow starthistle adult populations remained relatively constant (approximately 70 plants per square foot) in years with moderately dry and wet springs (Sheley and Larson 1994a).

In early fall, yellow starthistle plants lose their leaves and dry to a silvery-grey skeleton with cottony-white terminal seedheads. With the arrival of fall rains, seeds on or in the soil begin to germinate, and the cycle is repeated.

• Yellow Starthistle in the Plant Community

The competitive success of yellow starthistle is directly related to its rapid growth and resource capture (Sheley et al. 1993, Sheley and Larson 1994b). Yellow starthistle seedlings and rosettes require adequate light, water, nutrients, and space, and die easily under stressful conditions. With ample resources, however, yellow starthistle seedlings grow more rapidly than most perennial grass seedlings. This characteristic leads to poor grass stand establishment when grass seedlings are infested with yellow starthistle. Once established, however, vigorous stands of perennial grass can limit re-invasion by yellow starthistle (Larson and McInnis 1989a, 1989b). Perennial grasses that initiate growth in the fall, maintain some growth through the winter months, and continue growth into mid-summer have the best success of competing with yellow starthistle (Borman et al. 1991).

In deep-soiled rangelands dominated by annual grasses (for example, cheatgrass [*Bromus tectorum*]), the rapidly growing and deep, penetrating roots of yellow starthistle tend to avoid direct competition (Sheley and Larson 1995). In areas where cheatgrass is widely dispersed, yellow starthistle roots and shoots can grow twice as fast as cheatgrass. This results in deep soil penetration by starthistle roots, continued growth well into the latter part of the growing season, and increased starthistle seed production. In such circumstances, yellow starthistle can dominate the site. However, yellow starthistle growth rates tend to decline as plant density increases or soils become shallow (Sheley and Larson 1995). This shift in competitive ability under these conditions means that yellow starthistle will become a secondary or co-dominant species in the plant community.

Management

Successful management of yellow starthistle-infested rangelands requires integrating strategies to prevent further introductions, detecting and eradicating new introductions, containing existing infestations, and controlling large-scale infestations.

• *Prevention*

Because yellow starthistle is an annual that spreads by seeds, any action that prevents the production or spread of seeds is helpful in preventing an invasion. Actions might include thoroughly cleaning the undercarriage of vehicles that may have driven through patches of yellow starthistle, or warning the general public not to pick and redistribute "the pretty yellow flowers." Monitoring roadsides and rangeland for new infestations is essential. Prevention is always the most cost-effective weed management tool.

• *Mechanical Control*

Because yellow starthistle is a winter annual, hand-pulling individual plants can be effective. As much of the root as possible must be removed. Hand-pulling is especially effective on new introductions around homesites, in gardens, and in crops such as alfalfa. Hand-pulling is also especially useful along rivers and riparian zones when weeds are detected before they produce a crop of seeds. To prevent spreading the seeds of pulled plants, transport the plants in plastic bags to a location where the plants can be destroyed in a hot fire. Systematic surveys and repeated removal should be conducted every two to four weeks throughout the growing season.

412

On a larger scale, any tillage operation that severs the roots below the soil surface should provide complete control of those plants. Like hand-pulling, this must be conducted periodically throughout the growing season. As the seeds in the soil become depleted, the severity of re-invasion will lessen. Once re-invasion by yellow starthistle is under control, desirable plants should be established to occupy the site.

Mowing is generally ineffective for controlling yellow starthistle because the weeds react by growing more prostrate to escape cutting. In dry years, mowing can reduce seed production, but mowing followed by rain may actually prune the plants and produce more seeds. Mowing can be used to stress weeds once they grow above desirable grasses during revegetation.

• Burning

Burning is an ineffective method for controlling yellow starthistle. In most cases, burning increases plant size and seed production by releasing nutrients tied up in plant material. Rapidly growing yellow starthistle plants quickly capture these newly available resources.

• Revegetation

Following initial control, a perennial grass cover should be established on the site to interrupt the cycle of re-invasion. Grass stand establishment will increase the competition for resources faced by yellow starthistle seedlings and rosettes, limiting their survival and the rate of re-invasion. Research plots using the "Oahe" intermediate wheatgrass (*Agropyron intermedium*), "Tualatin" tall oatgrass (*Arrhenatherum elatius*), "Paiute" orchardgrass (*Dactylis glomerata*), "Covar" sheep fescue (*Festuca ovina*), "Critana" thickspike wheatgrass (*Agropyron dasystachyum*), and "Sherman" big bluegrass (*Poa ampla*) have successfully controlled or reduced the rate of starthistle re-invasion (Larson and McInnis 1989a, 1989b). In southwestern Oregon, "Berber" and "Palestine" orchardgrass and sheep fescue have effectively suppressed yellow starthistle (Borman et al. 1991). The degree of success or failure of any seeding will depend on the selection of a grass species suited to the site, the density of the established stand of grass, and the land manager's ability to maintain grass vigor. Land managers should refrain from fertilizing new grass seedlings that are infested with yellow starthistle, because that practice has been shown to increase starthistle production (Larson and McInnis 1989a).

• Chemical Control

Yellow starthistle can be controlled using various herbicides, including picloram, dicamba, 2,4-D, clopyralid, and glyphosate. Chemical control is an appropriate tool to use: 1) on large infestations, especially when desirable plants are abundant in the understory; 2) in highly productive soils; and 3) around the perimeter of infestations to contain their spread.

Picloram applied at a rate of 0.25 lb active ingredient per acre (0.28 kg ai/ha) to seedlings, rosettes, or to plants beginning to bolt has provided selective control for about three years (Callihan et al. 1989). Well-established grasses are unaffected by picloram at this rate, but grass plants with less than four leaves may be killed.

Dicamba and/or 2,4-D low volatile ester applied at 0.5 lb ai/ac (0.56 kg ai/ha) and/or 2,4-D amine can be used in sensitive areas such as riparian zones or around homesites. Control lasts about one year.

Clopyralid at 1.2 lb ai/ac (1.35 kg ai/ha) applied before the bud stage can be effectively used to control yellow starthistle in areas with a high water table.

Glyphosate applied at one pint per acre is a non-selective, contact herbicide that is used in most revegetation projects to prepare an initial seedbed for seeding desirable grasses.

• *Grazing*

Grazing and grazing management are critical to the successful management of yellow starthistle. Full recovery of desirable rangeland plants should be allowed before the area is regrazed. This will help keep the plants vigorous and competitive, and minimize re-invasion by yellow starthistle.

Before it has spines, yellow starthistle can be grazed by livestock. In California, grazing by cattle or sheep suppressed the weed (Thompson et al. 1990). In this case, proper timing was critical. The first grazing should be timed to the bolting, pre-spiny growth stage. Crude protein ranges from 5% to 10% during this stage. Repeated grazing is required to suppress yellow starthistle. The intensity and frequency of subsequent grazing depends on the environment. Those areas receiving moisture after grazing are likely to require additional defoliation for weed suppression.

• *Biological Control*

The long-term effectiveness of biological control of yellow starthistle is under investigation. Plant pathogens are being researched at universities and the U.S. Department of Agriculture. A number of yellow starthistle-attacking insects have been introduced into the United States.

As of 1996, three weevils (*Bangasternus orientalis, Eustenopus villosus, Larinus curtus*) and three flies (*Chaetorellia australis, Urophora sirunaseva, U. jaculata*) had been released on yellow starthistle in Canada and the Pacific Northwest (Turner et al. 1995, Rees et al. 1996). All of these biological control agents attack the flower head. The goal is to reduce seed production and thereby reduce plant colonization.

Bangasternus orientalis was introduced into the United States in 1985. Ten years later it was established in California, Idaho, Oregon, and Washington. The weevil is readily available in these states. Larvae tunnel through leaves and stalks to the flower head where they feed on developing seeds.

In 1996, *Eustenopus villosus* was readily available in Arizona, California, Idaho, Oregon, and Washington. Larvae feed within flower heads, while adult weevils feed on small flower buds.

Larinus curtus lays eggs at the base of flowers. After the eggs hatch, larvae feed on developing seeds within the flower head, then pupate within the damaged area. This weevil was available in limited numbers in California, Idaho, Oregon, and Washington in 1996.

Larvae of the fly *Chaetorellia australis* feed within the flower heads of yellow starthistle and cornflower (*Centaurea cyanus*). In 1996, the fly was readily available in Oregon and Washington, and could also be found in Idaho.

Urophora sirunaseva larvae feed on developing seeds, then build woody galls within the flower heads. The flies pupate within the galls, then emerge in the spring. By 1996, this species was established and readily available in California, Oregon, and Washington. (A similar fly, *U. jaculata*, was introduced, but did not become established in the United States.)

• Integrated Weed Management

Yellow starthistle's fast growth and high seed viability necessitate a land manager's long-term commitment to control programs. The commitment will likely include an initial control program using herbicides and grazing, along with the establishment of desirable vegetation. This may be followed by a program of vegetation management and monitoring, with periodic chemical applications to control localized infestations. Insectaries for biological control agents may be established. Eventually, biological control agents may be effective enough to provide long-term, sustainable control of this weed, but in the meantime, land managers must integrate all the weed management tools available to contain and reduce yellow starthistle.

Literature Cited

Borman, M.M., W.C. Krueger, and D.E. Johnson. 1991. Effects of perennial grasses on yields of associated annual weeds. J. Range Manage. 44:318-26.

Callihan, R.H., F.E. Northam, J.B. Johnson, E.L. Michalson, and T.S. Prather. 1989. Yellow starthistle biology and management in pasture and rangeland. Moscow, ID: University of Idaho. Curr. Inf. Ser. No. 634.

Callihan, R.H., T.S. Prather, and F.E. Norman. 1993. Longevity of yellow starthistle (*Centaurea solstitialis*) achenes in soil. Weed Technol. 7:33-35.

Lacey, J.R., C.B. Marlow, and J.R. Lane. 1989. Influence of spotted knapweed (*Centaurea maculosa*) on surface water runoff and sediment yield. Weed Technol. 3:627-31.

Larson, L., and M. McInnis. 1989a. Response of yellow starthistle and grass biomass to grass, picloram, and fertilizer combinations. Weed Technol. 3:497-500.

Larson, L., and M. McInnis. 1989b. Impact of grass seedings on establishment and density of diffuse knapweed and yellow starthistle. Northwest Sci. 62:162-66.

Maddox, D.M., A. Mayfield, and N.H. Poritz. 1985. Distribution of yellow starthistle (*Centaurea solstitialis*) and Russian knapweed (*Centaurea repens*). Weed Sci. 33:315-27.

Rees, N.E., P.C. Quimby, G.L. Piper, E.M. Coombs, C.E. Turner, N.R. Spencer, and L.V. Knutson (eds.). 1996. Biological Control of Weeds in the West. Western Soc. of Weed Sci., USDA-ARS, Montana St. Univ., Bozeman, MT.

Roché, B.F., Jr. 1991. Achene dispersal in yellow starthistle (*Centaurea solstitialis* L.). Northwest Sci 66:62-65.

Roché, C.T., and B.F. Roché, Jr. 1988. Distribution and amount of four knapweed (*Centaurea* L.) species in eastern Washington. Northwest Sci. 62:242-51.

Roché, B.F., Jr., and C.J. Talbott. 1986. The collection history of *Centaureas* found in Washington state. Agric. Res. Cent. Res. Bull. XB0978, Washington State Univ. Coop. Ext., Pullman, WA.

Sheley, R., and L. Larson. 1994a. Observation: comparative life-histories of cheatgrass and yellow starthistle. J. Range Manage. 47:450-56.

Sheley, R., and L. Larson. 1994b. Comparative growth and interference between cheatgrass and yellow starthistle seedlings. J. Range Manage. 47:470-74.

Sheley, R., and L. Larson. 1995. Interference between cheatgrass and yellow starthistle at three soil depths. J. Range Manage. 48:392-97.

Sheley, R., L. Larson, and D. Johnson. 1993. Germination and root dynamics of range weeds and forage species. Weed Technol. 7:234-37.

Thompson, C.D., W.A. Williams, M.R. George, W.B. Henry, F.L. Ball, and R.S. Knight. 1990. Managing yellow starthistle on rangeland. Calif. Agric. 44:4-7.

Turner, C.E., J.B. Johnson, and J.P. McCaffrey. 1995. 72. Yellow starthistle, *Centaurea solstitialis* L. (Asteraceae). *In:* Nechols, J.R., L.A. Andrews, J.W. Beardsley, R.D. Goeden and C.G. Jackson (eds.), Biological Control in the Western United States: Accomplishments and Benefits of Regional Research Project W-84, 1964-1989. Univ. Calif. Div. Agric. Nat. Res. Pub. 3361. Oakland, CA, 270-75.

Contributors

K. George Beck is Associate Professor of Weed Science, Dept. of Bioag Science and Pest Management, Colorado State University, Ft. Collins, CO 80523.

Michael M. Borman is Associate Professor, Dept. Rangeland Resources, Oregon State University, Corvallis, OR 97331.

Stephen C. Bunting is Professor, Dept. of Range Resources, University of Idaho, Moscow, ID 83844.

Alvin J. Bussan is Assistant Professor, Dept. of Land Resources and Environmental Sciences, Montana State University, Bozeman, MT 59717.

Robert H. Callihan is Weed Extension Professor Emeritus, Dept. Plant, Soil and Entomological Sciences, University of Idaho, Moscow 83844-2339.

Michael F. Carpinelli is Graduate Research Assistant, Dept. Land Resources and Environmental Sciences, Montana State University, Bozeman, MT 59717.

David Clausnitzer is a Graduate Research Assistant, Dept. of Crop and Soil Science, Oregon State University, Corvallis, OR 97331.

Eric M. Coombs is Biological Control Entomologist, Oregon Department of Agriculture, 635 Capitol St. NE, Salem, OR 97301.

Steven A. Dewey is Extension Weed Specialist, Utah State University, Logan UT 84322.

Celestine Duncan is an independent consultant, Weed Management Services, Helena, MT, 59604.

Joseph M. DiTomaso is Non-Cropland Weed Specialist, Weed Research and Information Center, University of California, Davis, CA 95616-8746.

William E. Dyer is Associate Professor, Dept. of Plant Sciences, Montana State University, Bozeman, MT 59717.

John O. Evans is Professor, Dept. of Plants, Soils, and Biometeorology, Utah State University, Logan, UT 84322.

Duane Griffith is Extension Farm Management Specialist, Montana State University, Bozeman, MT 59717.

Robert T. Grubb was Research Associate, Dept. Plant, Soil, and Environmental Sciences, Montana State University, Bozeman, MT 59717.

Joseph M. Hudak is Lincoln County Vegetation Program Manager, Libby, MT 59923.

James S. Jacobs is Post-Doctoral Research Associate, Dept. Land Resources and Environmental Sciences, Montana State University, Bozeman, MT 59717.

Douglas E. Johnson is Professor, Dept. Rangeland Resources, Oregon State University, Corvallis, OR 97331-2218.

Susan Kedzie-Webb is Graduate Research Assistant, Dept. Land Resources & Environmental Sciences, Montana State University, Bozeman, MT, 59717.

Sherry E. Lajeunesse is an independent pest management consultant with Biological Alternatives™, Newbury Park, CA 91320.

Larry L. Larson is Associate Professor, Dept. of Rangeland Resources, Oregon State University, Corvallis, OR, stationed at OSU-Eastern Oregon State University Agriculture Program, LaGrande, OR 97850.

Rodney Lym is Professor of Weed Science, North Dakota State University, Fargo, ND 58105.

Mark E. Manoukian is Phillips County Extension Agent, Malta, MT 59538.

Gerald Marks is Missoula County Extension Agent, Missoula, MT 59851.

Bruce D. Maxwell is Assistant Professor, Dept. Land Resources & Environmental Sciences, Montana State University, Bozeman, MT 59717.

Joseph P. McCaffrey is Professor of Entomology, Department of Plant, Soil and Entomological Sciences, University of Idaho, Moscow 83844-2339.

Erin G. McConnell is Weed Management Specialist, Bureau of Land Management, Lakeview, OR 97630.

Kirk C. McDaniel is Professor, Dept. of Entomology, Plant Pathology and Weed Science, New Mexico State University, Las Cruces, NM 88003.

Peter B. McEvoy is Professor of Ecology and Biological Control, Department of Entomology, Oregon State University, Corvallis, OR 97331.

Heather C. Miller is a Graduate Teaching Assistant, Dept. of Biology, University of Northern Colorado, Greeley, CO 80631.

Don W. Morishita is Associate Professor of Weed Science and Extension Specialist, Div. of Plant Science, University of Idaho, Twin Falls, ID 83303.

Jeffrey C. Mosley is Associate Professor and Extension Range Management Specialist, Dept. of Animal and Range Sciences, Montana State University, Bozeman, MT 59717.

Barbra Mullin is State Weed Coordinator, Montana Dept. of Agriculture, Helena, MT 59620.

Bret E. Olson is Associate Professor, Dept. Animal and Range Sciences, Montana State University, Bozeman, MT 59717.

Robert Parker is Extension Weed Specialist, Washington State University, Prosser, WA 99350.

Janet K. Petroff is an agricultural communications consultant, WriteOn Communications, Bozeman, MT 59715.

Gary L. Piper is an Associate Professor, Dept. of Entomology, Washington State University, Pullman, WA 99164-6382.

Peter Rice is a Botanist in the Division of Biological Sciences, University of Montana, Missoula, MT 59812.

Ben F. Roché, Jr. passed away during preparation of the manuscript. He was Extension Range Management Specialist, Department of Natural Resource Sciences, Washington State University, Pullman, WA 99164-6410.

Cynthia T. Roché is Rangeland Weed Ecologist, P.O. Box 897, Asotin, WA 99402.

Roger L. Sheley is Assistant Professor, Dept. Land Resources and Environmental Sciences, and Extension Noxious Weed Specialist, Montana State University, Bozeman, MT 59717.

Tracy M. Sterling is Associate Professor, Dept. of Entomology, Plant Pathology and Weed Science, New Mexico State University, Las Cruces, NM 88003.

Jack Stivers is Lake County Extension Agent, Ronan, MT 59864.

Tony Svejcar is the Supervisory Range Scientist, USDA-ARS, Eastern Oregon Agricultural Research Center, HC 71 4.51 Hwy 205, Burns, OR 97720. The Eastern Oregon Agricultural Research Center is operated jointly by USDA-ARS and the Oregon State University Experiment Station.

Donald C. Thill is Professor of Weed Science, Department of Plant, Soil, and Entomological Sciences, University of Idaho, Moscow, ID 83844.

David C. Thompson is Assistant Professor, Dept. of Entomology, Plant Pathology and Weed Science, New Mexico State University, Las Cruces, NM 88003.

419

Charles E. Turner passed away during preparation of the manuscript. He was Research Scientist, USDA Agricultural Research Service, 800 Buchanan St., Albany, CA 94710.

Roseann T. Wallander is Research Associate, Dept. Animal and Range Sciences, Montana State University, Bozeman, MT 59717.

Tom D. Whitson is Extension Weed Specialist, Dept. of Plant, Soil and Insect Sciences, University of Wyoming, Laramie, WY 82070.

Linda M. Wilson is Research Support Scientist, Dept. Plant, Soil and Entomological Sciences, University of Idaho, Moscow 83844-2339.

David L. Zamora is Pesticide Specialist, Washington State Dept. of Agriculture, Wenatchee, WA 98807.

Richard K. Zollinger is Extension Weed Specalist, North Dakota State University, Fargo, ND 58105.

Index